THE SUPER-STRATEGISTS

Also by Col. John R. Elting:

A Dictionary of Soldier Talk, 1984 (coauthor)
American Army Life, 1982
Military Uniforms in America: Long Endure: The Civil War Period, 1852–1867, 1982 (coeditor)
Battles for Scandinavia, 1981
Military Uniforms in America: Years of Growth, 1796–1851, 1977 (editor)
The Battles of Saratoga, 1977
The Battle of Bunker's Hill, 1975
Military Uniforms in America: The Era of the American Revolution, 1755–1795, 1974 (editor)
A Military History and Atlas of the Napoleonic Wars, 1964 (coauthor)
The West Point Atlas of American Wars, 1959 (associate editor)

THE SUPER-STRATEGISTS

Great Captains, Theorists, and Fighting Men Who Have Shaped the History of Warfare

COL. JOHN R. ELTING
U.S. Army, Ret.

CHARLES SCRIBNER'S SONS
NEW YORK

Library of Congress Cataloging in Publication Data

Elting, John Robert.
The superstrategists.

Bibliography: p.
Includes index.
1. Strategy—History. 2. Military art and science—History.
I. Title.
U162.E57 1985 355′.02 85-10792
ISBN 0-684-18353-6

Published simultaneously in Canada by Collier Macmillan Canada, Inc.

1 3 5 7 9 11 13 15 17 19 F/C 20 18 16 14 12 10 8 6 4 2

Printed in the United States of America.

To Ann

"Strategy is horse sense."
(unknown American)

". . . when practiced by Indians, it is called treachery."
(an equally unknown American)[1]

TABLE OF CONTENTS

INTRODUCTION

Alfred, with his great bow,
Roland with his horn.
Men have heard their horses' hoofs
Many a scarlet morn.[2]

Because war "is fate knocking at the door . . . nature deciding the life and death of nations,"[3] the question of how wars may be won is still a major preoccupation of mankind. Men have given their lives to that study; a certain few, whether by example or written word, have taught generations of soldiers after them and have hard lessons for us yet today. Exemplars and teachers, they are the "superstrategists," the men who—for good or for ill—have shaped the nature of this world's wars.

This book stands as a brief introduction to such men. They include great captains, world acclaimed as masters of strategy, and carping pedants "that never set a squadron in the field,"[4] and carefully avoided any opportunity of getting mixed up in such bloody doings. Some were not, in the modern sense, strategists. Maurice de Saxe made no pretense of being one; Frederick the Great was a strategic bumbler, yet both influenced future strategists. Flavius Vegetius and Dennis H. Mahan were ivory-tower theorists; Jomini followed the drums for years without really becoming a soldier. Henry Lloyd was an obscure mercenary; Napoleon changed the currents of world history. Julius Caesar, Liddell Hart, and Douglas MacArthur owe much of their fame to their genius for self-glorification, but Caesar and MacArthur also were mighty men of war. Genghis Khan understood little of the world beyond his steppes; unless dissuaded by trusted counselors, he instinctively destroyed what he did not comprehend (see chapter ix).

It is worth noting that this assortment of great captains, hard-luck soldiers of fortune, scholars, and pontificators were, on the average, not an especially bloodthirsty lot. Some were hard men, born into a harsh age, but most could be merciful—by instinct rather than merely for policy's sake. They considered war a natural thing; they were proud of the skill with which they waged it. But only Genghis Khan and his like felt that man's greatest joy lay in large-scale throat cutting, rape, and looting. To most of the Westerners it was a serious business, not to be lightly entered upon. Most waged it—at least until World War II loosed the strategic "bomber barons"—with a certain moderation toward civilians and a professional respect for a worthy foe. Some of the superstrategists were builders and lawgivers; most were patriots. All of them left us their legacy of martial skills.

The superstrategists are rare. One and all, they are products of their own times, shaped by its wars, trained up in arms by veterans of past campaigns. Consequently, any superstrategists can be best understood, and his contributions to the art of war best appreciated, in the context of the men who trained him, the authors he studied, and the soldiers who served with him and against him. Many of those men were competent strategists or combat leaders, with impressive accomplishments to their names. William of Orange and Henry of Navarre were not great generals, but their courage and obstinacy baffled all the power of Spain. Young Turenne's literally Dutch uncles put him into the ranks to prove himself with common pikemen and musketeers before they would make him an officer. Napoleon remembered with lifelong gratitude the training the du Teil brothers gave him when he was a mere artillery shavetail. During the American Civil War, Ulysses S. Grant and William T. Sherman won the reputation of superstrategists. Henry W. Halleck and George H. Thomas, a desk soldier and a mighty battle captain, respectively, did not. However, but for Halleck and Thomas, it is unlikely that Grant and Sherman would have achieved any such distinction. Only a few specialist historians remember men like Arrain, Folard, Montecuccoli, and Fox Connor today, but their deeds and writings inspired generations of soldiers, superstrategists included.

This is a book about superstrategists, and so, necessarily, also about those men who developed the armies, fleets, and weapons that the superstrategists employed—the men who campaigned with and against them, and sometimes even defeated them in battle.

Therefore, the reader will meet many warriors who long ago vanished from most histories: Captains courageous, roaring sea dogs, drillmasters, theorists, "old fighting men broke in the wars,"[5] they all contributed to the art of war, even if only as horrible examples of how *not* to wage it.

In this study of strategists and strategy, I make no apology for stressing American adaptations and contributions. We are not a military people, but we are a warlike nation. Otherwise, the United States would not exist today. And we have given the world a good many of its most potent military lessons.

I

STRATEGY AND STRATEGISTS

. . . so few empire-makers have looked the part.
Fate has a way of picking unlikely material,
Greasy-haired second lieutenants of French artillery,
And bald-headed, dubious, Roman rake-politicians.[1]

STEHEN VINCENT BENET

Strategy is the art of winning wars. It appears simple, since it is nothing more than the application of common sense. Given the available information on a military situation, the average citizen feels perfectly capable of roughing out an acceptable strategic solution. Unfortunately, common sense is somewhat uncommon. And strategy involves not only the development of a plan of operations but the carrying through of that plan, which is entirely something else again! As Napoleon wrote: "Anyone can plan a campaign, but few are capable of waging war, because only a true military genius can handle the developments and circumstances."[2] Even competent generals have found the transition from planning to execution beyond their abilities. Maj. Gen. Joseph Hooker—"Fighting Joe"—had rare personal bravery under fire; he had proved himself an aggressive division and corps commander; he understood the American volunteer soldier and took care of his men. In early 1863 he commanded the Army of the Potomac, which he had brought to a strength of 134,000 and its highest state of efficiency. Opposing him, dug in along the south bank of the Rappahannock River around Fredericksburg, were Gen. Robert E. Lee's 60,000 Confederates. Knowing that a direct attack on Lee's position would be suicidal, Hooker developed an excellent plan: fixing Lee in place with a threatened frontal attack, he would swing most of his army widely to the west in a turning movement that would bring him into Lee's rear. Launched as planned, his offensive moved like clockwork; Lee was baffled long enough for Hooker to come down upon him with over 70,000 men, while 40,000 more crossed the Rappahannock south of Fredericksburg. Then, as the first shots crackled, Hooker—so he later would wryly confess—"just lost confidence in Joe Hooker."[3] He went over to the defensive. Lee quickly seized the initiative and handed him a humiliating defeat at Chancellorsville.

Early writers on the art of war often made no clear difference between *tactics,* which is the art of handling troops in combat, and *strategy,* the science and art of planning and executing large-scale military operations. (During the nineteenth century the term "grand tactics" might be employed to cover at least some aspects of strategy.) The great German military philosopher Carl von Clausewitz succinctly defined the relation of the two: "Tactics is the art of using troops in battle; strategy is the art of using battles to win the war."[4]

In its most comprehensive meaning, military strategy is the art and science of employing a nation's armed forces to secure its national objectives either by force, the threat of force, or the deterrent effect of force available for immediate use. (There will be no deterrence unless—comes worst to worst—you are willing to *use* that force and your enemy knows that you will.) The fundamental law of military strategy is: Be stronger at the decisive point.

Military strategy, however, should be only an integral part of national strategy, which is the sum of a nation's desires, policies, plans, and programs to secure its present and future existence. National strategy combines the nation's political, economic, and scientific strengths with its military strength; it also may exploit a favorable geographic location and such spiritual, intellectual, or ideological leadership as it possesses. Though military considerations should definitely influence the development of national strategy, military strategy itself must be subordinated to the national strategy in peace or war. Clausewitz laid down this dictum with clarity and force in his massive work *Vom Krieg* (*On War*): "War is no pastime. . . . It is a serious means to a serious end . . . a continuation of policy [national strategy] by other means . . . always an instrument of policy . . . in no sense autonomous. . . . Policy is the guiding intelligence and war only the instrument." Otherwise, war becomes "something pointless and devoid of sense." It was proper for a commander to request that his government avoid wars he thought his forces were too weak to win, but that was all.[5]

By contrast, the national strategy of the United States during World War II was completely militarized—pared down to achieving the unconditional surrender of Germany and Japan. Every possible propaganda trick was used to stimulate the American people into total war against them. The German resistance movement was given no encouragement in its attempts to overthrow Hitler. But there was next to no solid study on what postwar developments would be most beneficial to the United States. After victory, so President Franklin D. Roosevelt assumed, the world's problems would be handled peacefully by his great design for a United Nations organization. To achieve both absolute military victory and a functioning United Nations, Roosevelt considered the cooperation of the Soviet Union essential. He wooed Russian Premier Joseph Stalin with all-out, no-strings-attached aid, much praise, and acquiescence to Russian territorial expansion in the Baltic, Eastern Europe, and the Far East. In return, he trusted that Stalin, out of a sense of *noblesse oblige,* would work with him for a peaceful, democratic world.

That strategic myopia may be the death of many of us yet.

A carefully thought out national strategy could have supported the German underground. Certainly it would have avoided Roosevelt's demand for unconditional surrender and Secretary of the Treasury Henry Morgenthau's plan to strip a conquered Germany of its industries—threats that merely increased the desperation with which Germans fought. Aid to Russia could have been carefully rationed. There would have been vast clamor, but the Russians—always tough and wily bargainers—would have understood Americans acting out of self-interest.

A nation without a definite national strategy approved by a majority of its people—an unfortunate predicament in which the United States frequently finds itself—can hardly develop a military strategy appropriate to its international problems. There are questions as to the relative emphasis to be given nuclear and conventional weapons, defensive and offensive weapons systems, and different types of ships, aircraft, and ground force units. There are problems as to the necessity and character of overseas bases and military alliances, and half a hundred others—and all dependent on a decision by the American people as to what they want in this present disorderly world. ''If the trumpet give an uncertain sound, who shall prepare himself to the battle?''[6]

Strategy takes many forms, which have been given many names. There is the ''Strategy of Annihilation,'' or ''Overthrow,'' which seeks a quick victory by forcing the enemy's armed forces to give battle and there destroying them. (Such destruction normally consists of destroying the opponents' ability and/or will to continue to offer resistance, not their literal annihilation.) Most of Napoleon's campaigns exemplify this strategy, as did Hitler's conquest of Poland in 1939 and of France, Holland, and Belgium in 1940. To be successful, however, the attacker must achieve *surprise* (see appendix) and be able to *mass* superior combat power at the decisive point. If he fumbles the operation, he lays himself open to a dangerous counteroffensive.

The ''Strategy of Attrition,'' or ''Exhaustion,'' attempts to wear out the enemy by comparatively limited operations over a considerable length of time. It may be employed either offensively or defensively and frequently proves more costly in lives, money, and social strain on the combatants' homelands than a quick run of major, decisive battles. One anonymous American officer defined it as ''being nibbled to death by ducks.'' On occasion, such a strategy may evolve when a campaign of annihilation fails. In World War I both sides had planned and prepared for a swift-moving, violent campaign that would achieve victory in a few months at most. On the western front the armies proved too big, the means of controlling them too ineffective, the area of operations too small, the generalship generally too amateur, to carry through the planned offensives. Four years of grinding trench warfare—attrition at its roughest—followed.

Conversely, attrition can develop into annihilation when one side builds up sufficient strength to break the stalemate. In World War I, the arrival of some two million Americans supplied the needed combat power. After World War II, Vietnamese rebels began guerrilla operations against the French in Indochina with varying success. However, at the end of fighting in Korea in 1953, Communist China was able to divert great amounts of weapons and supplies to the Vietnamese, who quickly changed over to a strategy of annihilation against the almost-exhausted French. This culminated in the French defeat at Dien Bien Phu in 1954.

The term "Strategy of Containment" is sometimes applied to a national strategy of "containing" (walling off) a hostile aggressive power by establishing alliances with nations around its borders.

The basic rules of strategy do not change. They would govern an all-out nuclear war—so long as it lasted—just as they have all wars past. These lessons, wrung from war and study by great strategists through the centuries, have a stern similarity and continuity. In approximately 500 B.C. the legendary Sun Tzu Wu of eastern China wrote, "The natural formation of the country is a soldier's best ally." Some nine hundred years later and half the world away Vegetius recorded much the same thought: "The nature of the ground is often of more consequence than courage."[7] Throughout World War II, American soldiers heard the slogan "Take the high ground." During the past century the substance of these lessons has been codified into our modern "Principles of War" (see appendix). The ability to parrot them does not make an officer a superstrategist; the ability to apply them intelligently to strategic problems very well may.

In one regard our general concepts of war and strategy differ from those of earlier soldiers—our Principles of War contain no mention of "luck" or "Fortune." Today we have computers, simulation games, and other electronic marvels that are popularly presumed to make all knowledge exact and shield us from errors those ancient unreliables might introduce. Possibly Henry Lloyd's observation that whatever passes through men's hands will reflect their imperfections should be engraved across the front of all computers. A computer's "intelligence" can be no greater than that of the men who program and use it, which on occasion puts the poor machine under an impossible handicap.

Fortune, luck, chance, divine intervention—all were very real to our military forefathers. The Greek military historian Thucydides (ca. 460–400 B.C.) noted: "War proceeds only in the smallest way in accordance with definite laws; its principal part it creates for itself by itself."[8] Caesar acknowledged Fortune's influence; Vegetius cautioned that "fortune has often a greater share than valor in a battle's outcome."[9] Napoleon asserted that "War is composed of nothing but accidents."[10] and that "luck" was really the ability to exploit them. Clausewitz considered such mischances part of the "friction" that made the efficient conduct of war so difficult. Even today, few combat soldiers do not believe in luck, if only from the almost whimsical irrationality with which Death takes or passes by.

Officially acknowledged or not, Dame Fortune still joggles the elbows of modern strategists. Besides her old-fashioned gifts of unexpected bad weather, collapsed bridges, mislaid orders, wrong roads taken, inaccurate maps, the chance shot that kills the only guide, she now bestows modern tribulations—the glitch in the vital computer program, sudden outbursts of solar flares to scramble high-frequency radio communications, the sophisticated and costly weapons system that performs perfectly on the proving grounds but only occasionally in the field. The lady is too lively and fickle to be quantified or predicted; she comes and goes at no man's pleasure, like death and the wind and rain.

Like great captains gone before, the modern strategist must recognize and accept this irrational element in war and prepare himself to meet and master it. One of Napoleon's light cavalry officers understood the matter: "Let nothing surprise you. Depend mostly on yourself and a little on luck, and always go forward."[11] And Thomas J. ("Stonewall") Jackson advised, "Never take counsel of your fears."[12]

II

BY PHALANX AND LEGION

Qui procul hinc, qui ante diem periit,
sed miles, sed pro patria.[1]

Herodotus (ca. 484–430 B.C.) was neither a captain nor a student of things military. If he ever did any soldiering, it probably was a brief spell as a yardbird spearman, second class, in the rear ranks of his hometown phalanx. And yet the study of strategy in the Western world undoubtedly begins with him.

Originally a native of the Persian-ruled Greek city of Halicarnassus on the western coast of what now is Turkey, Herodotus traveled widely around the eastern Mediterranean. Probably he was a trader, possibly a political exile. He had sufficient intellect and education to lecture successfully on history in Athens; finally, he settled in the

Athenian colony of Thurii in southern Italy. Somewhere in his life he began writing sensible accounts of the various nations and cities through which he had journeyed. Since most of these had been conquered by the growing Persian empire, his *Histories* was expanded to include the story of that empire and then the tale of the wars between Greece and Persia.

Most of his work must have been based on oral history—tales told by eyewitnesses and participants or by keepers of legends—since there were few documentary archives during this period. He seems to have attempted to get straight accounts; where facts were in dispute, he would record the differing views without comment. He had no military savvy; his descriptions of battles and campaigns seem exact but, when studied, are full of puzzles. He grossly exaggerated the strength of the Persian army that Xerxes led into Greece in 481 B.C., possibly tenfold. But he wrote well, and he loved stories of brave men, whatever their race. His account of the Persian Wars is an epic of free Westerners against the might of the despot-ruled East, of self-sacrificing patriotism and high valor, whether he tells of the death stand of Leonidas and his 300 Spartans at Thermopylae or of the final victories. His title of "father of history" was well and honestly earned. Soldiers of many states and nations read his works and pondered the wars he described.

Herodotus' successor, Thucydides, was a different personality altogether. With Persia defeated, the Greek states resumed their complex feuds, which usually involved hostilities between Athens and Sparta and their shifting allies and vassals (Peloponnesian Wars, 458–404 B.C.). Sparta's strength was her grimly trained citizen soldiers; her habitual strategy was to invade and ravage an enemy's territory, forcing a decisive stand-up battle in which the red-tuniced Spartans almost invariably were victorious. Athens was a commercial city; her soldiers were good enough, but her power was her ships and skilled seamen. She built up the League of Delos from smaller Greek states in the Aegean Sea and along the coast of Asia Minor, reaching through the Dardanelles and Bosphorus into the Black Sea. As years passed, Athens reduced these allies to mere vassals, establishing a potent maritime empire. Neither side was completely self-supporting: Athens drew much of its grain from areas north of the Black Sea and timber for shipbuilding from Thrace and other states in northern Greece; Sparta and its allies imported grain and olive oil from Sicily and southern Italy. War between these two powers went by fits and starts, replete with the double-crossing diplomacy, treachery, and fifth-column operations common in Greek history. Pericles, the leading Athenian statesman, evolved a strategy of standing on the defensive in Athens' home territories while sending the Athenian fleet to raid the coastal areas of Sparta and its allies in hope of forcing them into a decisive sea battle. This strategic stalemate slowly worked itself out. Athens was crowded with refugees from the devastated countryside; an unknown type of plague, brought in by ship from Egypt or Asia Minor, burned through the city in 430–429 B.C. and again in 427–426. Pericles was an early victim. His less competent successors attempted to conquer Sicily to stop its grain exports to Sparta and were badly beaten in 413. Meanwhile, the Spartans, with money from Persia and ships and men from anywhere, built up a navy and struck at Athens' jugular—the Dardanelles. After a run of failures and half successes, they broke the Athenian navy there in 405. With a Spartan army at its gates and a Spartan fleet offshore, starving and helpless, Athens capitulated.

Thucydides, born 460 B.C., an experienced officer from a wealthy, aristocratic Athenian family, studied this war from the sidelines. In 424 B.C. he had been sent to nail down Athenian influence in Thrace, but his luck had run out. Brasidas, one of those imaginative fighting men that supposedly stolid Sparta did sometimes produce, had gotten there before him by a remarkable overland march from southern Greece, and promptly added Thucydides to his list of defeated Athenian commanders. Athens punished Thucydides by exile as long as the war might last. (At that, he was lucky; in one case, Athens executed six victorious admirals who had failed to recover the bodies of drowned Athenian sailors from a stormy sea after a desperate battle.)

With plenty of time on his hands, Thucydides proceeded to write a history of those wars, working from his own experience and from eyewitnesses whose reliability he checked as thoroughly as possible. His viewpoint was that of an intelligent Athenian soldier, yet admirably impartial. As an exile, he could pass anywhere and seek information from Spartans and other enemies of Athens. His style was restrained and straightforward; he omitted some details that we would like to have—for example, the terrain over which major battles were fought. Probably he thought most of his readers would be familiar with it. Unlike Herodotus, he occasionally offered his own opinion of events but apparently only where he had personal knowledge of them.

In one respect he really let himself go: almost a fourth of his work consists of long speeches given by various Greek leaders on important occasions. Such talk *was* very important in determining Greek public policy, but Thucydides naturally could not furnish the exact text of these speeches. He did try to determine their main points, which he then draped with logical oratorical accompaniments. Possibly some of these reconstructed speeches are more effective than the original versions.

Naturally, Thucydides was particularly interested in the uses and limitations of sea power, which in his opinion grew out of a combination of domestic industry and overseas trade. A wise maritime power, he concluded, would avoid major wars, since war harmed commercial activity, but it would use its fleet unhesitatingly to keep trade routes open and safe for its merchant shipping. In case of war, it should protect its home territory, keep its fleet strong and use it to maintain constant pressure on the enemy until they sued for peace, but not risk large-scale conquests of foreign territory. Much of this doctrine would reappear some two thousand years later in the writings of an American naval officer, Albert Thayer Mahan.

War to Thucydides was an accustomed thing, to be accepted and endured. He noted that "opportunities in war do not wait" and that unexpected developments were to be anticipated. In the words of King Archidamus of Sparta, "the course of war is hidden, and much comes about from very small things, and passion brings about accomplishment."[2] In a less speculative mood he quoted Pericles' view that success in war "is most often achieved by good judgment and abundance of money."[3] Thucydides also was much interested in man as a political animal and in his appetite for power and his use or misuse of it. In that sense, his history is a keening over the fall of Athens, the story of how a powerful free state (Athens was never a "democracy" in our sense of the word, but its citizens did enjoy considerable liberty for that period) went down in ruin. Athenians would not deal justly with their allies; they chose leaders who promised them further conquests and increased commercial profits—and got incompetent demagogues and slippery adventurers who brought them defeat abroad and subversion at home.

Thucydides came home briefly to Athens after the war's end in 404 B.C. He died three or four years later in far-off Thrace. By then he had completed his history of the war from 458 to 411 B.C. Possibly because of ill health, his writing style had weakened, but he still was master of his subject.

Herodotus had labored to depict the brave days of old; Thucydides wrote mostly of contemporary events. Neither provided a complete picture of Greek history—such would have been beyond the ability of mortal man—but they worked skillfully and honestly, and Thucydides ranks as one of the great military historians of all times.

The best known of those historians who continued Thucydides' work was Xenophon (approximately 430–350 B.C.), another Athenian exile. He had served in the Athenian cavalry during the last years of the Peloponnesian War but thereafter found himself at loose ends. His study under the skeptical philosopher Socrates may well have left him in a restless, questioning mood, and Athens' daily politics were a messy turmoil. In 401 B.C. he became a junior officer in a force of 12,000 assorted Greek adventurers recruited by Cyrus the Younger, pretender to the Persian throne.[4] Cyrus met the Persian king's army at Cunaxa, near ancient Babylon (fifty-five miles south of modern Baghdad). At the day's ending the Greeks still held the field, but Cyrus had gotten himself killed. The Spartan officer commanding the Greeks accordingly requested safe passage for his troops over the 1,000 miles back to the seacoast. The Persian commander invited the senior Greek officers to parley in his tent and there had them murdered. If his intent had been to demoralize the Greek rank and file, it failed. Seemingly trapped in the center of a hostile nation, those stubborn spearmen constituted themselves a temporary Greek city-state and chose new leaders, including Xenophon. Then they burned their tents and wagons and moved out, up the Tigris River and into the mountains, standing off a halfhearted pursuit. For eight months and 1,700 miles they moved northward, living off the country, fighting off the wild Carduchi (modern Kurds) and other unlicked hill tribes. Half of them came through to Trebizond, a Greek settlement on the Black Sea. Xenophon and his fellow officers had displayed an amazing versatility in mountain warfare and flank and rearguard tactics and in maintaining morale and discipline through constant hardship and danger. During this time Athens had declared Xenophon an exile, possibly because he had taken service with the Spartans and Persians. Thus cast out, Xenophon (along with the survivors of Cyrus's Greeks) joined Spartan forces that had begun operating against the Persians in Asia Minor. He seems to have continued to serve efficiently with them after they were recalled at the outbreak of the Corinthian War (395–387 B.C.) in Greece. (Subsidized by Persia, Athens had formed a new alliance against Sparta.) Again, the Spartans won. They gave Xenophon a small country estate where he turned to writing, agriculture, and hunting and followed the wars no more, though he continued to study the military art. Later, he seems to have lived both in Corinth and in Athens, which had restored his citizenship.

Xenophon's continuation of Thucydides' history, his *Hellenica,* covers 411–362 B.C. It is a more uneven work that rather neglects events in central and northern Greece, and it has something of a pro-Sparta bias. At the same time it contains much useful information, unrecorded elsewhere. Xenophon was a prolific author whose subjects included Persia, Socrates, Spartan institutions, management of country estates, and dogs and hunting. More important were two short manuals he produced on cavalry and horsemanship—subjects still foreign to many Greek soldiers. His triumph is his *Anabasis* (*March Upcountry*), the story of Cyrus's campaign and the retreat of his Greek contingent; written in a concisely clear style, it brims with information on the country and peoples the Greeks encountered, as well as their battles. Always the practical soldier, Xenophon concerned himself with rations, baggage animals, camp followers, guides, and roads (subjects hitherto of no particular interest to Greek commanders with their short-range wars), as well as tactics and strategy. Simply and modestly, he tells a proud tale of what soldiers of free nations can bring themselves to endure. Every soldier should study *Anabasis* at least once, but there is much in it for every reader. Through it and Xenophon's other military works run his matter-of-fact concepts of strategy: "Wise generalship consists in attacking where the enemy is weakest, even if the point be . . . distant." "If you attack . . . do it in full strength, because a surplus of victory never caused any conqueror one pang of remorse."[5]

Anabasis proved one of the most popular military works ever written. In 1758, Gen. James Wolfe, who next year would die at the taking of Quebec, claimed to have trained his newly organized light infantry after Xenophon's description of the tactics used by the Carduchi. But the effects of Xenophon's book on his Greek contemporaries were immediate, confirming a growing suspicion that the wealthy Persian empire had gone soft and was ready for conquest and looting. Among its fascinated readers must have been young Alexander, bookworm son of Philip II, king of Macedon.

Philip II badly needs a competent biographer. He built the disorderly, rustic, north-country principality of Macedon into a potent kingdom. Courageous, swift in thought and execution, a good king, and an unscrupulous foe apt at fraud and bribery, one-eyed and crippled in a hand and leg from his wars, his limping presence usually is considered only a darker background for Alexander's spectacular career. Yet it was Philip who organized and trained the amazing Macedonian army; broke Athens, Thebes, and their allies—more dangerous foes than any Alexander would face—at Chaeronea in 338 B.C. and united all Greece (Sparta excepted) into a Hellenic League. This league made him commander in chief of its armed forces and authorized him to proceed with the conquest of Persia. Before he could take the field, he was assassinated—probably at the instigation of his estranged wife, Olympias, Alexander's mother and a strange, vehement creature.

Alexander (356–323 B.C.) thus came to the throne at the age of twenty. Philip had trained him carefully in every apsect of kingship; for his tutor Philip secured the famous Aristotle, a universal genius of that time, who confirmed Alexander's love of reading and gave him the available smatterings of knowledge of natural science and geography. We know that Alexander was fascinated by Homer's *Iliad;* it is only logical that considering Philip's long-planned invasion of Persia, he would have studied Herodotus, Xenophon, and probably Thucydides.

He would live for not quite thirteen more years, and crowded years they were. On Philip's death his status was doubtful even within Macedon. Fortunately, Philip's generals stayed loyal to him. A few over-ambitious relatives and hostile nobles were quickly disposed of (Olympias took care of Philip's second wife and her infant), and Alexander swept suddenly south to restore Greece to obedience. That done, he spun northward to put the fear of Macedon into the tribes between it and the Danube, then drove south again. Aware of Alexander's plans, Persia had bribed Athens, Thebes, and other states to revolt anew. Thebes was leveled and its 30,000 surviving inhabitants sold into slavery. Athens crawled in fear and abasement and was forgiven.

Early in 334 B.C. Alexander crossed into Asia Minor. He destroyed three Persian armies, broke up the Persian fleet, occupied the Persian capital, and pursued King Darius of Persia into the wilderness where he was murdered by his own men. Thereafter, Alexander set about mopping up the interior of Persia, parts of which never had been subdued by any king. One tribe of mountaineers was won over when Alexander captured and married their chief's daughter, Roxane, reportedly the most beautiful woman in Persia. (Unfortunately for modern soldiers, such opportunities no longer occur in guerrilla wars of this century.) In 327 B.C. he moved on into India, penetrating beyond the Indus River. There his army balked. They had made great conquests, won famous victories, but India was hot, and the rain seemed endless. They had marched 17,000 miles, and always there were new enemies before them, new sicknesses, new hardships. Alexander gave in and returned westward.

Alexander could call himself "King of Asia." His rule ran, more or less, from the Danube and the Adriatic south through Greece, and from Egypt eastward across Palestine, into modern Turkey and Iran, and much of Afghanistan, Pakistan, and the southern provinces of Soviet Russia. Its borders were vague and not all its inhabitants reconciled, but no other king or people challenged him. Back in Persia he found this new empire, as might be expected, in considerable disarray and worked energetically at putting it to rights. He had great visions of a kingdom in which all races were equal and of further explorations and campaigns. But he was worn thin with exhaustion and eight wounds, and getting Macedonians and Persians to work together was more than difficult. A fever carried him off suddenly as he toiled in Babylon in 323 B.C. Months later, Roxane bore him a son (and murdered his second wife, a daughter of Darius), but both were killed by one of his former officers. His lieutenants, the Successors, split up his empire and promptly came to blows. Several of them proved skilled strategists and excellent tacticians, but the story of their wars is now relegated to one of history's back attics.

Alexander was a true warrior king, leading his troops in person regardless of odds or dangers, generous and openhanded with his soldiers, yet iron hard in matters of discipline. He looked the part—golden-haired, handsome, athletic, a skilled horseman. His campaigns covered every type of warfare—sieges, naval engagements, mountain and desert fighting, major river crossings, guerrilla operations and pitched battles—and he was victorious in all of them. He moved with a swiftness that caught experienced generals and wily savages alike by sheer surprise. Totally aggressive, always striking first, endlessly inventive when confronted by difficulties, his natural qualities and his victories gave him a presence that was more than kingly and made him a legend to many peoples while he still lived.

Yet for all his striking bravery in battle, he was a cautious strategist, always careful of his base of operations and his communications. Before he invaded Persia, he had brought Greece and the northern frontier tribes to heel, so that a relatively small home army could keep them in order. Once in Asia Minor, he found his communications endangered by the Persian fleet, manned by Greeks from Asia Minor and sea-roving Phoenicians from the coast of Palestine. His own fleet was comparatively weak, his Greek allies reluctant to support him. His solution was to move down the coast of Asia Minor and seize the Persian naval bases one by one. Most of the Greek city-states were won over by diplomacy; others joined him after his first major victory over Darius at Issus, 333 B.C. Those that resisted—particularly Tyre—were battered into submission. Even on his long march to the Indus, Alexander operated by methodical stages, building cities to protect his communications and reconnoitering the country ahead. There is little exact information on his logistic (supply and evacuation) system, but it must have been highly efficient. Recruits, remounts, and supplies went forward (reportedly, replacement armor for 25,000 men reached him in India), and worn-out veterans, captured elephants, and other booty went back to Persia or Greece.[6] For the first time western armies learned to deal effectively with great distances and prolonged campaigns.

Alexander's strategy was based on a careful study of potential enemies, including the supplies their territories could furnish an invading army. Plundering was held to a minimum because Alexander's objective was conquest and enlightened conquerors do not deliberately wreck new additions to their realm; also, it wasted supplies. Useful institutions—for example, the Persian courier service—were left intact; local governments were told to continue bringing in taxes and supplies and maintaining the roads. Restless hill tribes that might take advantage of the war's turmoil to get in a little quick raiding and looting were routinely squashed. When possible, Alexander's favorite antiguerrilla strategy was to move against them in winter, when deep snows would have forced them down out of the high country or immobilized them in their villages.

Alexander was, in truth, a multitalented genius. Unfortunately, nothing written about him during or just after his lifetime has survived. What histories we have of him—such as Arrianus's—were written three centuries or so after his death, mostly from sources also lost to us. Even after much painstaking research in this century, in many ways he remains a legend and a mystery, a warrior comet that flashed across our history and passed on into the darkness. Other great captains could only hope to equal him.

Westward, in the center of the Italian peninsula, the small city of Rome grew slowly into a major power. Its first major enemy was Carthage, a Phoenician state on the North African coast (near modern Tunis). During the First Punic War, Rome improvised a fleet and wrested naval command of the western Mediterranean from Carthage. A city of merchants and commerce, Carthage depended largely on mercenaries but did have one brilliant military family, the Barcas. During the Second Punic War (219–202 B.C.) Hannibal Barca canceled Roman naval superiority by leading a Carthaginian army out of Spain, across southern Gaul, over the Alps, and into northern Italy. This march had been as carefully planned and prepared as any of Alexander's; Hannibal's excellent intelligence service had cleared the way. In Italy he won victory after victory, maintaining his army off the country and attempting to win over Rome's allies. One of his most spectacular victories was at Cannae (216 B.C.) where he outflanked, surrounded, and butchered a Roman army almost twice the strength of his own; his maneuvers here would haunt later strategists (see chapter VIII).

Hannibal finally was checked by one Quintus Fabius, who chose not to risk raw Roman farm boys against Hannibal's veterans. Instead, exploiting the rugged Italian terrain and the Roman knack for building field fortifications, he instituted a strategy of blocking and harassment, cutting off Carthaginian detachments and making foraging hazardous. When some cities, such as Capua, went over to Hannibal, Roman armies promptly dug in around them, one line of works facing the city, another facing outward to hold off Hannibal's attempts to relieve his allies. Hannibal could never break through, and the cities were starved into submission. Ironically, though Fabius's careful strategy won Rome time to train troops and develop competent commanders, it was not too popular with his fellow Romans, who hoped for a quick riddance of the Carthaginian invaders. It is doubtful that the nickname ''Cunctator'' (''Delayer'') they gave him was entirely inspired by admiration. (Possibly the same spirit was behind it as moved citizens of Vienna to toss nightcaps into the carriage of Field Marshal von Daun, who had employed a somewhat similar strategy against Frederick the Great; see chapter V.)

Hannibal held out in Italy under increasing pressure and with little help from Carthage for fifteen years, until he was summoned back to Africa. The Romans, having decided that the best way to extract Hannibal from their territory was to invade Carthage, had sent out an expedition under Publius Scipio, later dubbed ''Africanus.'' Scipio beat Hannibal in a knock-down, drag-out battle at Zama in 202 B.C., and Carthage capitulated.

Hannibal lived twenty more years in exile. He had been one of Rome's best military instructors, and Romans feared him as they did no other enemy. He finally committed suicide to escape their vengeance. Practically everything we know about him comes from his enemies, but it is plain that he was a soldier's soldier, able to win the personal loyalty of tough mercenaries of many nations, as well as a master strategist and tactician. Napoleon considered him the most audacious of generals, daring, firm, and great in all things. The Duke of Wellington thought him the greatest soldier in history.

The military history of Rome covers a reach of more than one thousand years. Its earlier centuries are mostly myth, and historians still dispute the rest of it, since Roman military organization, weapons, equipment, and tactics were constantly evolving to meet new circumstances. There was as much difference between the Roman home-guard militiaman of 500 B.C. and his long-service-Regular descendant of A.D. 100 as there is between the part-time soldier of a New England trainband in King Philip's War (1675–76) and today's American paratrooper.

The Roman art of war was a homegrown product, owing little to Greece and less to Macedonia. Alexander the Great might be a semidivine hero to Romans (one of them, Flavius Arrianus, a soldier-administrator-military historian of Greek parentage, wrote the most complete account of Alexander's campaigns we have), but few Romans attempted to emulate Alexander.[7] One of the few who showed a touch of the Macedonian's genius was Gnaeus Pompeius, better known as Pompey the Great. Combining careful perparation with swiftness of concept and execution, between 83 and 61 B.C. he won an uninterrupted series of victories from Spain to the eastern end of the Black Sea. The most impressive was his elimination of the swarms of organized pirates who had literally seized command of the Mediterranean, cut Rome off from its overseas sources of grain, and even attacked the few Roman warships in their harbors. Given command against them in 67 B.C., he set up a coordinated offensive by sea and land across the whole length of the Mediterranean and utterly broke them in three months, resettling their survivors in inland towns. Moreover, he was an honorable man, the only general of his time to break with the Roman tradition of ceremonially murdering defeated enemy leaders. Unfortunately, he also was vainglorious and a political innocent, with neither Julius Caesar's cutthroat will to supreme power nor the vision to form his own plan to rescue the wobbling Roman Republic. Also, he has had no friendly biographer.

Certain characteristics marked the Roman art of war, even into Rome's last imperial years. The Roman was practical. He fought not for glory but to win; he might lose battles, but he would win the last battle—and the war. He was willing to learn from his enemies. His strategy was cautious, military force working together with an aggressive diplomacy; usually, his conquests went methodically, after careful preparation. (It was Julius Caesar's contrary style, his unpredictable, start-the-campaign-on-a-broken-shoestring recklessness that baffled other Roman generals.)

Where Alexander's way had been dash and drama, the surge of armored horsemen, and swift conquests, the Roman way was the steady, disciplined march of its legions—heavy infantry, who were as handy with their entrenching tools as with short sword and javelin, who could do any sort of combat engineer work, and built roads and canals and drained swamps in times of peace. The trace of those roads can be seen today; their nation endured for centuries. The average Roman officer had a career that alternated between military and civil offices. (George Washington, Andrew Jackson, and William Henry Harrison are examples of the early American application of that same rough practice.) This had the advantage of ensuring that some statesmen would understand military problems and some generals would have the diplomatic skills needed to deal with allies—highly desirable, since Rome apparently never had anything resembling a general staff or war department.[8] (Cynics have attributed its centuries of military successes to that fact!) Even so, the comparative shortage of keen professional generals tended to make Roman strategy slow and improvised. Each new enemy was a fresh problem of unknown potentialities; therefore, the average middle-aged Roman politician-general tended to go carefully until he was certain he understood the situation confronting him. Many apparent Roman wars of conquest actually were a reaction to enemy activity, if only the necessity to civilize wild mountain tribes that insisted on raiding fertile Roman-ruled lowlands. At the same time, the efficiency of veteran Roman armies was such that any reasonably competent general could handle the average enemy, and a fair number of complete blockheads suddenly became conquering heroes.

The Roman influence on future generations was large and generally beyond easy description. In specifically military matters, their historians were long read. Flavius Arrianus left works on cavalry training and his defeat of invading Alans in 134, as well as on Alexander. Publius Cornelius Tacitus (A.D. 55–110) wrote of Germany, operations in Britain, and general Roman history. Probably an aristocrat with some military service, Tacitus had a polished style and described military matters with a certain expertise but without much attention to geography and other vulgar details. He also made no effort to be either consistent or impartial and decorated his works with long and often improbable speeches that various commanders supposedly made before joining battle. Rome's outstanding historian was Polybius (ca. 205–125 B.C.), a Greek soldier-statesman who was taken to Rome as a hostage and there achieved fame as a writer. While not completely impartial, he used the best sources available and wrote competently from personal knowledge on geography and wars. His comparison of the Roman legion and the Macedonian phalanx is famous; Maurice de Saxe would later incorporate it into his *Reveries* in 1740. Another captive who passed into the Roman service was Flavius Josephus (ca. A.D. 37–100), a Jewish scholar whose *History of the Jewish War* gives a good picture of the Roman army that captured him.

Julius Caesar and Flavius Vegetius Renatus, the two Romans, who most influenced future strategists, were not true historians in the strict sense of that discipline, though each produced one seemingly authoritative work. Caesar ranks with history's "Great Captains," and Vegetius is only a casual name, yet it would be hard to determine which has been the more widely read.

Julius Caesar, "our bald whoremonger . . ." to his hard-used, hard-bitten veterans, grew up in a republic that was collapsing internally. A tall, dark-eyed man of great physical stamina and personal charm, deeply intelligent, he was a splendid orator and a slippery politician. Rome of the last century B.C. was not an environment to nourish the traditional virtues, but even when judged by its raunchy standards, Caesar emerges very much a solipsist, careless of any standard of good or evil. Supreme power was his basic desire, and he did whatever might seem necessary to acquire it.

Like most Roman aristocrats, he had a smattering of military service in his younger years and had read Xenophon's works and descriptions of Alexander's campaigns, but his actual generalship began only in 58 B.C. when, at the age of forty-two, he became governor of both Cisalpine and Transalpine Gaul (today, respectively, northernmost Italy and France). Eight years sufficed to conquer all Transalpine Gaul (with forays into England and Germany) in campaigns marked by swift maneuvers and coldly systematic atrocities. (An example was Uxellodunum, where an especially stubborn Gallic garrison finally surrendered on Caesar's promise their lives would be spared. Caesar kept his word—and cut off their hands.)

From 51 through 45 B.C. he fought Romans (the Civil War) from Spain to Asia Minor, emerging supreme ruler of Rome, proclaimed a living god over all men.[9] This time he pardoned and rewarded his enemies and began a rebuilding of the shattered Roman state. But his growing arrogance angered old enemies and friends alike. They assassinated him in early 44 B.C.

Caesar's legacy to future soldiers was his *Commentaries,* covering his Gallic War and most of the subsequent Civil War by which he won control of Rome. Probably these originated as a series of annual reports to the Roman Senate on his operations as governor of Gaul and included the reports he had received from his subordinate commanders. Written (or rewritten) in a starkly elegant style that has endured as a model of Latin literature, they actually were expert propaganda, aimed at influential elements of the Roman public. It may be assumed that they are fairly accurate—Caesar hardly could have written major lies about current battles and campaigns in which thousands of Romans (not all of them his friends) had just taken part. It is even more certain that they were given the "wax job" traditionally applied to reports to higher headquarters, with enemy numbers and casualties upgraded, embarrassing incidents discreetly minimized. Also, Caesar contrived to give his accounts a certain deceptive aspect of modesty by always referring to himself in the third person as "he" or "Caesar," as if he were reporting on the doings of another general.

The more experienced reader, however, eventually discovers that these *Commentaries* are little more than a bare-bones summary of events and not really military history. There is far too little information on logistics, tactics, and the terrain. Moreover, Caesar seldom bothered with explanations or reasons, leaving the military student little more than the description of his actions to consider. Because of this, Frederick the Great doubted that a soldier could learn anything from the *Commentaries*. Napoleon considered them worth study but complained that Caesar's battles "had no names"[10] and so could not be studied effectively for lack of clear geographic detail. The average reader, however, was swept along by Caesar's terse, persuasive story of mighty deeds and victories won out of desperate peril. And since this average reader had little or no acquaintance with other Roman historical sources from that period, he tended to accept Caesar's propaganda image of himself as the perfect general and wise, all-seeing leader.

Rarely, Caesar might include some observation on the importance of troop morale or the importance of chance or accident in war. "But fortune [chance] which influences many a thing . . . especially in war, produces the greatest changes with small impulses."[11] Attention to both might be expected from a commander who lived by his skill, wits, and ability to persuade other men to get killed "that Caesar might be first in Rome."[12]

In truth, Caesar remained a "bald-headed, dubious Roman rake-politician"[13] throughout his life, and the combination of those skills with his military ability made him formidable indeed. He could smoothly set various Gallic tribes at each others' throats—or arrange a conference with the leaders of two German tribes, seize them despite a declared truce, and then surprise and slaughter their leaderless folk. His preparations for the Civil War included bribing as many of his abler opponents as possible. (Reportedly, he also crossed the palms of certain slaves who were believed to have considerable influence over their masters.) At the same time he won over the surviving Gallic chieftains with calculated rewards and mercies so that they would remain quiescent while he waged the Civil War against his fellow Romans instead of seizing the opportunity to revolt.

As a military commander, Caesar made no real innovations in the art of war but became extremely expert in the employment of the existing Roman military organization, especially in exploiting the traditional Roman skill in the construction of field fortifications. He was a superb battle captain, able to grasp the critical place and moment of an engagement and to snatch victory out of defeat by personal bravery and example. And he managed to hold the loyalty of his troops through all hardships and dangers.

His strongest characteristics were speed, aggressiveness, and audacity. Believing that "the most potent thing in war is the unexpected,"[14] he hit first and—whenever possible—hit from an unlikely direction. At the outbreak of the Civil War in 49 B.C. Caesar had only one legion in Cisalpine Gaul and eight more north of the Alps in Gaul proper. His opponents (in general terms, the aristocrats of the Senate, with Pompey as their general) had seven legions in Spain and three in Italy. With such odds against him, the Senatorial party was certain that Caesar would pause until he could bring more legions out of Gaul; meanwhile, they could raise additional troops. Instead, Caesar struck immediately southward across the Rubicon.[15] The Senate went into panic; Pompey feared that two of his available legions, having previously served under Caesar, were unreliable. He therefore withdrew eastward across the Adriatic Sea into northern Greece and began concentrating a large army while his fleet held the Adriatic. Briefly pausing to establish his authority in Rome, Caesar plunged westward into Spain; in three months of able marches and maneuvers, he forced the surrender of the seven legions there, almost without fighting. Returning swiftly to Rome, he found the Adriatic raked by winter storms and all the seaports along its eastern shore held by Pompey's troops and ships. Again, his enemies thought he would not risk a crossing until the spring calms, but Caesar crossed anyway, eventually—after several defeats—routing Pompey at Pharsalus (48 B.C.).

On several occasions this audacity landed Caesar in deep trouble, requiring his full resources of courage, cunning, and plain dumb luck to escape—not to mention much unnecessary hard fighting and dying by his soldiers.[16] After he defeated Pompey, he seems to have assumed that his name and presence alone were worth several legions. A series of tight scrapes in Egypt immediately after Pompey's death there in 48 B.C. did not enlighten him. The next year he landed in North Africa with insufficient forces and came very close to being rubbed out or starved out before his lieutenants could get reinforcements and supplies to him. His emphasis on speed was not matched—at least so far as the *Commentaries'* scanty descriptions can inform us—with any logistic preplanning to make certain that his men would be fed. Frequently, they weren't. Finally, he could be irresponsible. In 48 B.C., with Rome in chaos behind him, the eastern Mediterranean seething with small wars, and the Senatorial party's adherents rallying successfully in North Africa and Spain, Caesar went on a two-month spree up the Nile with Cleopatra, his newly confirmed Queen of Egypt. This side trip undoubtedly was high-seasoned fun, but it also prolonged the war into late 45 B.C. He was assassinated a few months later as he was about to leave a still-unhealed Rome for a three-year campaign against the Parthian Empire (roughly, modern Iran) that promised to be a desperate affair. Upon his death, his tacked-together, one-man government collapsed into renewed civil war. It was 31 B.C. before his grandnephew and heir Octavius finally could end it.

Caesar's influence on European soldiers, however, was definitely second to that of an obscure late-Roman aristocrat, Flavius Vegetius Renatus, the author of one short, simply worded book, *De Re Militari* (*Concerning Military Affairs*).[17]

Vegetius remains thoroughly obscure. From his book it is obvious that he was neither a field soldier nor a historian. Probably he wrote it as a concerned citizen, troubled by the declining efficiency of the Imperial Army. (Just possibly, he may have been some high-level official who was told to work up a staff study on that problem.) Even the approximate date of his work is uncertain. He dedicated it fulsomely "To the Emperor Valentinian," but Rome had three emperors by that name. It now is generally agreed that he meant Valentinian II (A.D. 372–92), a hapless boy "Emperor of the West," murdered by his commander in chief before he had opportunity to show whatever qualities he possessed.

By land and sea Rome's frontiers were sagging under seemingly endless barbarian inroads, ranging from minor cattle lifting to major tribal invasions. The strain of these continuous wars had weakened the empire's economy; there was a shortage of suitable native-born recruits and an increasing reliance on barbarian mercenaries. Discipline and morale frequently were not what they had been in the brave days of old. Moreover, the need for mobile troops, capable of intercepting fast-moving raiders, had led Roman commanders to increase the strength of their cavalry and light infantry, arms in which native Romans had never shown particular interest or skill. The traditional Roman legions were neglected; the new units tended to get the pick of the officers and recruits.

To Vegetius the solution was a return to the military organization, training, and stern discipline of "the ancients" and the restoration of the old-style legions. He recounted the established Roman methods of raising, training, and disciplining new troops; the organization of the Roman armed forces; tactics and strategy; the art of besieging and defending fortified places; and naval operations. He patently mistook the reasons for the legion's decline, and his dredgings through earlier Roman records and writings produced something of a pedantic jumble of material from different periods. All the same, his work (the one treatise on Roman military institutions that has come down to us intact) was honestly meant and contains a great deal of useful information and common sense, for example, his dry observation that "the expense of keeping up good or bad troops is the same."[18] He did not pretend that the Roman armies of the earlier days were always successful and was willing to admit that the "ancients" were less skilled in the employment of cavalry.

His version of strategy involved the careful use of relatively small armies of highly trained professional soldiers against enemies who probably would outnumber them and might well possess greater physical strength and better weapons but would lack the Roman training, discipline, and organization. His generals would be prudent and alert, careful to obtain all possible information concerning the enemy and the countryside, to keep their troops well supplied and in good health, and to seize any opportunity. Their wars would be basically defensive, fought to preserve the Roman Empire rather than to achieve new conquests. Their troops would be comparatively few and must not be squandered in unnecessary battles. Risks were to be avoided: "It is much better to overcome the enemy by famine, surprise or terror than by [large battles], for in the latter instance [luck] has often a greater share than valor."[19] This same system reappears throughout history, as in the eighteenth century, whenever armies of long-service regular troops must be employed in wars fought for limited objectives.

It is unlikely that Rome's last great generals—men such as Flavius Stilicho and Flavius Aetius, waging desperate wars of maneuver against the unending barbarian invasions with whatever troops they could scrape together—profited much from Vegetius's recapitulation of Rome's past glories. But his book certainly survived. Medieval rulers, constantly and royally thwarted in their attempts to build reliable armies out of their disorderly feudal vassals, found his precepts useful. A French translation, retitled *The Art of Knighthood,* appeared around 1300; a considerable number of manuscript copies in French, English, and even Bulgarian, dating from the seventh to the fifteenth centuries, still exist. *De Re Militari* was among the first books printed, being published in Utrecht in 1473. William Caxton brought out an English version in 1498. In the late seventeenth century Count Raimund Montecuccoli (see chapter IV) mentioned Vegetius as being widely read by aspiring young officers; a hundred years later, the Austrian field marshal Prince Charles Joseph von Ligne called it a "golden" book, divinely inspired.[20] (One British book dealer of that same period *did* refuse to reprint Vegetius, fearing that he would lose money because "very few of the British officers perused professional books, except a few of the Artillery Corps."[21])

Julius Caesar—and to a lesser degree, Vegetius—would be of service, however obliquely, in the establishment of the young United States. In eighteenth-century society a sound knowledge of Latin was the hallmark of educated, cultured gentlemen; its mastery was inculcated in their sons—if necessary, with the assistance of posterior corporeal stimulation. Undoubtedly, most Latin instructors were thoroughly unmilitary pedagogues, but Caesar's *Commentaries on the Gallic Wars* was considered an excellent beginners' text because of the purity of its style and grammar. Most young Americans probably found it sheer drudgery, but to some—as to this author—their first stumbling translation was the opening of a mighty gate into a new kingdom. So it must have been to young Henry Knox, reading through the stock of his Boston bookstore, never imagining that he would one day become Major General Knox, Chief of Artillery in Gen. George Washington's Continental Army, and then President Washington's secretary of war. And such it certainly was to restless Anthony Wayne, who took little interest in his schooling until introduced to Caesar. Then, as his uncle-schoolmaster wailed, "He has already distracted the brains of two-thirds of the boys . . . by rehearsals of battles, sieges, etc."[22]

Today Vegetius is practically forgotten. And we learn that Caesar must be avoided in teaching Latin because his "military tyranny is . . . infelicitous, intolerable, and deleterious."[23] Readings from poets, orators, and philosophers are substituted by well-meaning people to whom it would never occur that Virgil and Ovid are known today *only* because various bloody-minded proconsuls and their sweaty, hobnailed, tough-barked centurions built and kept the Roman state, thereby making Latin the universal language of the Western world. Their thumbprints are on the foundation stones of our republic.

Knox, Wayne, and officers like them had much to do with the winning of our Revolutionary War. But the real testing of Knox's and Wayne's Roman indoctrination came in the desperate days of 1792–95, with the new, half-hatched United States caught between a revenge-minded England and an aggressive Spain, both intent on penning it east of the Appalachian Mountains. Their active agents were the great Indian tribes of the Ohio and Mississippi valleys, skilled warriors compared to whom the Sioux and Comanche were amateur show-offs. When, in late 1791, the tribes north of the Ohio (considerably aided by British advisers) capped a series of victories by destroying half the tiny Regular Army and much of its attendant militia and volunteers, President Washington had considerable difficulty persuading Congress to support one more effort to hold our western territories.

It was then that the Legion of the United States, organized by Knox after the Roman model, went into the wilderness under Wayne's (now "Mad Anthony's") command. Vegetius would have applauded the care with which Wayne drilled, disciplined, and toughened his apprehensive recruits into competent woods fighters; Caesar, the campaign in which he led it to victory at Fallen Timbers, winning the United States unchallenged title to the Old Northwest.

Anthony Wayne died in 1796, aged fifty-one, while inspecting the little frontier fort at Presque Isle, New York. At his request, his soldiers buried him in full uniform at the foot of the fort's flag pole. "*Qui procul hinc, qui ante diem periit, sel miles, sed pro patria.*"

III

KNIGHTS, PEDANTS, VIRGIN QUEENS, SEA DOGS AND CAPTAINS COURAGEOUS

> . . . he that entereth the enemy's country without purpose to fight and hazard, let him henceforth keep his head warm at home and entertain ladies.
>
> MATTHEW SUTCLIFFE

Rome had fallen in the West, and there had been dark centuries of invasions and tumult, with no man's head safer than his own wits and right arm could keep it. Out of it, strong kings raised their banners, and the nations of the West took shape.

In the East, however, Rome still existed where the Greek-speaking Byzantine Empire kept high and magnificent state about its capital of Constantinople. It ruled much of the Balkans and Asia Minor until the late eleventh century and would drag out a cliff-hanger existence until 1453, when the Turks finally stormed Constantinople. It had able generals and emperors—men such as Basil II, ''Destroyer of the Bulgarians.'' (In 1014 Basil captured 14,000 Bulgarians and blinded ninety-nine out of every hundred of them—the hundredth man was left one eye to guide the others home. Their appearance literally broke their king's heart.) In its great days their army, based on armored mounted archers and lancers, was trained and organized on a practically modern basis, from companies to divisions. Every ''regiment'' had a medical detachment; there were specialized engineer and service units and a country-wide signal system.

The Byzantine rulers did not consider war a thing apart. Their realm was under pressure from all types of enemies—Lombards and Normans in the west; Bulgars, Slavs, Patzinaks, Magyars, and Russians from the north; Persians and Arabs from the east and south. Time after time, only hard and skillful fighting saved it from bloody extinction. The Byzantine commanders studied their enemies for their strengths and weaknesses and chose their strategy and tactics accordingly. Byzantine warfare was methodical and highly professional. Their troops were expensive to raise, equip, and pay; in contrast, most of their enemies were barbarian or semibarbarian hordes. Consequently, Byzantine commanders tended to fight only when circumstances favored them. Their training stressed maneuver, ambushes, faked retreats, deceitful negotiations, bribery, and stratagems of all descriptions.

Three works give an excellent account of their armed forces and military policy. Maurice (emperor, A.D. 582–602) seems to have written his *Strategicon* while he was still a general involved in reorganizing the Byzantine army. It has echoes of Vegetius and is a remarkably complete description of that army, including training, organization, supply, marches, camps, and tactics. Leo VI, ''the Wise'' (emperor, A.D. 886–912), was no soldier but an excellent ruler. His *Tactica* covered much the same subjects as the *Strategicon* but he improved Byzantine frontier defenses by organizing them in depth and increasing the mobility of the border garrisons. Strategy he rather passed over, his policy being basically defensive, but he did specify strategic counteroffensives: if the Arabs invaded Asia Minor by land, the Byzantine navy would raid their coastline, which probably had been left weakly defended; if they attacked by sea, the Byzantine army would advance overland. Leo's work is especially interesting in its analyses of Byzantium's various enemies: cold, wet weather soon took the fighting edge off Asiatic troops; western Europeans were extremely dangerous but lacked organization and were headlong and easy to lure into ambush. Finally, Constantine VII (emperor, A.D. 944–59) left his son a highly classified review (now titled *De administrando imperio*) of Byzantine's general situation, with much excellent advice as to how its various neighbors must be handled and how force and diplomacy could be applied by turns to wear away the strongest resistance.

Little of Byzantine military skill penetrated western Europe during the Middle Ages. The old imperial army, hamstrung by the imperial bureaucracy, had been destroyed by the Turks at Manzikert in 1071. The Crusaders saw only the wreckage of it. But the revival of interest in classical history during the sixteenth century brought Byzantine military doctrine, especially the *Tactica,* to the attention of several Dutch and English soldiers. Meanwhile, western Europe developed its own strategists.

Robert the Bruce, Earl of Carrick and one of the great nobles of the realm of Scotland, was crowned its king on March 27, 1306 at Scone Abbey, the traditional site of Scots coronations. The ceremony necessarily was somewhat improvised: the famous "Stone of Destiny"—by pious legend, the veritable pillow of the patriarch Jacob, miraculously floated from the Holy Land—upon which previous Scottish kings had sat to be crowned, had been carried off by Edward I of England ten years earlier. For Edward I, having battered the Welsh into submission, had set about doing likewise to the Scots. English garrisons held Scotland's important towns and castles; most Scots nobles, including Bruce, had pledged themselves his loyal vassals.

Edward I ("Longshanks") of England was a skilled warrior-king and an ill man to cross. He had begun his Scottish wars with a general massacre at Berwick when that seaport finally fell after a stubborn defense. Those Scots who rose against him in 1297 under William Wallace—a simple knight's son, fearless, stormy, and wry-humored—were hunted and harried. Wallace, finally betrayed and captured in 1305, was charged with treason and judicially murdered, though he never had sworn fealty to Edward—"hanged, cut down while still living, disemboweled, and then beheaded."[1] His heart and entrails were burned, his head placed on London Bridge, his body quartered and its parts displayed at Newcastle-upon-Tyne, Berwick, Stirling, and Perth. Bruce could expect no kinder fate if he lost.

Lose he did. His hastily raised army was surprised and routed. Almost alone, Bruce fled through winter's storms to refuge somewhere in the outer islands—Arran, the Hebrides, or the Orkneys. Edward used his captured followers and family cruelly. But the next spring Bruce was back, evading the tangle of English garrisons and personal enemies. (He had cleared his way to his coronation by putting a dagger into his major rival, Sir John "the Red" Comyn, during an angry parley before the high altar of Dumfries Church: The Comyns hoped to reciprocate.) Edward came grimly north again, but he was too old for Scottish wars. Wrenched by dysentery, he died while still on English soil, cheered—so tradition has it—by smoke from a burning Scottish village on the northern horizon. His son, Edward II, indolent, stubborn, and uninspired, was happy to leave the task of governing to his favorites but was unwise in his choice of them. His nobility was rebellious and uninterested in further Scottish adventures. For money, he borrowed heavily from Italian merchants, who in return got control of the English wool trade to the disgust of the Hanseatic and Flemish merchants who had handled it previously. Meanwhile, Bruce made himself King of Scotland, in fact as in title.

He was a king out of the ancient tradition of Arthur and Charlemagne—a knight of "strange, outrageous courage"[2] and deadly skill and hardihood in man-to-man combat against any odds. Behind these were intelligence, vigilance, and a strength and charm of character that drew able, willful men to serve him faithfully.

The strategic situation he faced was daunting: England was his next-door neighbor, infinitely richer and more powerful than Scotland, able to equip and field far larger armies. Constant wars in France and Wales had given the English forces a semiprofessional organization and expertise. With money from Italian loans and their improved system of national taxation, English kings could hire troops to supplement their feudal levies, which would serve only for the requisite forty days each year. Consequently, they could establish permanent garrisons in conquered areas and wage punitive campaigns through the winters. Add that Scotland was poor, with few industries. Weapons and armor had to be imported. With war smoldering across her fertile southern areas, food and even clothing might be scarce.

Bruce could count on few advantages beyond the loyalty of his commoners and minor gentry—hard men in a harsh land, much of which was still little known to Englishmen. He also had the backing of the Scottish clergy—a stiff-necked lot who wanted neither English bishops nor the Pope himself interfering with the Church of Scotland. The Pope might heap excommunications upon Bruce's head for his rebellion against Edward II, "our dearest son in Christ." Scots clergymen responded by preaching that killing Englishmen was as pleasing to God as going on crusade to free the Holy Land from Paynim rule.

Militarily, the odds against Bruce were heavy indeed. English nobility and gentry could afford complete armor and great warhorses; few Scots except their greater nobles could, and the Scottish nobility was the least reliable element among their kingdom's defenders, many of them maintaining open or covert English loyalties. Scottish archers were few and seldom as skilled as those from South Wales and the English shires who could "loose" five aimed arrows a minute and hit their man at well over two hundred yards. Scottish infantry, massed in dense "schiltrons" (blocks) behind their twelve-foot spears, might beat off charges of English heavy cavalry, but as Wallace's defeat at Falkirk (1298) showed, they were helpless targets for English archery which riddled their schiltrons, opening gaps through which English knights charged home, slashing down the unarmored Scots.

To risk another such pitched battle was obviously stupid. Bruce adopted a cautious strategy, striking at isolated British garrisons and detachments while simultaneously putting down the Comyns and other hostile Scots with a hard hand. When the English moved against him in strength, his troops faded back into their hills, woods, and mosses—to regroup swiftly and strike elsewhere. Lacking siege engines and the skilled men to operate them, Bruce reduced the English strongpoints by stratagem, starvation, or surprise. Thomas Randolph, Bruce's nephew and Earl of Moray, took Edinburgh Castle by a night escalade up the face of the cliff on which it stands, the garrison being distracted by a faked attack on its east gate. Bruce himself seized Perth in another night attack, leading the way through its moat in icy water up to his chin and mounting the wall before the bewildered English realized they were under attack. Another of Bruce's lieutenants, "The Black [James] Douglas," made himself so dreaded that English mothers quieted disobedient children with the threat that he would come for them. Rather than weaken his small army to provide garrisons for recaptured castles, Bruce dismantled many of them, thus also preventing their future use by his opponents. At the same time, he was consistently merciful to his prisoners, making it easier for trapped Englishmen to surrender with honor.

Strategy, however, needs more than sheer hard fighting. Organizing and ruling Scotland from his saddle as he liberated it, Bruce reached out for foreign aid. Food and some weapons could be secured in northern Ireland; to ensure their delivery, he improvised a navy of sorts (probably half piratical) from his western coasts and islands. Flemish and Hanseatic merchants happily brought better arms and armor and good clothing, bartering them for Scotland's wool, hides, deerskins, and timber. It must have been a profitable trade—English merchants were willing to turn smuggler and join in. Then, his forces larger, more experienced, and better equipped, he turned to "defending himself with the longest stick he had."[3] Keeping the few remaining English strongholds in Scotland under siege, he loosed raiding parties across northern England. These were increasingly well organized affairs, veteran troops mounted on small, hardy horses that could go as far as seventy miles a day. These raiders carried no supplies except a bag of oatmeal under one saddle flap and a small iron griddle under the other; otherwise, they lived off the country, boiling the meat of captured cattle in improvised cowhide caldrons. Too mobile to intercept, they generally avoided battle, concentrating on peeling the countryside of livestock. Bruce's navy raided English shipping down the Irish Sea to the coast of Wales. The Isle of Man was taken and retaken. An attempted English counterinvasion in 1309 reached Berwick but stalled there, unable to find food.

For Bruce had developed a unique strategy, which later would be put into rough verse as "Good King Robert's Testiment."

> On fut suld be all Scottis weire,
> (Scots should wage their wars as infantry)
> By hyll and mosse themselffs to reare.
> (Fighting from positions in swamps and hills)
> Lat woods for wallis be bow and speire,
> (Taking refuge in forests, rather than in castles)
> That innymeis do them na deire.

(So that their enemies may do them no great harm)
In strait placis gar keip all store,
(Keep your supplies securely cached in difficult country)
And byrnen ye planeland thaim before.
(And burn off the open country before an enemy's
advance)
Thane sall thai pass away in haist
(Then they will pass away in haste)
When that thai find na thing but waist.
(When they find the country wasted, and no food available)
With wyles and waykings of the nyght
(With night alarms and attacks)
And mekill noyis maid on hytht,
(And loud noises made from nearby hills)
Thaim sall ye turnen with gret affrai,
(You shall drive them off in as great affright)
As thai ware chassit with swerd away.
(As if you defeated them in pitched battle)
This is the consall and intent
Of gud King Robert's testiment.[4]

The system worked splendidly. The constant Scottish raids, driving ever deeper, made northern England a hungry place. When English armies did get into Scotland, the Scots methodically devastated their own farms; noncombatants and cattle were moved off deep into the hills. Empty-bellied Englishmen pitched their camps—frequently in the rain—amid gnat-swarms of Scots who knew the ground thoroughly. Some sat on the nearby hills all night, blowing horns (and possibly bagpipes), which at least made sleep difficult and might provide "noise camouflage" for horse lifting, sentry gutting, and (as the English wearied) large-scale night attacks.

This strategy was tested repeatedly. In 1312 Scots raiders temporarily occupied Durham and Hartlepool, exacting stiff ransoms. By 1314, England held only Berwick, Dunbar, and Stirling in Scotland, and Stirling was on the point of surrender. There was distress throughout the north of England. Edward II and his nobles shook themselves into temporary accord and mustered a splendid host from England, Wales, Gascony, and Ireland. They came north through empty country to finally find Bruce awaiting them in the woods south of Stirling near the village of Bannockburn, his front covered by obstacles, streams, and marshes. Early on June 24, Bruce caught the English just as they were forming up, his massed schiltrons—dismounted knights shoulder to shoulder with commoner spearmen—sweeping downhill against Edward's milling horsemen, his handful of cavalry outflanking and riding down the English archers. It was a red rout, a notable killing of proud barons such as Englishmen seldom have known. Survivors scurried home, pursued so closely by embittered border farmers they had no chance to "make water" safely.[5] Edward left the field somewhat early, losing all his royal baggage.

Yet Bannockburn merely intensified Edward's normal bullheadedness. With a great deal more resolution than is credited him in popular histories, he kept trying to reconquer Scotland—to the eventually fatal neglect of his domestic problems. In 1315 there was crop failure throughout the north from too much rain; Edward had to call off operations because "bread could scarcely be found for the sustenance of his family."[6] In 1318 an English army mustered around York, but its commanders quarreled, and nothing was done. Meanwhile, the Scots continued raiding; the battered northern counties paid bail and promised the raiders free passage. Some English borderers, "despairing of protection from their own King . . . became companions and guides of their incursions into England, and sharers with them of the spoils.[7] Bruce's power even spread briefly into Ireland, where his younger brother Edward led a successful revolt against English rule and was crowned king in 1316, only to be killed in battle two years later.

In 1318 Edward II finally got another strong army into the field and made a determined effort to recover the seaport of Berwick, which would give him a forward base for future operations—supplies and reinforcements could be brought there from southern England by ship, thus avoiding the long march through the ravaged northern counties. Much helped by one John Crab, a Flemish pirate and expert military engineer, the Scottish garrison beat off his attacks by land and sea but slowly weakened as their rations dwindled. Unable to raise enough troops to challenge Edward's army, Bruce sent Douglas and Moray into England; their raid carried clear to York, cutting Edward's communications, routing the Yorkshire levies at Myton and raising howls for help. Edward countermarched vainly, the Scots easily dodging his weary columns. A three-year truce followed. At its ending, the Scots once more "shook loose" the border. Edward had used the respite to build up another major offensive.

Once again he found northern England swept bare, the Scottish fields burned off before him. His foragers came back empty-handed—when they came back at all. One decrepit old bull, too lame to be driven off, was all they found. Supply wagons emptied, dysentery ate through the English. Heckled and starving though his host was, Edward pushed on to Edinburgh, but three days before its walls was as much as his troops could endure. The retreat was a disaster, with Bruce in close pursuit. At Byland, north of York, he surprised the English camp with a dawn attack on its front and flanks. Edward barely escaped to York, again losing all his baggage, while Scottish raiders raked the countryside far to the south and east. Along the borderland only the fortress city of Carlisle held out against repeated Scottish forays.

This defeat encouraged revolt in England; Edward's estranged wife, Isabella, the "she-wolf of France," joined his enemies. He was captured, imprisoned, and murdered in 1327. That same year, while Bruce was carrying out a diversion in northern Ireland to secure grain, Douglas and Moray continued with their annual ravaging of northern England. By way of asserting its authority, the new English government—in theory, fifteen-year-old Edward III; in fact, Isabella and her lover, Roger Mortimer, Baron of Wigmore—determined to destroy them. Concentrating a powerful army, including 2,500 expensive German men-at-arms, Edward III took the field. It was cold, wet weather; food soon ran low, even on English soil; the Scots flitted just beyond reach, making the nights clamorous with their horns. Worn by hunger and exposure, difficult marches and fireless camps, horses dropping from exhaustion and short feed the English expedition bogged down—and Douglas struck suddenly out of the night. His thrust was shrewdly directed at the king's pavilion; Edward escaped only by the devotion of some attendants, who stood to be killed to cover his flight. And the year went out in a pelting of more Scottish raids. Baffled, the British made peace in 1328, acknowledging Scotland's independence and Bruce's kingship. Scotland had been enriched by ransoms and plunder, and her fighting men appeared invincible.

Bruce died in 1329, leaving his throne to his five-year-old son, David II; Douglas, on pilgrimage to the Holy Land, was killed in Spain the next year; Moray died in 1332. That same year war broke out again, and it quickly became obvious that few Scottish leaders had taken "Gud King Robert's testiment" seriously. Proud from years of victory, they attempted to beat the English in open battle and were defeated, repeatedly and thoroughly. Again and again Scotland seemed almost conquered, but the innate toughness of its folk somehow always won out. There never was another Bruce, but there were always leaders who moved swiftly and attacked unexpectedly out of the night and waste until the English yet again ebbed back across the border.

For some months in 1650, during the unncessary war between Scotland and the English Commonwealth, the Scots commander, David Leslie, applied Bruce's strategy brilliantly to block Oliver Cromwell's drive on Edinburgh, while foul weather, short rations, and dysentery riddled Cromwell's army. But Leslie maneuvered with newly organized troops and the clammy hand of the Scottish clergy and politicians on his neck. (Set on making war after the precepts of the Old Testament, these fanatics had been quick to purge Leslie's forces of any officer who was not an abiding Presbyterian, replacing them with "ministers' sons, clerks, and such other sanctified creatures."[8] Leslie finally moved out to offer battle, to be caught half-ready by Cromwell's veterans and routed.

It is quite possible that Bruce's example influenced another forgotten medieval strategist, Charles V of France. There were Scots enough in the French service who could have explained Bruce's strategy to the king. Coming to the throne in 1364, Charles was faced with desperate circumstances: Edward III controlled roughly a third of France; the rest was afflicted by *routiers* (out-of-work mercenaries) who plundered almost unchecked. Edward had much improved his grandfather's military system, utilizing masses of archers and men-at-arms who fought dismounted "in the Scottish fashion."[9] This combination, standing on the defensive, was certain disaster to any enemy foolish enough to attack head-on, as the French *noblesse* repeatedly insisted on doing. Courageous enough but too sickly to command his battered armies in person, Charles was also determined and longheaded. To him it was plain that the English were overextended, that many of their French subjects were resentful and ripe for revolt. He managed to divert large numbers of *routiers* to wars in Spain, where their casualties were satisfactorily heavy. Fortifications of French-held towns and castles were strengthened; French commanders were ordered to avoid pitched battles. As his Constable (commander in chief) Charles selected a Breton captain of free lances, Bertrand Du Guesclin, who was no great general but brave, energetic, and willing to obey orders. Du Guesclin had the knack of irregular warfare, wearing the English down by small-scale surprises, ambuscades, and raids, while Charles pushed a shadowy campaign of subversion among the lesser nobles of the English-ruled provinces. At the same time, with help from his Spanish allies, Charles created a navy that raided the English Channel and its seaports and also made shipping between England and Bordeaux, the major English-held seaport in southeastern France, extremely risky. Finding the French unwilling to fight large-scale battles, the English attempted to force them to action by a series of great plundering raids across France—one of these in 1373 left Calais in early July and ended fagged out in Bordeaux on Christmas. Such raids inflicted great damage but achieved nothing. The newly strengthened fortresses held out against them, while cold, hunger, disease, and minor actions caused increasing English losses. At the last the war ran down from mutual exhaustion, but the English had lost approximately three-fourths of their holdings in France.

In fact, Bruce's strategy has seldom been copied. One reason, undoubtedly, is that devastating your own territory to deny an enemy food and shelter demands a determined patriotism from all concerned. Contrary to popular fable, the Russians did not use this strategy either in 1812 or 1941. In 1810 the Duke of Wellington imposed it on Portugal during his withdrawal into the Lines of Torres Verdes (a fortified position around Lisbon) before a stronger French army, but since it was his Portuguese allies who did the suffering, the Duke's case-hardened heart was not troubled. A better example of strategy similar to Bruce's is Gen. Philip Schuyler's devastation of the upper Hudson Valley in 1777, destroying or removing crops and cattle, breaking bridges, and ruining roads to slow the advance of Gen. John Burgoyne's invading English army while striking successfully at Burgoyne's communications and detachments. In both cases, their strategy was highly successful—the enemy armies were worn down by hunger and irregular warfare—but there is no evidence that either Schuyler or the duke ever studied Bruce's campaigns!

Bruce was a king and undoubtedly a superstrategist. His story is one long lesson in the uses of *surprise,* the *offensive,* and *maneuver* to exhaust and crush stronger enemies. Yet his fame is minor compared to that of an Italian intellectual who wrote on war and kingship without having experienced either one.

Niccolò Machiavelli (1469–1527) began in obscurity; he may have come from a poverty-pinched noble family (Italy was full of them) and certainly must have had a good classical education. He comes into our knowledge as an official of the Florentine Republic, a smart, imaginative young man who could handle all sorts of administrative and diplomatic errands.

Florence was one of the more powerful of the assortment of duchies, republics, and kingdoms that jostled each other throughout Italy. In 1494 it had expelled the Medici family, which had ruled it for several generations, and finally had managed to establish a republican-type government, precariously balanced between its aristocrat and burgher factions. Italy as a whole was prosperous and the fountainhead of European art and scholarship, but it was not exactly peaceful, its small states being continually engaged in small bickerings. These, however, were waged by the *condottieri,* mercenary leaders who peddled their services to any principality that might require them, and had little impact on the average Italian. Machiavelli raged against the *condottieri* in his writings as unreliable and incompetent but obviously—probably because of his lack of practical experience as a soldier—failed to understand the rationale of their way of making war. They insisted on being paid. Once paid, they generally served faithfully; many of them were skilled practitioners of a sort of trade-union war game, the object of which was to win without any avoidable discomfort, exertion, or bloodshed.

Into this military Utopia in 1494 came cloudy-witted Charles VIII of France, with a vague claim on the sovereignty of Naples and Constantinople and the dream of leading Christendom on a new crusade to redeem the Holy Land. He also had his Scottish guardsmen, French *gendarmerie* (regular heavy cavalry), Swiss pikemen, and newfangled mobile artillery. The *condottieri* and Italy's fortresses went down before them in an almost-comic-opera pratfall—these northern barbarians intended to win in a hurry and considered their enemies completely expendable.

The rest of Europe chose not to share Charles's vision, and so began some sixty years of Italian Wars, in which Italy was trampled down by every possible species of foreign invader, including Barbary pirates. The tangible results were the failure of the French attempt to conquer Italy and the accelerated spread of syphilis across Europe as poxy survivors of many armies straggled homeward. Florence's rule had been harsh to its subject towns and territories; the proud old seaport of Pisa took advantage of the confusion following the French invasion to reassert its independence. The best *condottieri* Florence could hire were baffled by the Pisans' defense—chiefly, it would seem, because they knocked off work during the winter months, allowing the Pisans to rebuild their fortifications and lay in supplies. At last, in 1506, some Florentine—quite possibly Machiavelli himself—suggested raising a militia among the inhabitants of Florence's rural areas to supplement the *condottieri*. Machiavelli drew up the necessary regulations, supervised recruiting, selected officers, and handled supplies. Some two thousand militiamen finally were added to the besieging force, the siege was kept up through the winter, and Pisa was starved into submission in 1509.

This success, plus the fact that the Medici were planning a comeback with Spanish support, led to the formation of a militia in Florence itself in 1511. However, its organization—designed to prevent any of its officers from staging a coup d'etat—must have made effective operations almost impossible. No captain could be elected from the district in which his company was raised, and once a year all captains would be shifted to other companies. The whole setup was to be supervised by several "councils," the members of which could serve for one year only. When the Spanish veterans arrived in 1512, there was a quick stampede and an unnecessarily enthusiastic slaughter.

With the Medici back in power, Machiavelli's public career was over. After days of farm work he turned to study and writing, producing *The Prince* (1513), *The Discourses* (1519), *The Art of War* (1520), and *The History of Florence* (1525). All contain his views on war and armies, but *The Prince* remains the best known of them. It was written in a barefaced but unsuccessful attempt to curry favor with Lorenzo Medici, to whom Machiavelli dedicated it. In it, Machiavelli dealt with practical politics in much the same fashion that Clausewitz would later write on war. His screed held that politics had its own morality; power was the decisive factor, and the wise ruler used any means to gain and keep it. He must be flexible and prepared to employ either magnanimity or cruelty as the situation might require. The only rule was that the method he chose must be successful. But it was safer to be feared than to be loved, and a wise prince would never keep his pledged word when such honesty might be to his disadvantage. As a model, Machiavelli presented that gaudy sixteenth-century *mafioso*, Caesar Borgia (1476–1507), who had built up an ephemeral state in central Italy. (Machiavelli completely missed the fact that Caesar succeeded only because he was the beloved bastard son of that old rip Pope Alexander VI and that he was a poor soldier, even if sufficiently cruel and devious for Machiavelli's taste. Once the Pope died, Caesar was a lost dog.) For military service, Machiavelli urged "his" prince to rely on native-born troops, chiefly infantry, raised by a type of conscription. And the prince should unite with his fellow Italian rulers to expel the "barbarians" from Italy.

The Art of War was directed primarily at his fellow Italians, hoping to arouse a new national spirit among them. His description of the general nature of war, here and in his *Discourses,* has omens of Clausewitz and Mao Tse-tung:

> Many are now of the opinion that no two things are more discordant and incongruous than a civil and military life. But if we consider the nature of government, we shall find a very strict and intimate relation between the two conditions; and that they are not only compatible and consistent with each other, but necessarily connected and united together. . . . Where the very safety of the counry depends upon the resolution to be taken, no consideration of justice or injustice, humanity or cruelty, nor of glory or of shame, should be allowed to prevail. . . . That cannot be called war where men do not kill each other, cities are not sacked, nor territories laid waste.[10]

He applauded one *condottiere* who utilized fraud in preference to force. But beyond these general proclamations, Machiavelli's military ideas are flimsy, the best of them a repetition of Vegetius's insistence on training and discipline. His concept of tactics (and of strategy insofar as he comprehended it) would have pleased the most conservative *condottieri*: battles are to be avoided unless all conditions are in your favor or the situation leaves you no alternative.

There is much attention to retreats, evasive maneuvers, entrenched camps, and stratagems but very little on generalship. He approved highly of Fabius Cunctator.

The army Machiavelli proposed was a re-creation of the Roman legion—mostly infantry, half of whom would be armed only with swords and bucklers, one-sixth with firearms or crossbows, the rest with pikes. There would be a few cavalrymen for scouting and raiding and a few pieces of artillery. (Machiavelli considered artillery useless against brave troops.) The soldiers were to be Italians, raised by conscription. This proposed organization ignored every lesson so far taught by actual warfare. His legion would have been run over and massacred by any of the forces then in Italy, but he closed his book with an enthusiastic account of an imaginary battle in which they were victorious.

Since his death, Machiavelli has been seen in many different lights—an Italian patriot, an unscrupulous opportunist, a mere pedant. At least one academic historian has proclaimed him "the first military thinker in modern Europe" and advanced the somewhat shattering claim that "military thought has proceeded ever since on the foundations which Machiavelli laid."[11] Both *The Prince* and *The Art of War* were widely read throughout Europe during the sixteenth century. A good many Protestants seem to have detested the former: Francois La Noue, a formidable Huguenot cavalry officer, noted that reading Machiavelli "saps all fundamental ideas of honor and justice."[12] Matthew Sutcliffe, Judge Advocate General of the British forces in the Low Countries, considered that setting up "that impious Atheist Caesar Borgia for a pattern . . . was detested even of the barbarous nations."[13] As an interesting comparison, Clausewitz, who dealt with theories of the nature of war, thought that Machiavelli "had a very sound judgment in military matters";[14] Napoleon, who dealt with warfare's realities, snorted that "Machiavelli wrote of artillery as a blind man [would] of colors."[15] His fellow Italian, Montecuccolli, studied Machiavelli approvingly, as did Laurent de Gouvion St. Cyr, the odd man out among Napoleon's marshals, who made war like a chess player. By the nineteenth century interest in Machiavelli was fading, though an American translation of *The Art of War,* "to which is Added Hints Relative to Warfare by a Gentleman of the State of New York," was published in Albany in 1815.

Machiavelli was a man of his own time, with a low opinion of his fellow man. He sought to write realistically of the actual facts of life in sixteenth-century Italy—frequently a messy, dangerous place, full of disrespectful foreigners whom he longed hopelessly to chastise and drive forth. He had had great expectations and had been cast into obscurity while still young and active. We know practically nothing of him as an individual, but his works suggest a cynical would-be operator with a well-trained conscience and no religion. His descriptions of the nature of war had value, especially for his most unmilitary countrymen, but held nothing new for the barbarians across the Alps to whom warfare certainly was "connected and united together" with their daily life. As for his views on armies, weapons, and warfare, Humphry Barwick, veteran captain to Queen Elizabeth, gave a flat summation: Machiavelli never had been a soldier—and never had the least desire to acquire the experience of being one.[16]

The Italian Wars ended, and Frenchmen turned viciously against one another in their Wars of Religion (1562–98), a Catholic-Huguenot (Protestant) struggle that saw no end of bitter fighting and treachery but very little effective strategy. The best fighting general on either side undoubtedly was Henry of Navarre (later King Henry IV of France), the Huguenot commander, who—even when king—led his cavalry in person. Hear him at the crucial battle of Ivry (1590): "Comrades, God is with us; there are his enemies and ours; here is your King—have at them! If you miss my pennon, rally round my white plume."[17] (Henry's pennon bearer was killed almost at once, and that white plume was the point of the slashing, pistoling wedge of cuirassiers that split the enemy's line.) After Ivry he could have moved immediately on Paris, then held by the extremist pro-Spanish Catholic League; instead, he sat for two weeks, allowing the Parisians to get over their panic, and then marched southward to besiege a minor town. Three years earlier, Henry had celebrated an earlier victory at Coutras, somewhat like Julius Caesar in Egypt, letting the war slide while he rode off to present twenty-two captured flags to his current mistress. This apparent lack of strategic sense was much bemoaned by his followers, yet we cannot be sure of his seemingly trifling motives. He *did* enjoy the hack-and-scramble of a cavalry charge; the Duke of Parma (see page 49) considered him nothing more than a captain of light horse. Yet he took his kingship seriously; his Huguenots were only a militant minority in France, and he could not hope to reunite that nation by military conquest alone. Unlike too many commanders in these wars, he was merciful in victory, a difficult task when most Huguenots remembered butcherings of families and friends. After their victory at Coutras the cornered Catholic commander, who hanged Huguenot prisoners, cried that he would pay 100,000 ecus (possibly that many dollars today) ransom for his own life. He was instantly shot down. It is quite possible that much of Henry's apparent dilatoriness came from reluctance to drive moderate Catholics to desperation, which meant into the service of Philip II of Spain, paymaster and protector of the Catholic League.

Western Europe was twisted by a major power struggle, a head-on collision of national strategies, as Philip's program of subversion and conquest met the French and Dutch determination to survive as independent states and England's newly conceived dream of free adventure across the high seas. These wars produced few great strategists but many expert generals and lessons in plenty on the art of war.

Philip was engaged in restoring the Catholic faith throughout Europe. This would also include the imposition of Spanish rule, direct or indirect, so far as it might be stretched. A ruler of method and conscience, self-possessed, stubborn, if dilatory, Philip intended to bring all his empire directly under his thumb, wiping out all local freedoms and customs that opposed his centralizing directives. He could equally provide the men and ships that checked the westward Turkish advance at Lepanto (1571), shake half Spain apart to wipe out the last pockets of Moorish culture, and be an affectionate father. Above all, he would not be a ruler of heretics, and he loosed the Inquisition against Jews and Protestants everywhere he ruled, chillingly refusing them even the right of private worship according to their faiths. Philip was utterly sincere in his belief that all this was "what a Christian prince fearing God ought to do in His service, [and for] the preservation of the Catholic faith and the honor of the [Papacy]."[18] To accomplish his ends, he possessed the wealth of the Americas and the eastern Indies, the backing of the Pope and the whole apparatus of the Catholic church (especially its shock troops, the Jesuits), and the military resources of Spain, Portugal, Naples, Sicily, Lombardy, Burgundy, and Flanders. His agents filtered throughout Europe, spying, bribing, suborning, propagandizing, proselytizing, murdering on occasion, with the certain feeling that they did God's own work and a versatility that would make the KGB envious.

Philip's problem was threefold: first, Henry of Navarre must not rule a reunited France; Philip wanted a bobtailed, dependent, all-Catholic France with either one of his French Quislings, the Guise brothers, or his own half-French daughter as its ruler. Second, his rebellious Netherlands provinces must be restored to unquestioning obedience and Catholicism. And finally, the infuriating English nation must be brought to heel. Philip had sacrified himself once to that end, marrying homely Queen Mary I of England, eleven years older than he, who had loved him deeply, found money and men for his French wars, and set about restoring the Catholic faith in England through the godly incineration of almost three hundred Protestants. After Mary's death he gladly would have wedded her skinny, red-headed, quicksilver younger sister, Elizabeth, but the wench proved first elusive, then dangerous.

The three tasks interlocked, and it was a long, awkward, complex business. In hope of simplifying its history, we shall deal with each of Philip's antagonists more or less separately. Philip built up a mighty Spanish army in Flanders—actually a mix of Spaniards, Italians, Burgundians, and Walloons (modern Belgians) from his empire, with German mercenaries, Irish refugees, and English renegades. They had—especially the Spanish veterans—the reputation of being the finest soldiers in Europe. From its position in the Netherlands this Army of Flanders could operate against the Dutch rebels, threaten England with invasion, or strike directly at the heart of France. (France has suffered for centuries under the strategic vulnerability of having its capital, Paris, located so close to its northwest frontier, and that frontier was much closer in the late sixteenth century.)

But Philip's Army of Flanders needed constant supplies, money, and new recruits. These could come either by sea from Spain into the few Flanders ports, such as Dunkirk, under Spanish control, or overland up the Spanish Road (actually a number of routes) from northern Italy through Switzerland, Lorraine, and Luxembourg. (This "Road" also was used to reinforce the French Catholic League forces in their struggle against the Huguenots.)

Instinctively, it would seem, Philip's enemies took the strategy of striking at these two lines of communication. The sea route went first. Dutchmen might make unsteady pikemen, but they were natural seamen. Their "Sea Beggers" literally savaged Spanish shipping in the English Channel and on out across the oceans; fewer, but equally unkind, Huguenot raiders shipped out of La Rochelle to sweep the Bay of Biscay and the Caribbean. And more and more Englishmen joined the game, at first almost as pirates, later with Queen Elizabeth as a silent partner. Year by year these raiders lengthened their reach into Philip's overseas empire. In 1580 Francis Drake brought his *Golden Hind* into Plymouth harbor, crammed with Spanish treasure, after a three-year voyage around the world, and became Sir Francis and a great man. In 1628 the Dutch would capture the entire "treasure fleet" bound for Spain from Vera Cruz. And all through these years of wars, Dutch, English, and French would rake the Spanish settlements in the Western Hemisphere; Dutchmen would sweep the East Indies. Drake and others would "singe the King of Spain's beard" by raids along the coast of Spain itself. Fortunes were gathered in by bold and lucky seamen and by the merchants who outfitted them—the most delightful combination of patriotism and pleasure since Alexander wedded Roxane.

The losses Spain suffered from this strategy were incalculable, and they snowballed. Fleet after fleet went down to destruction and had to be replaced. Spain's colonies needed new fortifications for their seaports. Money lost to freebooters had to be made up by increased taxation at home and additional borrowings from Italian financiers. Interest rates climbed steadily; Spain went through a series of bankruptcies.

Left unpaid for months, the Army of Flanders broke out in mutiny after mutiny, repeatedly crippling promising campaigns. Its elite Spanish units were foremost in such actions; they would move to the rear, seize some fat town within their own lines, and live off the country. In 1576 they selected Antwerp, the great commercial center of western Europe and largely Catholic. In three days of "Spanish fury" they burned almost half the city, butchered some 7,000 citizens, and robbed the place—even the Catholic churches—bone bare. Such affairs increased public hostility to Spain and sapped the Army of Flanders' morale, especially when Spanish commanders frequently broke their pledged pardons to mutineers and endeavored to swindle them out of their back pay. Eventually, Spanish officers would surrender their posts to the Dutch on condition the latter would make up the arrears in pay long due them.

The Spanish Road was less vulnerable. But Henry of Navarre, maneuvering far more carefully politically than he did on the battlefield, became King of France in 1589, turned Catholic in 1593, and reunified France, guaranteeing his Huguenots their rights. At the same time he chipped away the Road through conquests on his eastern frontier and pressure on Spain's north-Italian allies. By 1601 he had a chokehold on it, forcing Spain to seek less satisfactory routes farther to the east. Reconciling and rebuilding with the intent that every farmhand in his realm might have a chicken in his pot on Sundays, Henry remained a *vert galant* among the ladies to the last. A Catholic fanatic assassinated him in 1610. By then, Holland had won its independence.

Flanders—modern Holland, Belgium, and much of northwest France—had been the most prosperous portion of Philip's domains, but each of its thirteen provinces had its own laws and customs and stood stoutly by them. Worse, they tolerated men of all nations and creeds. To make them dutiful taxpayers and Catholics, in 1567 Philip dispatched the flint-souled Duke of Alva, who systematically killed off the provinces' natural leaders, Catholic and Protestant alike, and confiscated their estates. The provinces revolted. Even when a later Spanish commander, the Duke of Parma, brought the Catholic southern provinces back under a somewhat gentled rule, the seven northern seacoast provinces remained defiant. Their leader was William "the Silent" (properly, "the Wily"), Prince of Orange and Count of Nassau, a statesman of courage, skill, and character but a mediocre general. His first campaign in 1568, waged largely with German mercenaries, was a sad failure; he had to flee the country, which Alva then set himself to harrow into submission.

One unconquered bit of free Holland remained, however—a score or so of the little ships of the "Sea Beggars." Operating out of any Protestant port that would tolerate them, they waged war to the death on the seas and along the coast of Flanders against the Spaniards and the Dutch "glippers" who were loyal to Philip. William of Orange gave them letters of marque so that they might claim the legal status of privateers and organized them into an effective fleet. His admiral was one William de la Marck, a Flemish nobleman known as "Longnails." (When Alva executed his cousin, Count Egmont—a Catholic and a loyal soldier in Philip's earlier wars—Marck had taken oath never to cut his hair, nails, or beard until Egmont was avenged.) But no amount of vengeance sated Marck; his hobby seems to have been the devising of painful exits for captured monks and priests. In 1572

Spanish pressure on Elizabeth forced her to close English ports to the Beggars. Desperate, Marck led his sea wolves to seize the small Dutch port of Brill and then wipe out a Spanish force sent to recover it. He hoisted William of Orange's banner over his new base, and the northern provinces again exploded in revolt.

William of Orange won no battles, but his United Provinces possessed an endless stubbornness. Also, they seemed intended to bog down hostile armies; their terrain was cut up by rivers, canals, woodlands, and marshes. Most of their towns were fortified, an increasing number in the latest fashion, with low, thick walls and projecting bastions. Their climate gnawed at invaders from sunny southern Europe. And despite the usual ham-handedness of the raw Dutch soldier in pitched battles, the average Hollander, soldier or citizen, at bay behind his town walls, proved a hard man indeed. The wholehearted bestiality of Spanish troops in captured towns, the merciless arrogance with which Spanish commanders could break their word and hang a whole garrison that had surrendered on promise of quarter after a stout defense, made it better to starve or die fighting rather than yield. Women would fight beside their men at a disputed breach, screeching defiance above the din, dashing fireballs into Spanish faces.

To Philip and his generals there seemed one certain strategy to end this ungrateful rebellion—cut the United Provinces' sea dikes and flood them. This, however, was finally rejected, since it might also damage areas loyal to Spain, probably permanently ruin the countryside, and shock the rest of Europe. As a more humane measure Philip was prepared to burn out the United Provinces' agricultural areas, but between the usual Spanish mutinies and Dutch resistance, his troops failed to reach them. In a desperate counterstrategy, William did not hesitate to cut his dikes on several occasions. This was most effective in 1574 at Leiden, which had been almost starved into submission before William flooded the countryside, drowned the panicked Spaniards out of their siege works, and sent a fleet of shallow-draft ships in to the rescue. A Catholic assassin shot William down in his Leiden home in 1584 (you still can see where the bullet chipped the plastered wall), but that brought England openly into the war.

For years this war dragged on, a desperate hole-and-corner struggle. The Army of Flanders won battles but thereafter had to turn again to besieging stubbornly defended towns in mud, rain, and floods, with half-starved peasants waylaying its stragglers and outlying detachments. Between the towns a shifting, irregular war surged back and forth—led sometimes by freebooter captains, with little loyalty to either side, seeking supplies or the chance to surprise an unwary garrison. Sir James Turner, an English captain, saw it as an affair of "onfalls and escaladoes"—sudden, savage clashes that required "proven swordsmen and masters of the whole art of warre" and was no place for the "homely wits fresh from the fields, nor tinkers from the hedgerows, nor lackeys, nor runaway 'prentices, nor out-of-time drawers newly from the stews"[19] who made up most of the recruits Queen Elizabeth sent to Holland. Spanish veterans excelled in such cutthroat clashes. And in 1578 Philip finally had found an outstanding commander in his nephew, Alexander Farnese, Duke of Parma, probably the ablest general of that era. An unusual blend of soldier, politician, diplomat, and military engineer, Parma had great dash, personal courage, and the physical toughness to share his men's hardships. Against the Dutch he was a master of siege warfare, but Philip also expected him to support the failing Catholic League in France and prepare an invasion of England. He baffled Henry IV's cavalry tactics by marching his army in a hollow square and entrenching his camps Roman style, then striking suddenly with his own horsemen. Once or twice he almost bagged Henry, who was at his usual sport of beating up Parma's advanced guard. But his tasks were too many and his support from Spain insufficient. The army he massed for the invasion of England lost heavily from sickness and hardship during the winter of 1587–88. Unable to understand Parma's problems, Philip first interfered in his operations, then ordered him replaced. Before Parma could receive that expression of royal and avuncular disapproval, he was dead from a minor wound taken at the siege of a minor French town and the bungling of the surgeons who tended it. Philip, his sandy hair and beard gone gray, later found a talented general in Ambrosio Spinola, who also was a member of a Genoese banking family and so had a better credit rating than the King of Spain!

William of Orange's murder put the Dutch command into considerable confusion, but it soon came into the firm hands of William's son, Maurice of Nassau (1567–1625). A hard-nosed organizer and drillmaster, Maurice was also a bookworm who had found and studied the Emperor Leo's *Tactica*. Combining Leo's teachings with others culled from Polybius, Caesar, and like antique writers, he rebuilt the hapless Dutch army. His was a polyglot creation—Dutch, English, Scots, Huguenots, Germans, and Swiss—tautly drilled and disciplined, capable of swift maneuver under all conditions. When older officers protested Maurice's emphasis on small, flexible units as the product of mere book learning, he gave them a cold lesson: "He who does not know the ancient art of war, I say to him roundly that he is a mere recruit, and does not deserve the name of soldier."[20] Then he proved his statement on the battlefield: At Turnhout (1597), with only 800 Dutch and English troopers, he rode down some 5,500 veteran Spanish, Germans, and Walloons; at Nieuport (1600), he broke the main Spanish field army. But, for lack of manpower, his strategy necessarily was cautious. Building up the Dutch frontier fortifications in depth, to give him a secure base, Maurice set about a masterful campaign of sieges, massing men and guns against stronghold after stronghold to force their speedy surrender before a Spanish relief force could concentrate against him, then shifting quickly to another target. Thus, in 1591, he took Arnhem in western Holland, then shuttled his little army westward in barges along canals and rivers to capture a town near Antwerp, then darted eastward again to take Nijmegen, completely confusing the Spanish command. One of his junior officers was his nephew, the Vicomte de Turenne. By 1609 the exhausted Spanish agreed to a twelve-year truce. Somewhat recovered and emboldened by successes in Germany, Spain tried again in 1621 in a series of campaigns that got entangled with the Thirty Years' War (1618–48) in Germany. Finally, fought to a standstill in Flanders and repeatedly defeated at sea, Spain reluctantly made peace and recognized Dutch independence in 1648.

For all of Henry's dash, William's endurance, and Maurice's skill, Philip undoubtedly found Elizabeth of England his most infuriating opponent. Inaccessible and unpredictable, she presented a strategic problem he would have lief left alone until he had finished with France and Holland. Also, there was for years the possibility that she might be dealt with by a strategy short of open war. Mary Stuart, Queen of Scots, a woman of such allure and presence that her stupidity usually went unnoticed, was next in succession to the English throne. Run out of Scotland for willful incompetence in 1568, she had taken refuge in England with her cousin Elizabeth. She had been well treated and soon had joined in the plottings, foreign and domestic, against her protector. (Philip had competition here. Mary's allies included the murderous Guise clan, who served Philip but plotted to seize the French crown, the Pope, and English Catholics, especially those in exile overseas.) Assassination might remove Elizabeth, or the Catholic nobility of England's northern marches might rise in rebellion and depose her. Waves of devoted priests would infiltrate England to win Englishmen back to the Catholic creed and impress on them that Pope Pius V had excommunicated Elizabeth, declared her no longer Queen of England, and released her subjects from their duties to her. In the meantime, it also would be well to stir up discontent in Ireland and to consider the possibility of an actual invasion of England itself. But Philip's habit of thinking twice on any subject—and then a third time—and of waiting for heavenly guidance would delay any such drastic measures.

Elizabeth was as given to evasion and procrastination as Philip. She had come young to the throne of an impoverished, divided England, and she would give her life to restoring and strengthening it. She had no standing army beyond a few guards and garrisons; her navy was ancient and weak. She detested war. It was uncertain, expensive, and the one aspect of ruling she could not manage in person. But Philip obviously must be faced if England were to survive, and she too developed a strategy of delay and attrition. Philip got honeyed words and small concessions, but aid in money, men, and supplies trickled out from England to Huguenots and Dutch; more and more English sea captains brought home treasure (of which she claimed a share) from the Spanish Main. Meanwhile, she put that old sea dog and slaver John Hawkins to building warships. Hawkins turned out splendid new galleons, big, tight, swift, and heavily gunned, and rebuilt her older ships in that same image.

Elizabeth's aid was never generous, but it proved enough to thwart Philip repeatedly. And, one by one, his hopes failed. None of the assassinations planned for Elizabeth ever came off; her secret service was efficient, if informal. In 1569, with the Huguenots twice beaten in France and Alva victorious in Flanders, the northern nobility rose in arms for Mary Stuart and the old religion under "Simple Tom" Percy, Lord of Northumberland. They had money from Rome and promises of aid from Alva; Elizabeth's officers had much trouble raising and equipping troops, but her ships held the coast, and tough troopers from her border garrisons rode out against all odds. Aided by a hard winter, the rebel lords made a rare hash of their operations. Simple Tom fled across the Scottish border for refuge with the reiving Armstrongs of Liddesdale, who sold him to Elizabeth's Scots friends. There were revolts in Ireland; in 1580 a Papal/Spanish expedition put 700 troops, mostly Italians, ashore with arms for 5,000 Irish at Smerwick. English reinforcements poured in, and the Italians were unceremoniously knocked on the head. The Catholic missionaries were hunted across England like wild game, and many of them died horribly.

Matters approached the breaking point in 1583–84 with the Spanish ambassador being implicated in the latest plot to murder Elizabeth, the assassination—after five unsuccessful attempts—of William of Orange, and the dispatch of English troops to Holland. (In a typical bit of obfuscation, Elizabeth assured Philip she wasn't attacking Spain, merely assisting her friends.) Philip requested a large shipment of grain from English merchants under pledged safe conduct. As the ships arrived in Bilbao harbor, he seized them, but one fought its way out and home with the news and a captured Spanish official. Elizabeth sent Drake with some of her ships and soldiers on a long raid against northwest Spain, the Cape Verde Islands, and the Spanish Main. Philip began to plan the invasion of England and collect ships.

What really broke any remaining pretense of peace was the execution of Mary Stuart, practically forced on a still-reluctant Elizabeth by her angry subjects and counselors in 1587. Mary had been too involved in too many plots: She died gallantly. Philip gave the word that his "Armada" must sail in 1588.

This "Most Happy Armada" was a mighty fleet, officered by gallant gentlemen but collected from all around the Mediterranean. It had few ships and men who had experienced the brutality of a North Atlantic gale, and it sailed for strange waters. It was a crusade against the heretic, with a standard the Pope had consecrated, and thoroughly confessed and absolved sailors who were told to refrain from profanity. Parma waited in Flanders; when the Armada had cleared the Channel, he was to get his army into barges and invade England. Once ashore, the Army of Flanders should make short and bloody sport of Elizabeth's raw trainbands. Parma, however, seems not to have approved of the business. He had no good seaports, and those he had were watched by the Sea Beggars, whose light warships operated freely in the shallow coastal waters where the Armada's heavy vessels dared not venture. In his hurry to dispatch his Armada, Philip had overlooked many things, especially the matter of ensuring tight cooperation between his sailors and soldiers.

Elizabeth dithered. In early 1587 she again had sent Drake to raid the Spanish coast (with specific instructions as to her cut of any booty), then rushed off an urgent order that he might raid Spanish shipping but not the Spanish mainland. Drake, knowing his queen, had put to sea before the second message could reach him. He sank some thirty ships in Cadiz harbor and raked the Portuguese coast, destroying valuable ships' stores the Armada would need. To Drake *this* was the way to deal with the Armada—destroy it piecemeal along its own coasts while it still was fitting out. He wrote Elizabeth that she should not "fear any invasion of her own country, but . . . seek God's enemies and her Majesty's wherever they may be found."[21] But Elizabeth, always hopeful that a cheap way out might suddenly appear, negotiated with Parma—who spun out the talks while the Armada thrashed northward—and kept her ships close in English ports. She cut their crews, provisioned them stingily, and would not replace beer brewed too sour for even sailors to get down. Once she wanted to demobilize the whole fleet. When the Armada appeared, her ships had to claw out to sea short of ammunition and with food for only three or four days. By sheer luck the available powder and shot (pieced out with what officers could scrape up) lasted until the Armada was shattered and driven north. The Spaniards had fought bravely, but vainly, against better ships and seamen and heavier guns. Parma did not move, having no intention of committing suicide.

Elizabeth's strategy of delay and shipbuilding had won. Her holding her ships in their home waters through the winter of 1587–88 instead of letting them risk storm and fortune along the Spanish coast, as Drake wished, has been praised as a farsighted intent to have them ready for action the next spring. This, however, overlooks her failure to keep them in fighting trim. Really, considering her complex, mercurial character, it is impossible to know what combination of pride, parsimony, and calculation inspired her decisions. At least her seamen won a great and famous victory. Moreover, she and her advisers had subsidized an army of Germans and Swiss to support Henry of Navarre's Huguenots and so keep Catholic France too busy to aid Philip. (Their operations proved a fiasco, which redounded to the reputation of the Guises, but that Elizabeth could not help.)

The ending, however, was wry. Bit by misfortune, Philip assessed his failures, built a new and potent navy, and strengthened the defenses of his colonies. Elizabeth let her navy slide. When Drake and Hawkins went back to the Spanish Main in 1595, they could accomplish little. Both died while at sea and fittingly were buried there. Drake had been a marvel to his age. His wonderful instinct for intercepting treasure ships in mid-ocean led Spaniards to believe him in league with the devil and to possess thereby a magic mirror that showed all the ships on all the world's seas. He was a father of English naval strategy and tradition—seize the initiative and hit first, pick the time and place that your blow will be most effective and remember that "continuing to the end, until it be thoroughly finished, yields the true glory."[22] When the British assault went in at the Falkland Islands, Drake's drum must have rumbled once again across the waters.

As noted, Elizabeth began with no standing army. Interest in military affairs had waned in England; the warlike and adventurous went to sea. Such military organization as did exist was primitive; armies were improvised out of temporary collections of independent companies. Most English captains swindled their Queen and cheated their men.[23] But the war in Holland brought Englishmen back to soldiering. That the high command was, as in all armies, reserved for the high nobility was not necessarily an evil: Charles, Lord Howard of Effingham, was Lord Admiral of the English fleet in 1588 and did remarkably well—and also spent his own scant estate to take care of starving seamen his queen dismissed unpaid. The troops sent to Holland in 1585–86, however, were commanded by Robert Dudley, Earl of Leicester, Elizabeth's aging favorite. Leicester was ambitious and greedy, with no military experience and a large ego. He wanted to succeed and was not without care for his soldiers, but his outstanding ability as a commander was that of quickly reducing any simple and promising military problem to absolute confusion, uproar, and despair. There can be no doubt that he inspired the quotation, given at the opening of this chapter, from Matthew Sutcliffe, once Leicester's judge advocate general.

These wars naturally stimulated interest in past military history. Caesar seems to have been especially favored because of his early expeditions to England. A Sir Clement Edmonds published an analytical version of the *Commentaries* in 1600, stating that any soldier, however experienced, could still learn from the experiences of others through reading because of the endless variety of human actions. Vegetius undoubtedly was the most influential, but a Peter Whitehorne also translated from the Italian the works of Onasander, with their insubordinate suggestion that generals should be selected for their intrinsic ability, not social rank or ancestry.[24] Machiavelli was translated into English in 1560, as was "Iron-Arm" La Noue's *Discours politiques et militaires* in 1587. (The latter work, with its blend of military expertise and Puritan ethic, seems to have attracted Oliver Cromwell's notice in later years.)

Moreover, like soldiers returning from the wars everywhere, Englishmen began to write their own books. Sutcliffe was thoroughly modern minded, wanting tighter army organization and administration, faster operations, and an improved intelligence system. His words are a landlubber's version of Drake's: "Opportunity to do great matters seldom offereth itself a second time. . . . Nor is anything more hurtful than delays."[25] Barnabe Rich, the most prolific of Elizabethan military writers, embodied his experiences in a book charmingly titled *Faultes, Faultes, and Nothing Else but Faultes* (1606). His rambling works show indebtedness to Machiavelli for ideas, but he frequently substitutes his own greater experience (he claimed forty years' combat service) and reasoning for the Italian's. His *Allarme to England* (1578) includes an immortal definition of pacifists—"right bastards to their country."[26] Thomas Digges was an unusual sort. A skilled mathematician with a strong practical streak in his character, he wrote on fortification, castrametation, harbors, and ordnance, and served as a muster-master (a species of combined paymaster and inspector general) in Holland, where his honesty made him completely unpopular.

The sorest head among these soldier-authors was Sir John Smythe, distant relative of Queen Elizabeth, diplomat, traveler, and soldier with the Hungarians against the Turks and in other foreign armies. Choleric and conservative, he also was something of a scholar—an intellectual Colonel Blimp. He had read Leo's *Tactica* and wanted to apply it to a badly needed reform of the English militia. But he scorned service in Holland—English soldiers learned nothing there except to drink too much and abandon the longbow. For Smythe lived in the past: Musketeers never could replace the traditional English archer, this new fad of discarding armor must be stopped, the fashionable long rapiers and daggers were only silly toys, and English soldiers had fallen so low as to use foreign jargon—for example to say "beleaguer" instead of the good English "besiege." To be a soldier was a proud thing; "all other arts do rest in safety under the shadow and protection of the Art and science military,"[27] but the profession needed purging. There were too many officers from the lower classes; also, there were too many corrupt, insubordinate officers of all classes. Smythe's first book, *Certain Discourses Concerning the Formes and Effects of Divers Sorts of Weapons* (1590), was supposedly intended as a demonstration of the superiority of the longbow to firearms, but it was so spiked with scalding descriptions of Elizabeth's generals that it was suppressed by royal warrant. In 1596 Smythe was imprisoned in the Tower for telling militiamen the Queen had no authority to send them overseas but soon was sent home, another broken soldier, to die obscurely.

Smythe's fondness for the longbow was countered by Humphry Barwick, the contemner of Machiavelli, and especially by Sir Roger Williams, who was reputed to be the model for Captain Fluellen in Shakespeare's *Henry V.* Williams was something of a nonpareil—an English professional soldier. From poor Welsh gentry but somehow educated, he was choleric, blunt, and always ready for deeds beyond any call of duty. A short man, he wore a gilt helmet with a high plume so that foe and friend alike could mark him. He studied his profession, seeking to combine the best of the new and the old; before England entered the Dutch wars, he served in the Spanish Army, which he considered as the finest possible school for soldiers, especially in the raising and training of troops. He also had a high opinion of La Noue, and ended his active service under Henry of Navarre. His *The Actions in the Low Countries* (1618) is lively and accurate. Captured at Sluys after a famous defense (Leicester completely bollixed the effort to relieve the town) in 1587, Williams refused Parma's admiring offer of a Spanish commission and went home, wounded in body and sick at heart and too poor to buy a horse. He considered marrying a rich merchant's widow and hanging up his sword, but though Elizabeth spared him no favors, he soon was back at the wars. Of like kidney was Sir Francis Vere, who looked after his men, became a general in the Dutch service, and so modeled himself on Julius Caesar that he entitled his memoirs *Commentaries*.

However much these captains courageous differed over weapons and equipment, they were much agreed that the English army needed better discipline, training, and organization from general to private. Rations, pay, and clothing must be handled honestly. England should maintain a sufficient force of trained men ready for emergencies and take proper care of its veterans. They considered tactics and strategy all together, English wars (except in Ireland) having been too small to give any real experience in land strategy. But they stressed the need in all operations for "expedition."

When the Armada came in 1588, England faced an immediate problem in defensive strategy. Besides Parma's army, supposedly ready to cross over from Flanders, the Armada itself carried over 18,000 excellent troops—enough to seize a port or beachhead along the Channel coast. The problem of whether to hold every possible landing place in strength or to strip the coastal area of food and concentrate troops further inland for a counteroffensive was put to a group of experienced officers, including Sir Roger Williams and Thomas Digges. Their final decision rather suggests that which followed in England in 1940 after Dunkirk, when England once more faced invasion, this time by Hitler's triumphant armies. Since there were too many possible landing places to fortify all of them, a screen of local troops would watch the coast, the seaports would be held in strength, plans made to move all cattle into the interior, and two mobile forces assembled as reserves.

There remained Ireland, as unpredictable as Elizabeth herself, a stew of tribal groups that frequently hated each other almost as much as they resented their English overlords. It was a large country which must be held by a comparatively small army, and the English commander there must be a strategist and something of a diplomat as well as a competent tactician.

After the Smerwick episode in 1580, the island had been quiet—so quiet that Spaniards, escaping from wrecked ships as the storm-hounded Armada fled around Scotland and Ireland for home, could find little help or mercy. But in 1595 Hugh O'Neill, Earl of Tyrone, the greatest chief in Ulster, rebelled. He was an able operator in tribal politics and an excellent soldier. Philip sent him money, weapons, and drill sergeants. His own genius enabled him to evolve a strategy of attrition, skirmishing, raiding, and overwhelming isolated English posts. At the same time, his levies learned enough drill and discipline to mount sizable actions, as at the Yellow Ford in 1598 where he caught an English column on the march and killed, wounded, or captured 2,000 out of 4,200 men. He moved swiftly, struck fiercely, and vanished.

The English soldier did not like these Irish wars. He usually was a "pressed" (drafted) man, which meant that his friends and neighbors considered shipping him overseas a definite public improvement. Ireland was a damp, rough place of few roads and much swamp and forest where even the few honest supply officers had trouble keeping him fed and clothed. Malaria, dysentery, and other "agues" might kill four men out of ten and weaken the rest. His basic weapons for chasing wild Irishmen were a heavy eighteen-foot pike or a matchlock musket, weighing between twelve and twenty pounds, which had to be fired from a rest and seemed designed for use by soldiers with three hands. Worse, there was little chance of valuable loot. (Sir Walter Raleigh, who served in both places, noted that in the West Indies English soldiers "run upon the Spaniards headlong in hope of [seizing] their gold plate and money, whereas had they been ordered to attack against the same odds in Ireland . . . they would have turned their muskets and pikes against their commanders, arguing that they had been senselessly exposed to butchery and slaughter.")[28] Like Americans in the Far East in the 1950s who remembered the United States as the "Land of the Big PX," the English soldier in Ireland thought of home as the "land of good meat and clean linen."[29] To keep up his morale and avoid sickness from drinking foul water, the private's rations included a half pint of "sack" (sherry) a day and a quart of beer and a quarter pint of "aqua vitae" (brandy or aquavit) every other day.

To deal with Tyrone, Elizabeth dispatched an unusually strong army of 17,000 men but entrusted it to her current favorite, Robert Devereux, Earl of Essex, thirty-three years her junior. Essex was personally brave, popular, obsessed with his personal honor, and ambitious beyond his capabilities and understanding. He might conceive sound military plans, but in their execution he always went off into some knight-errant's cloud-cuckoo-land. In Ireland he disobeyed his orders to move immediately against Tyrone and wasted men and time on various sideshows. When he did move, his forces were too weak; Tyrone practically dictated terms of a six-week truce to him. Essex bolted back to England, failed to win Elizabeth's approval, stumbled stupidly into rebellion, and was—in one of Elizabeth's favorite terms—shortened by a head.

Elizabeth's choice as Essex's successor—Charles Blount, Lord Mountjoy, an old friend of Essex—amazed everyone. (Essex's wildly charming sister, Penelope, was his open mistress and had given him five children; she also had had seven by her legal husband.) Mountjoy had served bravely and been wounded in Holland in 1586, but Elizabeth had thereafter recalled him. He now was either something of an invalid or a hypochondriac, cosseting himself with naps, careful diet, and warm clothing. Unlike other noblemen, however, he was an incessant reader of military books, both the classical sources and the practical comtemporary works of soliders like Williams, Smythe, Sutcliffe, and Rich—so much so that Essex earlier had advised the Queen against giving him a command in Ireland because "he was too much given to studies."[30]

Once in Ireland, he continued to be careful of his health and comfort. He also nursed the remnants of Essex's army back into fighting trim and went after Tyrone. His strategy was carefully calculated, his execution dazzingly swift. So well did he conceal his intentions that it was said nobody knew what he planned to do until it was actually done and that "surprise" might well have been his motto. Counterguerrilla warfare is a delicate and difficult branch of the military art. Historically, Mountjoy ranks as one of its most successful practitioners—a superstrategist in that one field.

Closing off Ulster with a line of mutually supporting strongholds, he moved into it, destroying crops and driving off cattle. His reorganized supply system enabled him to keep up this pressure through the winter of 1600–1601—the bare woods gave the Irish no shelter; their stored grain was easy to burn. Thrust off-balance, Tyrone could find no weak point in Mountjoy's advance. Moreover, exploiting his control of the sea, Mountjoy built up a base around present-day Londonderry in Tyrone's rear.

His advance was checked in 1601 when 3,500 Spaniards landed at Kinsale in southeastern Ireland. Apparently caught between them and Tyrone, Mountjoy went southward by forced marches to pen the Spanish in Kinsale before they could move inland. Tyrone pursued, but this brought him into open country. Mountjoy suddenly turned on him. There were 1,500 English against more than twice their number of Irish, but Tyrone's men lacked the steadiness for a stand-up battle and were routed. Mountjoy then forced the Spaniards to surrender, and hounded Tyrone until the Irish begged for peace. His duty done, he went home and married his Penelope, who had been newly left a widow. Court society, which had chuckled approvingly over their liaison, was deeply shocked and distressed.

IV

SNOW KING, BOY KING, SUN KING, AND BASTARD

You must love soldiers in order to understand them, and understand them in order to lead them.''

TURENNE

It is good to exercise caution, but not to such an extent that all [opportunities] become lost. . . . Hungry dogs bite best.[1]

CHARLES XII

Through the sixteenth and seventeenth centuries the nations of modern Europe were still in their making. Germany and Italy were clutters of independent principalities; even established states such as France and Spain were crazy quilts of differing provinces, each with its own traditions and dialect. Roads were little better than cart tracks; agriculture was still primitive. The Turks held the Balkans, North Africa, and the eastern Mediterranean; in 1571 raiding Krim Tartars ran Czar Ivan IV, ''the Terrible,'' out of Moscow and burned it to the ground.

Armies tended to live off the countryside and so seldom exceeded

20,000–30,000 men; larger ones would have starved, and even small ones might if held too long in one picked-to-the-bone district. They moved ponderously and might spend an entire summer campaign in an elaborate siege of some major town. It was good generalship to ravage the countryside, both to feed your own troops and deny the enemy supplies. With cold weather at hand, the wise general would maneuver his troops into winter quarters, if possible somewhere in untouched enemy territory, to rest until the next campaign. Civilians could hope for little more than bare survival, robbed and brutalized, at the hands of unpaid, half-disciplined soldiers and irresponsible officers, whether enemies, allies, or their own. The famous engraver, Jacques Callot (1592–1635) pictured this starkly in his *Miseries and Disasters of War* as he witnessed it in his native Lorraine.[2]

Off in northeastern Europe the nations around the Baltic Sea were fighting more or less private wars over control of the Baltic trade routes and their conflicting schemes for territorial aggrandizement. These were rough affairs, marked by shifting alliances and peppered with domestic revolts, but out of them came a new strategic tradition.

Denmark long had been the primary Baltic power, but in 1520–23 young Gustavus Vasa (1496–1560) led a successful revolt in Sweden and painstakingly built that disorderly collection of backwoods provinces into a prosperous Protestant kingdom. His descendants, the Vasa dynasty, were men of talent and energy, high-strung and conquest minded. They gradually developed a national army, raised by conscription and carefully trained. One of their first generals, Pontus de la Gardie, a wanderer from the south of France, became a legend for his successful winter campaigns, sweeping in amid a storm across a frozen arm of the Baltic or bursting out of snow-choked forests to surprise and destroy far larger Russian armies. His son, Jakob de la Gardie, led a Swedish army into Moscow in 1610 and brought it back.

The Vasa clan's outstanding soldier, Gustavus Adolphus (1594–1632), grandson of Gustavus Vasa, came to the throne at seventeen. Sweden was fighting a losing war against Denmark and a desperate one against Russia; the Swedish nobility was intent on regaining its old-time privileges. Gustavus, however, was no raw boy. He had been educated for kingship and was a linguist and student of history who particularly treasured Xenophon's *Anabasis*. Tall, commanding, gifted with a powerful personality, absolute courage, and a unique combination of high intelligence and shrewd common sense, he already had proved himself a thorough soldier and a born leader. Always, he shared his soldiers' hardships and dangers, leading in person like a Viking chieftain, with a berserker love for danger and battle. Withall, he was a man of imagination and method. Like his grandfather, he tended Sweden as if it had been his own farm, modernizing its government, schools, and industries. Dutch businessmen and Walloon smiths improved the country's iron and copper mines and set up an efficient armament industry so that Sweden could provide its own weapons. He began his reign by making the best peace he could with Denmark (1613), then moving eastward to smash the Russians. Thereafter came a series of campaigns in Poland (1617–29) to strengthen Sweden's hold on the eastern shore of the Baltic. At first Gustavus lost several battles to the famous Polish general Stanislaus Koniecpolski and the excellent Polish cavalry but, modifying his tactics and his weapons, he decisively gained the upper hand during 1626–29.

Gustavus had modeled his army on that of Maurice of Nassau, aided by officers who had been in the Dutch service. (This was a period when professional soldiers might pass freely from one army to another in search of adventure, on-the-job training, or higher pay.) His Polish wars taught him to use even lighter, faster-moving, and more flexible formations. He went beyond Maurice in developing light field guns and in teaching his infantry, cavalry, and artillery to work together. His cavalry was retrained to rely on shock action—the impact of troopers charging all out—and cold steel rather than their pistols. To increase his infantry's speed of maneuver and rate of fire, he equipped them with lighter muskets which could be fired without a rest and with paper cartridges in place of cumbersome bandoliers and powder flasks.[3]

Swedish national strategy was consistently aggressive, as expressed in the folk saying "It is better that we tether our horses to our enemy's fence, than he to ours."[4] Their tactics and strategy were inspired by the same spirit, which came to be expressed in Swedish slang as *gā pä*—literally "go on!"—in intent, probably something like "Get those bastards!"[5] Gustavus did not waste time in long sieges; if he could not take a place in "above five or six days" of bombardment and storming, he would "rise and to another."[6] At the same time, he was a "great spademan" who made his soldiers dig in whenever the situation required it, "not only to secure his Souldiers from the enemy, but also to keepe them from idlenesse." Robert Monro, a Scots officer, admitted that "the spade and the shovell are ever good companions in danger"—though Gustavus always had trouble getting his Scottish units to dig themselves in as thoroughly as his Swedes and Germans.[7]

Before the ending of his Polish campaign, Gustavus found himself involved in the crisis of the Thirty Years' War (1618–48). Having defeated the Protestant German princes and Denmark, the Catholic armies of the Holy Roman Empire and Spain were pushing northward to establish bases on the Baltic coast and create a fleet there. This, the Swedes could not permit; also, as good Protestants, they felt an obligation to succor their fellow "Gentlemen of the Religion." Necessary financial support for their offensive came from Catholic France: The deadly-devious Cardinal Richelieu, prime minister to Louis XIII (1601–43), intended to strengthen France's position in Europe by weakening Spain and the Holy Roman Empire and considered Protestant Sweden an essential ally to that end.

During 1630–31 Gustavus moved into Germany with approximately 23,000 men, mostly Swedish but including a considerable number of German mercenaries, several crack regiments of Scots volunteers, and a sprinkling of Englishmen. Each regiment marched with its two chaplains at its head and its provost marshal and his hangman behind it. Marching and camping was done with the greatest order and regularity; the army was always screened by outposts and patrols, and its supply trains were as thoroughly organized as the combat troops. The whole army had morning and evening prayers and sermons on Sundays; whenever possible, a special prayer was offered in line of battle, just before the shooting started. Looting, dueling, and similar offenses against good order and military decency were punished by hanging. Troops so disciplined in mind and body marched and maneuvered faster, held together better in action than their opponents. And their behavior to civilians greatly eased Gustavus's advance into Germany.

The Swedes won major battles at Werben and Breintenfeld in 1631 over the army of the Holy Roman Empire and another one in 1632 at Rain. Setting up headquarters in Nurnberg, Gustavus began organizing a league of German Protestant princes. His prestige was tremendous—enough to worry Richelieu, who had his own plans for

Germany. Catholics called him the "Snow King," but to the superstitious he was the foretold "Lion of the North" who was to bring peace and safety. His own political plans remain obscure, but they could have reached as far as the establishment of a new German empire. Holy Roman Emperor Ferdinand II recalled Albert of Valdstejin (better known by the German version of his name, Albrecht von Wallenstein), an immensely wealthy Bohemian nobleman who had been sent home from his army two years previously because his greed and arrogance had angered Ferdinand's allies.

Wallenstein had no religion and fought for his own hand. He believed in astrology and lived in more personal splendor than impoverished Ferdinand II could manage. He had raised an army for the emperor's service out of his own means in 1626; now he did it again. Veterans flocked to his standards, Wallenstein having the reputation for winning battles and letting his men have a free hand across the countryside between them. Also, he obviously knew one basic way to a soldier's heart—even his enemies admitted him to be a "good provisioner" with a "singular good catering wit of his own."[8] (Since Wallenstein held that "war should feed war,"[9] this "catering" undoubtedly consisted of squeezing the maximum amount of supplies out of an occupied area.) He had great, still-mysterious ambitions of his own; carried a grudge against Ferdinand for his earlier dismissal; and was up to his ears in assorted intrigues with Ferdinand's enemies. Oddly, he favored freedom of worship and an end to religious wars. Also, he was racked by the gout.

Wallenstein and Gustavus waged a war of maneuver and minor actions, Wallenstein having slightly the better of it. But in November 1632 Gustavus remained inactive long enough to convince Wallenstein that he was going into winter quarters. Wallenstein therefore begun dividing his own forces—and Gustavus came down upon him at Lutzen, just southwest of Leipzig. A savage day-long battle in mist and low-clinging powder smoke ended with Wallenstein's broken army in full retreat. (Wallenstein would be murdered by order of Ferdinand in 1634.) But, caught almost alone in the fog and shifting cavalry action, Gustavus was dead. Colonel Monro would pen a fitting elegy: "Such a General would I gladly serve; but [another] such a General I shall hardly see."[10] His death was compared to Alexander's—another great captain dead before his time, his work unfinished. Gustavus had been, and remained, a major hero to Protestant

Europe and was grudgingly admired by his enemies. Englishmen, including a country gentleman named Oliver Cromwell from the fen country of the eastern Midlands, read *The Swedish Intelligencer* and *The Swedish Discipline,* published in London. During the English Civil War (1642–51) both sides followed Swedish practice.

Gustavus's only child was his six-year-old daughter, Christina, but he also left behind gifted civilian ministers and outstanding generals whose names sounded a drum role of victories—Johan Baner, Lennart Torstenson, James King, Bernhard of Saxe-Weimar, Alexander Leslie, Gustaf Horn, and Karl Wrangel. As the war dragged on, the quality of the Swedish army declined until it became little more than another collection of mercenaries, however well trained, but when the Treaty of Westphalia ended the fighting in 1648, Sweden ruled a greater Baltic empire than ever.

It was, however, an empire under constant pressure from Poland, Russia, Denmark, and the north German state of Brandenburg (later Prussia). Christina abdicated in 1654, passing the throne to her cousin Charles X (1622–60). In 1657, while Charles was deeply committed in Poland, Frederick III of Denmark declared war on Sweden. Spinning on his heel, Charles came hotfoot across northern Germany, invaded Denmark from the south, and overran the Jutland peninsula. The Danish government still thought itself safe in Copenhagen on Zealand Island, protected by its navy, but an unusually severe winter froze much of the Baltic. When his engineer officer reported the ice thick enough to support troops, Charles announced, "Now, brother Frederick, we will talk in plain Swedish,"[11] and boldly led 20,000 men across frozen sea channels and small islands into Zealand. Caught unprepared, the Danes were forced to surrender and make extensive territorial concessions.

Charles's son, Charles XI (1655–97), began as a victorious general but turned to rebuilding Sweden's shaky economy and modernizing its army and fleet. He brought up his son, the future Charles XII, carefully but died before the boy was quite fifteen. Charles XII inherited an empire consisting (in modern times) of Sweden, Finland, the Leningrad area, and much of Estonia, Latvia, and northeastern Germany. The Baltic was practically a Swedish lake. Foreign diplomats, however, reported that Charles was immature and light-minded, and three of his neighbors saw an opportunity to collect large pieces of Swedish real estate. Denmark was the traditional enemy, dangerous chiefly because of her fleet. Augustus II, "the Strong," Elector of Saxony and King of Poland, was a new factor in European politics.[12] A big man, insatiable in all physical excesses, he was a Germanic version of Machiavelli's Prince—thoroughly shrewd, devious, and amoral. (He left 354 acknowledged bastards and one useless legitimate son.) Augustus intended to recover the territories Poland had lost to Sweden and to convert Poland itself into a centralized, hereditary monarchy. For this he had linked himself to Denmark and Peter I, "the Great" (1672–1725), Czar of Russia. Peter was another giant with wild appetites, half genius, half antic savage. He wanted control of a section of the Baltic coast; seeking a pretext for war, he could come up with nothing better than a charge that the Swedes "had not paid him sufficient honours when he went incognito to Riga" in 1697 on his way to visit Holland and England, thereby subjecting him to "indignities and humiliation."[13] In one of his whimseys, Peter had pretended to be a servant during the banquet offered his entourage; the "indignities" probably developed when the Swedes caught some of its members—in the best traditions of Russian diplomacy—trying to map Riga's new fortifications!

So began the Great Northern War (1700–21). It overlapped the War of the Spanish Succession (see page 89) in western Europe: The major European powers had little interest in the Baltic strife except to keep it from interfering with their greater wars or to secure Sweden as an ally. Both Peter and Augustus had professed their friendship for Sweden up to the last moment. Augustus did not bother to declare war before he struck. Almost immediately, the three allies were in the position of the sorcerer's apprentice who raised the devil.

Charles XII's father had insisted that his study of Latin should make him feel as if he had marched with Caesar. His childhood heroes were Alexander and Henry IV, and he was thoroughly trained in mathematics, fortification, and the strategy and tactics of Sweden's military history, as well as logic and ethics. Since his health was somewhat delicate, he had toughened himself by deliberate exposure and hardship. In personal conduct he proved a second Gustavus Adolphus—courageous in mind and body, quick-witted, careful of his soldiers, simple and Spartan in his daily life, openhanded, with the instinct for command. Like Gustavus, he had great technical skill in the development of new drills, tactics, weapons, and equipment. But he had been thrust, while still a boy, into a more dangerous situation than that which Gustavus had faced. His life gave him no chance to woo and wed, and he saw the world in a conscientious adolescent's stark terms of good and evil.

Charles had capable advisers, diplomatic and military. Securing naval help from Holland and England to overawe the Danish fleet, the Swedes swarmed ashore on Zealand Island and prepared to besiege Copenhagen. Denmark promptly begged peace. Augustus had made some small gains along the Baltic but now went over to the defensive. Learning that Peter was besieging the Swedish stronghold of Narva (in modern Estonia), Charles moved his small army across the Baltic through autumn storms, then across country to Narva by roads knee-deep in mud, to find 40,000 Russians within a strongly fortified camp bristling with 140 cannon. (Peter had departed hurriedly, ostensibly to locate reinforcements.) The Russians would not come out and fight, so—like de la Gardie, taking advantage of a sudden snowstorm—the 8,000 Swedes went in after them and broke them utterly.

In 1701 Charles cleared the Saxon invaders from the Baltic coastal areas but found himself in a strategic quandary. A direct stroke across Germany into Saxony would smash Augustus, but Charles needed the good will of England and Holland to restrain Denmark and keep the Baltic trade moving smoothly. England and Holland depended on German states such as Prussia and Hanover for auxiliary troops in their war against Louis XIV of France, and no German princeling would risk hiring out his soldiers when a Swedish army was rampaging across Germany. Therefore, barred from Saxony, Charles considered a campaign against Peter, to follow up his victory at Narva. But that could not be risked so long as Augustus could use Poland as a base for attacks against the flank and rear of a Swedish army invading Russia. So circumscribed, Charles tried to deal with Poland.

Poland was a moral, military, and political swamp in which cliques of ambitious nobles jockeyed for power. The only solution Charles could conceive was to unite Poland under a new, pro-Swedish king in place of Augustus. This held him through 1705; Augustus made repeated attempts to regain control of the country, with heavy support from Peter, giving Charles occasion to win a long series of battles that made him a legendary figure. Tall, lean, simply uniformed, high-booted, and laconic, he moved and hit like lightning unchained, always winning despite all odds of men, guns, and terrain. His men—Swedes and foreigners alike—revered him. Besides his open enemies, he had to deal with covert hostility from Holy Roman Emperor Leopold I and open opposition from the Pope, who opposed the spread of Protestant influence in Catholic Poland. In 1706, with the western war clearly going against France after the Duke of Marlborough's (see page 88) victories of Blenheim and Ramillies, Charles finally invaded Saxony. Augustus quickly gave up, with no intention of keeping his word once opportunity offered for breaking it with profit.

In the meantime, the Swedish position at the east end of the Baltic was gradually coming apart. Peter, who could afford to spend soldiers in lavish numbers, had adopted the strategy of mustering overwhelming forces against the Swedish towns and detachments in the area. If successful, he devastated the countryside, driving off thousands of civilians into literal slavery. If repulsed, he fell away, to come back in greater force the next year. Charles had thought to hold the area with small forces of second-line troops; convinced that a final victory would win back any of his losses there, he did not spare reinforcements to check Peter's advance. Peter made evident his intention to hold these conquests by creating a Baltic fleet and building a new capital, St. Petersburg, in a swampy area at the head of the Gulf of Finland. Its construction, which Peter pushed mercilessly, probably killed more men than the Great Northern War.

During the winter of 1706–1707 Charles rebuilt and reequipped his army from Saxon resources. The next spring, much to the general relief of Europe, he marched eastward again. Russian troops had shoved into Poland, where they sought to set up a pro-Russian king, and were pressing their offensive in Finland. During the winter of 1707–1708 Charles deftly maneuvered the Russians out of Poland and then pushed on into Russian territory.

Russia's surest defense has always been the land itself—forests, swamp, few roads, and those bad, mud, dust, big rivers to cross, a usual shortage of food and forage, and extremes of heat and cold. Add that Peter ordered the country ahead of the advancing Swedes burned off, and that the Russian army, trained by foreign officers, was improving. The Russians intercepted a vital supply train coming down from the north to join Charles's main body; expected help from the rebellious Cossacks proved mostly an illusion. After crossing the Dnieper River, Charles found good winter quarters in the Ukraine, but the winter of 1708–1709 was one of the coldest in generations and the Swedes suffered greatly. In the spring, hoping to provoke a fight with Peter's main army, Charles laid siege to the fortress town of Poltava. Peter moved to its relief as expected, but in the preliminary skirmishing a chance shot hit Charles's heel and ranged the length of his foot. The wound became infected; Charles was too weak to take command, and his generals—brave enough but lacking Charles's drive and split-second sense of a battle's decisive point and moment—bungled. Charles had followed the action in a stretcher; Russian fire smashed it. His reopened wound running blood, he somehow mounted a horse and rallied his disorganized troops to begin an orderly retreat. Again his generals, sick, weary, and discouraged, failed him, surrendering most of the remaining army to a smaller force of Russian cavalry. Charles, with possibly one thousand men, escaped into Turkish territory, halting at Bender (now Bendery in the Moldavian Republic of the Soviet Union).

Charles remained in Turkey until late 1714, largely, it would seem, because the Russians had reoccupied Poland, cutting off his direct route home. He hoped to arrange a combined offensive, with a new Swedish army driving south, to meet a Turkish army advancing north. The authorities in Sweden, however, were confused and often ineffective, though they managed to crush a Danish invasion, but Charles did incite the Turks into war against Russia. In 1711 Peter moved south to deal with this threat and was trapped on the Prut River by an army of Turks, Tartars, and Cossacks, advised by Swedish and Polish officers. Unfortunately, the grand vizier in command of the enveloping force was not gifted with precognition and had an itching palm. For a huge bribe and Peter's promise to evacuate both his previous conquests of Turkish territory and Poland, he let the Russians go. Naturally, Peter continued to hold on to Poland.

Charles had no more such luck. Eventually, the Turks (or some of their officials—the matter is worse than obscure) seized and held him as an honored prisoner; supposedly, they nicknamed him "Iron Head." But the situation in central Europe was changing drastically. With Charles away, Denmark, Saxony, Prussia, and Hanover (the Elector of Hanover was now King George I of England and so could use the Royal Navy for his petty thievery) had joined Russia in grabbing for Swedish territory. The prospect of a stronger Prussia and Denmark, along with Peter's advance through Poland, had shaken the Holy Roman Empire. Consequently, in late 1714, the way was cleared for Charles (and some 1,162 assorted comrades in misfortune) to go home across Hungary, Austria, Bavaria, and allied minor states. Riding ahead incognito with a single officer, Charles covered almost two thousand miles in a little over thirteen days to the fortress seaport of Stralsund on the Baltic.

He came back wiser and mellower but still sternly convinced of his duty to defend his realm. He could not hold the Swedish possessions in north Germany, but he deliberately set about rebuilding his armed forces and reforming his government. He was not beaten. Denmark and Russia planned a major invasion of Sweden in 1716 with British naval support. Charles forestalled this by a hastily prepared invasion of Norway (then a Danish dependency), penetrating to Kristiania (modern Oslo) before supply difficulties forced him to withdraw. The Danes had to divert much of their fleet to Norwegian waters, and the invasion was called off amid mutual hard feelings among the allies.

Having thus reasserted his repute as the most dangerous general in Europe, Charles prepared a counteroffensive. Sweden still could muster a powerful army without robbing either the cradle or the grave. Its growing navy was pugnacious, if awkward. Moreover, Charles had found the skilled statesman-assistant he hitherto had lacked. A sincere Christian, with a strait code of personal honor, Charles never had been able to match the diplomatic maneuverings of Augustus and Peter, with their specious offers of peace whenever he had them on the run. His unwillingness to try meeting fraud with guile had made him appear unreasonable and blindly belligerent. Now he acquired one-eyed Baron Georg von Gortz, intelligent, devoted, and as difficult to pin down as a nest of rattlesnakes. Gortz negotiated right and left, playing off each enemy against the others.

In late 1718 Charles moved into Norway again. This campaign would blood his new army with the least risk, and conquered Norwegian territory could be bargained against lost Swedish holdings in northern Germany and the Baltic states at future peace conferences. The invasion went swiftly; the Danish-Norwegians being maneuvered out of their border defenses. The Swedes soon were pounding the major fortress of Frederiksten into submission, and Charles went up into the head of his siegeworks to set an example to his pioneers. An apparently chance shot out of the night killed him. With him went Sweden's Baltic empire, the Vasa line, and Sweden's military tradition.

Charles had been set a task—to maintain Sweden against the enmity of most of Europe—that probably was beyond human competence. Few men's lives have been so filled with might-have-beens. He had the talent to be a successful peacetime ruler—one of the saddest of his eulogies was "He would have been a great King, had he lived."[14]

The Vasa lessons in strategy are simple: prepare carefully before risking war, then carry the war to your enemy. Hit swiftly with your concentrated forces from an unexpected direction; keep the initiative. Once again, the Vasa leaders proved the value of high-quality troops, armed with the best possible weapons.

But, in the end, the Vasas also proved unwittingly that sufficient *mass* can overwhelm better—but fewer—troops: Swedish armies and fortresses drowned in human seas of half-trained Russians. Charles XII added the lesson that the wise strategist must know when to stop or turn back.

Turenne—properly Henri de la Tour d'Auvergne, Vicomte de Turenne (1611–75)—was heir to both the Dutch and Swedish military traditions. He remains one of the greatest of French commanders, one of the extreme few to achieve the august grade of *maréchal général des armées*. He also was unique among the ramping, stamping French generals of his time, being as longheaded and practical in his planning as he was audacious in his strategic concepts and swift and shrewd in action. Innate ability aside, Turenne's generalship was shaped by ancestry and education. The first was only partly French; the second not French at all. He was a son of Duke Henri of Bouillon, a semi-independent principality on France's northeast frontier; his maternal grandfather was William the Silent, Prince of Orange. For most of his life he was a Huguenot.[15] A sickly, slow, nearsighted, stammering boy who loved to read about Caesar and Alexander, Turenne was trained up to war in that Roman-inspired army of his formidable uncles, Maurice and Frederick Henry of Orange-Nassau. Maurice put him into the ranks to begin as a private; Frederick Henry, who succeeded Maurice as *Stadholder* (chief magistrate) of the Dutch Republic, made him an officer. During five years of sieges and small battles he grew into a quiet, studious young captain, simple, direct, and eager to serve and to learn. Because his family had dabbled in intrigues hostile to France, he was required to transfer to the French army in 1630, almost as a hostage. Here he served with Bernhard of Saxe-Weimar, the last great German free-lance and former general in the Swedish service, who had brought a corps of Gustavus Adolphus's veterans into the French service.

Bernhard (1604–39) was praised as a general who "could make something out of nothing and not be puffed-up by his success . . . more intent on redeeming his mistakes than wasting his time in excuses . . . [trying] harder to make his men love him than to make them fear him."[16] There is an old story that Bernhard, dying in bed, called for his armor and had two of his officers hold him erect to die on his feet.

From his experience with Bernhard, his Dutch training, and his own studies, Turenne evolved a distinctive strategy of deception, speed, secrecy, and surprise. It was noted that he "forgets nothing that is of use and does nothing that is superfluous."[17] He had a knack for judging the terrain, whether for large-scale strategic operations or on the battlefield, even though his myopia made it necessary that he rely on keen-sighted staff officers while reconnoitering an enemy position. Confronted by a stronger army, he would maneuver so as to hang on its flank and threaten its communications, always camping in strong positions and covering himself with a screen of scouts. He employed this strategy successfully even after his worst defeat (Mergentheim, 1645) instead of retreating westward toward France.

Turenne's only French rival was Louis II, Prince of Conde, first prince of the blood[18] and a battle captain of rare personal courage and self-confidence. Conde got his first high command through intrigue and marrying a stunted hunchback niece of Cardinal Richelieu; he justified it brilliantly at the age of twenty-two by his victory at Rocroi (1643), which destroyed the almost invincible veterans of the Spanish Army of Flanders. He and Turenne were comrades against the armies of Spain and the Holy Roman Empire, but when Conde became a leader of rebellious French nobility in that series of internal squabbles called the "Fronde" (1648–54), Turenne defeated him. Conde then went over to the Spanish, and Turenne beat him again—with the aid of a contingent of Oliver Cromwell's redcoat ''Roundheads''—in 1658 in the once-famous battle of the Dunes, near Dunkirk. Comparing the two, Napoleon emphasized that Conde was never more brilliant than at Rocroi, whereas Turenne constantly improved and showed himself more expert in each new campaign, ''proving that Turenne had the intelligence to observe and profit by experience.''[19] Conde himself told Spanish generals that when facing Turenne, ''it was dangerous to make mistakes.''[20]

As he gained in experience and skill, Turenne showed increasing audacity, routing armies much larger than his own from strong positions by swift *maneuver,* expertly handled artillery, and his knowledge of the terrain of the Rhineland and Alsace-Lorraine. Once battle was joined, he went into the thick of the fighting, conspicuous on his piebald black-and-white mare, which his soldiers called ''the Magpie.'' Turenne's masterpiece—his 1674–75 winter campaign—was typical of his combination of painstaking preparation and driving execution. Withdrawing deep into Alsace when confronted by greatly superior numbers of Imperial and Brandenburg troops, Turenne began distributing his army into winter quarters. His weary enemies did likewise around Colmar in eastern Alsace. Suddenly regrouping, Turenne swept south through deep snow and bitter cold along the western slopes of the Vosges Mountains, which hid his march. He was around the southern end of the Vosges and in the midst of his snugly settled-in enemies before they realized he had moved.

At the same time, Turenne was seeking to modernize the French army. Having experienced the value of steady foot soldiers in the Netherlands, he began by building up his infantry. Native French infantry had been a neglected, often-scorned arm: French kings had preferred to hire Swiss or Germans. Possibly a turning point had come during the Wars of Religion at Coutras (1587; see pages 43 and 44) when some Huguenot arquebusiers, detached as *enfants perdus* to cover Henry of Navarre's left flank, went in against overwhelming numbers of enemy infantry with a howl of ''Il faut mourir dans le bataillon'' (a polite version of the old American ''Come on you sons-a-bitches! You want to live forever?''). Rolling under or smashing through the bristling front of pikes, they came up killing with sword, dagger, and gun butt.[21] Turenne improved his infantry's fire power and loosened their tactics so that they could move in ranks across broken terrain. More important, Turenne built up its morale. French cavalrymen considered themselves gentry, infantrymen mere peasants; the most junior cavalry officer could hardly be brought to salute an infantry colonel. Now Turenne's dogfaces made little jokes: Look at that funny animal with six feet, two hands, two heads, and only one hat. "Its brains are in the head that doesn't wear the hat."[22]

Among the Holy Roman Empire's generals, only Count Raimund Montecuccoli (1609–80) could match Turenne, and that rarely. Like Turenne, Montecuccoli had begun as an infantryman and had studied the Greek, Macedonian, and Roman art of war. He wanted lances for his cavalry, pikes and shields for infantrymen. He shared Turenne's appreciation of speed and surprise—"march energetically by night, along little-known, little-used roads." As a general of the chronically unready and impoverished Holy Roman Empire, he stressed that "It is necessary to make preparations ahead of time, while the nation is at peace" and that such preparations should include a full treasury. He won battles in Hungary against the Turks and wrote his memoirs, *Commentarii Bellici,* in Latin.[23] He sought to keep his army well concentrated, a moving fortress that could go where it pleased and repulse all attacks.

Turenne and Montecuccoli first clashed during the closing campaigns of the Thirty Years' War; their masters' contest of intricate maneuver and savage battle through the Rhineland during Louis XIV's Dutch War (1672–79) was much admired and studied. There was mutual respect between them. In 1675, when Turenne caught his wily opponent at Sasbach, he could remark, "I think Montecuccoli would approve of what we have done."[24] But—at the critical moment as his trap closed—Turenne took his death wound. His generals lost their wits and control of the action. Montecuccoli broke free and drove the French back across the Rhine. There the aged Conde, pardoned and once more a French marshal, rallied them.

Unemotional common sense was Turenne's governing characteristic. He lived simply in the field, "dressed all in rough woolens, bundled up in his cloak, recognizable only by his hard, satanic features."[25] He made himself unpopular with many nobly born officers who wished to drag along wagonloads of luxuries on campaign. He took excellent care of his men and had much to do with getting the French army put into standard uniforms. Personally generous and humane, he had to wage war in Germany under royal orders to eat up the country, raise money, and create disorder. The French ravaging of Rhine Palatinate in 1674 was especially ruthless. Bands of desperate peasants trapped isolated soldiers to hang them upside down, burn them over slow fires, or leave them blinded and crippled along the roads. The French retaliated with interest, but Turenne—almost alone among French commanders—did not indulge in private looting.

Somehow he remained more Dutch than French. He did not care for life at the royal court, had no charm, and was a frequent target of sharp-tongued courtiers. A strict disciplinarian in a most undisciplined army, he was resented by French generals, even after Louis XIV made him marshal-general in 1660. One of them heard his famous ultimatum "I speak harshly to no one, but I will have your head off the instant that you refuse to obey me."[26] His enlisted men considered him their "father." When a number of generals were promoted to marshal after his death, the army contemptuously dubbed them "Turenne's small change." At Sasbach, as his generals dithered, his soldiers shouted, "Let the Magpie go forward. She will lead us!"[27]

Louis XIV employed Turenne as a trusted adviser in military matters and as a diplomat but never rewarded him according to his true deserving. Never wealthy, Turenne several times found money out of his own pocket to pay his troops; before the battle of the Dunes he broke up his own silver plate to get needed funds. He did not use his position to enrich himself and died comparatively poor. One of the officers he trained was a handsome English volunteer named John Churchill, who would become Duke of Marlborough. Among his nonmilitary friends was Jean de la Fontaine, author of the famous *Fables,* which Louis XIV considered somewhat subversive.

Turenne's *Memoirs* began in 1643 and end abruptly as of 1658. Since he was an awkward writer, they are not easy reading,[28] but—with his correspondence—they show his concept of strategy. He believed that siege warfare consumed too much time, manpower, and money that could be more profitably expended in defeating the enemy's main army: "It is a great mistake to waste men in taking a town when the same expenditure of soldiers will gain a province." His use of deception resembled Sun Tzu's: "Seem sometimes to show fear, to give the enemy greater confidence in their own strength, and to make them more negligent and less distrustful of you." To "throw terror and consternation into the enemy's country," secretly separate your army into several task forces and carry out several operations at the same time, while exaggerating their strength. If you have to face an army made up of different nationalities, try to break up their alliance by raising discord and jealousy among their generals. By contrast, like Stonewall Jackson, Turenne felt that the best defense was to hit first. And he left one unique comment: "A blockhead has sometimes perplexed me more than an able general."[29]

Napoleon proclaimed Turenne "the greatest of the French generals: he is the only one who became bolder with age" and carefully studied his campaigns, as did Maurice de Saxe and Frederick the Great. And Nathanael Greene read his *Memoirs,* along with Euclid's geometry and Saxe's *Reveries* (see page 84), beside his forge while he yet was only a Quaker ironworker. Greene's daring campaign with a weak, almost-forgotten army through the Southern states during 1781–82 showed what Turenne had taught him.

Louis XIV, the "Sun King," loved military display possibly as much as he loved his mistress of the moment and his court ballets. He might go in stately processional, complete with current mistress and all possible luxury, to observe the siege of some large enemy fortress—*if* the weather were pleasant, the distance not too great, and the risk minimal. (Once, at least, the weather suddenly displayed an appalling lese majesty and subjected Louis to a few common discomforts.) However, though he had no personal interest in war, he did see it as the means of making himself and France (inseparable in his opinion) the greatest and most feared power in Europe. Almost half of his seventy-two-year reign was spent in predatory foreign wars; by way of variety, his revocation of Henry IV's Edict of Nantes, which had guaranteed Huguenots freedom of religion, provided the ferocious ten-year Camisard civil war in southern France.

After Turenne, most of Louis XIV's and Louis XV's (king, 1715–74) generals were rather small-bore warriors. Perhaps the ablest one was François Henri de Montmorency-Bouteville, Duke of Luxembourg (1628–95), a dwarfish hunchback who sent so many captured flags back to Paris for display in Notre Dame that he was nicknamed that cathedral's "tapestry maker." He had the wit to equip his troops with skates for a winter invasion of Holland, but a sudden thaw intervened, and the French had to wade, swim, and wallow to escape.

He was accused of being an amateur poisoner, and his treatment of enemy civilians would have drawn unfavorable comment from Genghis Khan. Withal, despite far stronger forces and numerous victories, he ended baffled by the fighting fury of William III of Orange, great-grandson of William the Silent.

Claude Louis Hector de Villars (1653–1734) had served with Turenne and learned from him to move rapidly and keep his plans to himself. Like Turenne, he was more popular with his soldiers than at court; unlike Turenne, he was brutal and a braggart. On occasion, he was very much law unto himself. When an aide asked why he would allow an enemy column to escape while he enjoyed a lavish breakfast, Villars reputedly replied:

> Certainly I could have taken them all, and put an end to the war today. But what would have been the consequence? I should have returned to Versailles, to be lost in the crowd, or, perhaps, to Villars, to die of ennui. Now, I like much better to stay here as Commander-in-Chief.[30]

The one French general of this era to have a really long term influence on military strategy was Sebastien le Prestre de Vauban (1633–1707). He originally had been an infantry officer with a knack for military engineering, which had to be handled by volunteers from line regiments in the pre-Louvois days. Through skill and nerve he rose to *Commissaire-General* of Fortifications in 1667 and marshal of France in 1703. He is credited with having strengthened over three-hundred fortified places and building thirty-three new ones, showing amazing ability in fitting his fortifications into any sort of terrain. But his great achievement was the development of a scientific method of capturing fortresses with the minimum loss of time and lives. Using Vauban's new techniques, attackers would literally dig their way up to the fortress's ramparts, constructing a carefully calculated system of trenches that were so angled that the defenders' artillery could not enfilade them. This was coupled with a more effective employment of siege artillery. Where previously the siege of a strong, well-defended town might require most of a campaign and, even if successful, leave the victorious army used up by casualties and sickness, Vauban's methods usually made it a matter of ''Place besieged; place taken." This definitely altered European strategy. An attacking army might be able to knock out the enemy nation's border fortifications and drive deep into its territory before adequate defensive forces could be collected. Thus Holland, caught unprepared in 1672 (Louvois had utilized a Dutch traitor to buy up most of the ammunition in the country), was almost overrun by the first surge of some 125,000 French—until future William III of Orange opened the dikes, as his grandfather had done to thwart the Spaniards. (Sieges could still be dangerous operations; after the French took Namur in 1692, one of them remembered, "There were sixty of us engineer officers at the siege: Twenty-two came out alive. The rest died in the trenches."[31] A particularly risky business was reconnoitering breaches made in the enemy's defenses by mines or artillery fire, to determine if they were large enough for attacking infantry to break through into the interior of the place. In each of the fifty-three sieges he directed, Vauban undertook that mission himself.)

Vauban was a large, gruff man, rather careless of his appearance, humane but shrewd. He held that laborers paid for piecework would work harder than those receiving a daily wage. As *Commissaire-General* of Fortifications, he roamed France, often in a roomy sedan chair (practically a traveling office) slung between two mules. An itch for work and knowledge kept him constantly busy. Between campaigns he toiled as a civil engineer, constructing roads, aqueducts, and harbor installations, and tried to collect data on France's resources, population, and topography—probably the first attempt at a national survey.

The barrier of fortresses he constructed and strengthened around France's frontiers had a definite strategic importance, especially the network along her northeastern border. Until the introduction of rifled artillery in the mid-nineteenth century, they could delay an invader or serve as bases for a French offensive. Even Marlborough found them a major problem in 1706–12. Also, besides his fortifications, Vauban left France a corps of expert, dedicated military engineers, acknowledged the finest in the world. French engineer officers, trained in that school, served in the Continental Army and aided in the development of American military engineering and the U.S. Military Academy thereafter.

Louis XIV's actual right-hand man in waging his wars was François Michel de Tellier, Marquis de Louvois (1641–91), the first modern secretary of war. A talented, energetic organizer and administrator, he rebuilt the French army, establishing the Engineers and Artillery as separate arms, creating the office of the Inspector General, developing a centralized supply system, and standardizing uniforms, weapons, and equipment. He put soldiers into barracks instead of quartering them in citizens' houses and pushed the fortification of France's frontiers. Most of this meant a continuing battle with the French nobility who provided the army's officers. Colonels had regarded their regiments, and captains their companies, as business enterprises which should show profit. Too many cheated their soldiers; others simply neglected them. Louvois painstakingly corrected such habits. He gave his king a far more efficient military machine, and a much stronger one. France could muster 400,000 men, properly equipped and supplied, and wage war in the Netherlands, Germany, Italy, and Spain at the same time.

Louvois, however, was a megalomaniac. Not content with creating this splendid army, he took it upon himself to command it from his Paris bureau. He had no real military experience or any grasp of tactics and strategy, but assisted by his favorite "whiz kid," the Marquis de Chamlay, an expert topographer but also no soldier, he issued detailed advance orders for generals in the field, down to what roads they should follow, how many miles they should march, and where they should camp. This may have furnished incompetent commanders an alibi when things went wrong, but it irked the capable ones (much to Chamlay's surprise) and hobbled their operations. Turenne disregarded such nagging but at the risk of being refused supplies and reinforcements.

Louvois is blamed for encouraging Louis to make war on any pretext. Personally, he seems to have been a somewhat nasty specimen. The 1674 ravaging of the Palatinate was by his specific order. And in Holland, watching Luxembourg burn a small, unfortified Dutch town and its inhabitants together—after having his troops rape the women—Louvois exulted, "We lit the village and grilled all the Hollanders in it."[32]

These great armies that Louvois raised—and Louis's enemies perforce had to equal—produced a whole new set of strategic and tactical problems. The wars of the eighteenth century avoided the total brutality of the Thirty Years' War, but they were messy enough, the major limit on their destructiveness being the general inefficiency of the armies and their command systems. At Blenheim (1704), Marlborough and Eugene would thrust their 52,000 English, Austrians, Prussians, Danes, Hessians, and Hanoverians against 60,000 French and Bavarians. Five years later, at Malplaquet, the respective numbers were 100,000 and 90,000. Only truly competent generals could handle such forces effectively. Their staffs were comparatively small and unspecialized; their staff officers commonly without specific training in the vulgar details of reconnaisance, ration returns, and march discipline. Also, there was no permanent military organization larger than a regiment. Regiments were grouped into brigades, the brigades lumped into "lines" or "wings," but unlike modern divisions or corps, these were temporary formations and so had very little ability to maneuver independently. Consequently, battlefield control was often difficult, especially when it was necessary to shift troops rapidly to meet an unexpected attack or organize a pursuit. An army that had been drilled into a clockwork machine that could march a few paces more an hour, fire a round more per minute, keep its ranks better aligned over rougher ground, endure getting killed a few minutes more than the average generally won its battles.

It took long, intensive training to bring soldiers to such a degree of efficiency. Generals therefore did not like to waste them in actions that did not promise success, preferring to employ what has been termed a "Strategy of Evasion." This involved gradually sapping the strength of the enemy army by frequent small engagements, using light troops to cover your maneuvers and raid his communications. Meanwhile, you keep your main army intact, choosing your positions carefully so the enemy would not dare attack you. Applied by real experts, such as Austrian Field Marshal Ludwig Khevenhuller, who had efficient forces of irregular troops (see page 99), this strategy of calculated maneuver and carefully applied force could gradually destroy hostile armies without major battles. (In fact, there is a great similarity between this strategy and that of Mao Tse-tung; see chapter IX.) Frequently, however, it ended as a drawn-out fumbling match that killed men to no purpose.

Sometimes defeating the enemy was less of a problem than feeding your own troops. "Understand," wrote Frederick the Great, "that the foundation of an army is the belly."[33] Since it was no longer possible to feed these new, immense armies off the countryside, generals had to pay far more attention to their supply systems or see their armies dissolve from desertion and sickness. Generally, the soldier was issued bread, with a little meat from cattle driven along with the army. But forage for its animals, straw for bedding, firewood, cash contributions, and anything edible or potable that the soldiers and their camp followers could snap up in passing still had to come from the countryside. Just to supply the troops with bread required long wagon trains, mobile baking units, and a series of well-stocked "magazines." Moreover, most European roads were bad and supply matters were normally handled by contractors of minimal virtue and sticky-fingered semicivilian commissaries. Thus hampered, a daily march often was only three to six miles; an army risked hunger if it got much over one hundred miles from its nearest magazine. (In the French army, nobly born officers were accompanied by gaggles of "valets, lackeys, cooks, hairdressers, whores, priests, and actors . . . whole chests full of perfumes and scented powders and great quantities of dressing gowns, hair nets, parasols, and parrots," which added to the traffic jams.[34] Strategy had to consider more than ever the control of major roads, rivers, and canals by which supplies might be moved.

Two French officers with unusual backgrounds—Jean-Charles, Chevalier de Folard, and Maurice de Saxe—proposed reforms to make these armies less unwieldy. Folard (1669–1752) was probably the most influential military writer of his day. He had run away from home at fifteen to see the wars and had quickly won a reputation as a daring officer of irregular troops. When France was at peace, he sought active service in foreign armies. During 1716–17 he was in Sweden where he saw little combat, but was a witness of Charles XII's last reorganization of the Swedish army and staff, designed to provide maximum flexibility and coordination. Moreover, he was able to discuss the art of war with Charles (whom he considered superior to Alexander) and his veteran officers. The Swedish *gā-pä* tactics, with their reliance on cold steel and silent, smooth-flowing maneuvers, impressed him deeply. Folard combined these experiences with study of classical warfare to produce two widely read books, *Nouvelles Decouverts sur la Guerre (New Discoveries Concerning War)* and his six-volume *Histoire de Polybe (History of Polybius).*[35] He held that both experience and study were needed to really master the art of war, and sought to teach by the use of historical examples. For battle formations he favored a heavy column, something like the Macedonian phalanx. It was never tested in combat, but the idea would appear in later French tactics. Over a century later, Folard's work would inspire one unusual student—Louis-Nicolas Davout, the sternest of Napoleon's marshals, who never lost a battle.

Because Folard's critical spirit irked many of Louis XV's powder-puff generals, he was never promoted beyond the grade of *mestre de camp* (colonel) of infantry, but he found an appreciative friend in Maurice de Saxe.

Saxe (1676–1750) was the eldest of Augustus II's bastards. His mother was Countess Aurora von Königsmarck, famous as one of the most beautiful women in Europe, who had made the mistake of visiting Saxony. Augustus had the boy christened "Moritz" (German for "Maurice") "in memory of the victory I gained over his mother at the castle of Moritzburg."[36] (Aurora seems, thereafter, to have remained in Saxony; legend has it that Augustus loosed her against Charles XII in the hope she could play Delilah to his Samson. Charles *is* known to have refused to see the "royal whore.")

After an education of sorts, Augustus made his son an ensign (second lieutenant) of infantry at the age of twelve and sent him hiking from Dresden to Flanders to serve under Marlborough. The next year he was at "murdering Malplaquet," where Marlborough lost a fourth of his army in winning his last and most desperate battle. Saxe grew up thus in camps, battles, sieges, and long marches. At seventeen he was an experienced cavalry colonel, widely known for daring and competence, and the father of at least one bastard of his own. At eighteen, Augustus forced him into marriage with an extremely wealthy fourteen-year-old heiress. (It was not a success; Saxe soon got rid of her fortune, and the marriage was annulled.) When the wars in northern Europe were done, he fought under the famous Prince Eugene of Savoy against the Turks. In 1720 he went into the French service as a *marechal de camp* (roughly, a major general) and fought his way up to the grade of Marshall of France by 1743.

Saxe was almost a split personality. A big, immensely strong, fearless man, possessed by sweeping ambition and consuming energy, he was wild and irregular in his private life. His stable of mistresses was described by Madame de Pompadour, official mistress to Louis XV, as "a train of street-walkers."[37] He dreamed of gaining a kingdom of his own—location of no particular importance—and indulged in amazing intrigues to that end. The other Saxe was a sincere, careful, industrious soldier who kept his troops under firm discipline, studied all aspects of his profession, and taught that "The man who devotes himself to war should regard it as a religious order into which he enters. He should have nothing, know no other home than his [regiment], and should hold himself honored in his profession."[38]

Saxe's *Mes Reveries (My Reveries on the Art of War)*[39] were written in December 1732 in thirteen nights. He was sick and sought to amuse himself and probably, in modern slang, to get his ducks lined up. He had been unable to determine any definite rules governing warfare and was certain only that contemporary armies showed a sad falling off from the days of Gustavus Adolphus. Though somewhat jumbled, his book ticks off how troops were raised, supplied, trained, and fought and how all that should be changed to provide a really effective army. The work also contains elements of fantasy: He wanted to put his heavy cavalry into complete armor (he had invented a new type that would weigh not over thirty-five pounds); half of the infantry should have pikes and all of them oval shields, as Montecuccoli had advised. But the rest was either common sense or prophecy.

Saxe wanted universal military service; every man, "whatever his condition in life," should serve for five years, sometime "between the ages of twenty and thirty . . . the years of libertinage, when youth . . . is of little comfort to parents."[40] He advocated sensible uniforms, proper rations, decent pay, and promotion according to merit instead of social status. He had rediscovered the Roman habit of mixing vinegar with drinking water to purify it and the art of marching troops in cadence to military music, which made it possible to maneuver far more rapidly and exactly. He wanted soldiers and horses kept in good physical condition—"all the mystery of maneuvers and combats is in the legs"; practical training was much more important than barrack-yard formal drill, but this training must be uniform throughout the army. Soldiers must not be left in idleness (like the American occupation forces in Japan before the Korean War): "They murmur at every trifling inconvenience, and their souls soften in their emasculated bodies."[41]

He wanted to reorganize the French infantry into "legions"—

self-contained forces of light and heavy infantry, cavalry, and light artillery, much like our modern infantry divisions. The light infantry were to be trained as skirmishers and snipers, and all infantrymen as marksmen. To make their fire more effective, he had invented a breech-loading musket.[42] Being of an empirical turn of mind, Saxe had tested Folard's columns, apparently on the drill field, and concluded that they were too big and clumsy, though he honored Folard for having the courage to introduce new theories.

As for strategy, Saxe gave it little specific attention. He discussed river crossings, irregular warfare, sieges, fortifications, and the technique of feeding an army in winter quarters off the countryside with the least harm to the inhabitants. In covering mountain warfare, he noted the odd point that there usually were passes that the local people never used and might not know of; these could be found by a careful search and used to surprise an enemy on the far side of the mountains.

In theory, Saxe advocated the Strategy of Evasion. "I do not favor pitched battles, especially at the beginning of a war, and I am convinced that a skillful general could make war all his life without being forced into one."[43] In practice, he fought whenever there was a chance of crushing the enemy army, and he urged that a defeated enemy must be attacked, pushed, and pursued.

The most unexpected aspect of Saxe's doctrine is its humanity. He wanted the enlisted soldier treated fairly; he championed the "poor gentlemen who have nothing but their sword and their cape," yet made the best officers; he did not want officers who were "frank libertines," because they could stand neither hardship nor discipline.[44] But beyond that he puzzled over the question of soldiers' morale. Why would victorious troops suddenly give way to panic? Why were some regiments always successful when attacking but unreliable in defensive fighting? How could a general keep his men in good spirits?

> It lies in human hearts and one should search for it there. No one has written of this matter which is the most important, the most learned, and the most profound, of the profession of war. And without a knowledge of the human heart, one is dependent on the favor of fortune, which sometimes is very inconstant.[45]

Saxe's great period came during the War of the Austrian Succession (1740–48). Possibly through the good offices of Madame de Pompadour he secured command of the French army operating in the Austrian Netherlands (now largely Belgium) against Austrian-Anglo-Dutch forces. Like Turenne, he had many troubles from arrogant French noblemen who resented serving under a foreign bastard. Also, hard service and debauchery had ruined his health; he was so swollen with dropsy that he could seldom mount a horse and had to command at times from a small, padded wicker couch-on-wheels. Frequently in agony, needing to be periodically "tapped" by his doctor, Saxe nevertheless won battles at Fontenoy (1745), Rocourt (1746), and Lauffeld (1747). "It is not a question of living, but of acting," he told Voltaire.[46] Louis XV rewarded him generously. Saxe was given the vast castle of Chambord, with his own regiment of uhlans—men of all colors, religions, and races—as his guard. (Possibly out of native obstinancy, he had retained his mother's Lutheran religion, at least pro forma. But at Chambord he marched his roughnecks, Muslims and all, regularly to mass.) Saxe also had his own theater, having always admired that art—and especially various actresses. And he became Marshal-General of France.

His *Reveries* were printed posthumously in Amsterdam in 1757, and a badly botched English translation appeared in London that same year. They were the stuff to appeal to practical soldiers and so were very popular in America. They still are worth reading, not only for their unsparing picture of an eighteenth-century soldier's life, but also for their constant hammering at the importance of common sense and devotion to duty.

V

QUEEN OF PANDOURS, KING OF BRUMMERS

The army went to Flanders and swore horribly.
TRADITIONAL

But honour and dominion
Are not maintained so.
They're only got by sword and shot,
And this the Dutchmen know!
RUDYARD KIPLING[1]

Through the general period of the seventeenth and eighteenth centuries, France and England were alternately friends and foes. In comparison to France, English interest in the military art was sporadic. Lessons hard learned under Elizabeth were forgotten under James I. The English Civil War (1642–51) between Charles I and Parliament began as very much an amateur affair, guided by the few officers on each side who had served in Holland or with the Swedes. Out of it came Oliver Cromwell (1599–1658), who had no previous experience with soldiering but did possess resolution, common sense, and a natural talent. Cromwell had knowledge of Gustavus's system and probably also of La Noue; he followed the Swedish practice of carefully training and disciplining his soldiers and went beyond that in selecting (at least for his cavalry) men of character and zeal. Like Napoleon, he had no preconceived plan except to destroy the enemy's main armies, and he took whatever strategy was necessary to that end. Guided by an efficient intelligence service, he moved swiftly and had a particular ability to so move large bodies of troops. As a strategist, he maneuvered to pull his opponent into a trap or to strike from an unexpected direction at an unexpected time. His "New Model" army was one of the most efficient in history, but it vanished after his death.

Charles II, the "Merry Monarch," came back from exile in 1660. Louis XIV, who had been a most respectful ally to Cromwell against Spain, hired Charles—with cash payments and succulent Louise de Kerouallle—as an ally against Holland. (Charles made Louise duchess of Portsmouth. His loyal subjects called her the "French Whore" in contrast to popular Nell Gwyn, the "Protestant Whore.") This alliance brought England only disgrace in sea battles and home waters, Charles having let Cromwell's efficient navy go to pot. The Dutch sailed into the Thames in 1667 and burned or took unready warships there at anchor. Charles II was followed by his brother James II, a Catholic and unpopular. A revolt sent James fleeing to France; James's daughter Mary and her husband, William III of Orange, came over to rule in England.

One of the movers of that revolt was John Churchill (1650–1722), formerly an officer under Turenne. Very brave, charming, tactful, and astute, he was successful both as a soldier and a diplomat. He had been a protégé of James's, who relied on him. William and Mary made him Earl of Marlborough and entrusted him with the reorganization of the British army. Churchill served under William in Ireland and the Netherlands, concluded that he was not sufficiently rewarded, and was caught plotting with James. William gave him a few weeks in the Tower and six years "in Conventry," stripped of all his appointments. However, with war with France threatening, he finally was eased back into favor, and in 1701 he was dispatched to Holland to command the British army forming there and to serve as a special ambassador to the Dutch government.

William had created a "Grand Alliance" out of England, Holland, the Holy Roman Empire (Austria, Bavaria, minor German states), and Portugal to oppose France, Spain, and Savoy. (Savoy promptly sold out to the Grand Alliance.) The War of the Spanish Succession would rage from 1701 into 1714 across western Europe and around the world. American colonials called it "Queen Anne's War." In early 1702 William died; Mary having predeceased him, the throne passed to her younger sister, Anne. A woman of poor health and limited intellect, Anne nonetheless was a firm ruler. Marlborough's wife, the former Sarah Jennings, had been Anne's close friend since childhood and had great influence over her; Marlborough himself had Anne's trust because of his long association with her father. She made him Captain-General of her armies and later Duke of Marlborough. Sarah, unfortunately, involved herself in politics and had a biting tongue—and came to use it on both Queen and husband.

It is one of the oddities of military history that Marlborough's campaigns have received comparatively little attention. Seemingly they have been overshadowed by those of Frederick the Great (whom Marlborough would have eaten for a midmorning snack), Napoleon (who considered Marlborough an outstanding general), and the Duke of Wellington (a junior-league Marlborough). No general has had to carry such responsibilities for so long and discharged them with such grace and courtesy. His soldiers noted that he "was particularly happy in an invincible calmness of temper and serenity of mind; and had a surprising readiness of thought, even in the heat of Battle."[2] Only a third of his troops were English; the greater part were Dutch (including Germans and Scots in the Dutch service). The rest were Danes, Prussians, Hanoverians, and assorted German contingents paid by England or Holland. He had need of all his diplomatic skill and self-control to convert this polyglot assemblage into an effective army. It was said that Marlborough "possessed what so few men of genius are endowed with—ability to tolerate fools gladly."[3] That gift frequently must have been taxed to its utter limits in his constant dealings with the defense-minded Dutch, slippery German princelings, and corrupt English politicians who sometimes seemed determined to destroy their own army, and *were* determined to destroy Marlborough.

Some of his best English regiments were diverted to a fruitless invasion of Spain; in 1711 Queen Anne stripped him of five more for an expedition against Quebec. Since Sarah had proved too prickly,

Anne had found a new favorite, an Abigail Masham. (Mrs. Masham had a yardbird brother with the resounding name of John Hill—unofficially, ''two-bottle Jack'' to his unfortunate associates—and a thirst for glory. Hill, the Royal Navy, and greedy-gutted Massachusetts citizens collaborated to produce a major fiasco.) When not busy with military operations, Marlborough had to tend to such incidental chores as encouraging Charles XII to go fight the Russians instead of clobbering the Holy Roman Empire, which had been giving his enemies covert support; or persuading Prussia to send newly raised troops to his army rather than to Augustus II of Saxony; or urging the feckless Holy Roman Empire to do *something* useful; or, hardest of all, attempting to maintain his own political fences in England.

Marlborough waged ten campaigns. He took every town he besieged and won every battle he fought. He utterly broke the French army's reputation as the finest in Europe, though its troops fought gallantly and well and its generals were mostly competent. His strategy had nothing evasive about it. Like Cromwell, Napoleon, or U.S. Grant, he moved out to find his enemy, catch him at a disadvantage, and destroy him. Repeatedly, he lost excellent opportunities to really break France because the Dutch would refuse to risk an offensive battle, even with the odds all in their favor. Not the least of his abilities was that of keeping his army fed and cheerful. To his British soldiers and many of the foreigners who followed him for pay, he was ''Corporal John'' who took care of them, shared their dangers, and led them to victory. He was accused, with perhaps some justification, of being avaricious; most charges against him proved false or exaggerated, but he did not scorn to snap up an occasional windfall, such as charging a rich Jewish merchant £6,000 a year for the privilege of keeping an agent with his headquarters. (This agent reported the outcome of every battle by courier, enabling the merchant to make a killing on the stock exchange.)

Marlborough's great victories—Blenheim (1704), Ramillies (1706), Oudenarde (1708), and Malplaquet (1709)—are among the British army's proudest battle honors. His Blenheim campaign was particularly skillful: Bavaria had abandoned the Holy Roman Empire and joined France. There was pressing danger that a Franco-Bavarian offensive would overrun Austria, link up with a French-subsidized revolt in Hungary, and dismember the Holy Roman Empire. Marlborough's excellent spy service had learned the French plans; keeping his own intentions secret, he made a leisurely march up the

Rhine, threatening several different objectives to keep the French off-balance and devastating Bavaria in a fashion that Louvois would have found at least acceptable. Then he quickly linked up with the Imperial army of Prince Eugene and routed the Franco-Bavarian army at Blenheim. (Prince Eugene [1663–1736], *"der edle Ritter"* of Austrian song, had been born a French subject but was refused a French commission by Louis XIV, who found him unpleasing to the royal eye. Eugene therefore joined the army of the Holy Roman Empire to become a notable scourge to French marshals and Turkish pashas, and a loyal and gifted comrade-at-arms to Marlborough.)

By 1711 Marlborough's political opponents had secured control of Parliament, and England was war-weary. Louis XIV offered tempting terms for a separate peace. On December 31 Anne relieved Marlborough from active duty, and England crawled out of the Grand Alliance, abandoning her allies. Villars won several successes over Eugene and the Dutch, and the war ran down. Marlborough went into voluntary exile in Holland until Anne died in 1714. Her successor, George I, recalled him, but by then Marlborough was aging and sick. He did receive a splendid state funeral.

Like Robert E. Lee, a great captain of similar pugnacity, Marlborough left no memoirs and so must be judged from his deeds. His massive correspondence was far more concerned with bread contracts and the stubbornness of Dutch generals than with strategy. But his record as an undefeated general stands unmatched.

Through the seventeenth century, through all its wars, the north German principality of Brandenburg had steadily increased its size and influence, gaining territory from its neighbors by force or as concessions for the help of its oversized, efficient army. Its rulers—the House of Hohenzollern—were able and tough-minded; even their one frivolous member, Frederick I (1657–1713), who had a taste for the arts, silver furniture, and debts, got his status raised from "Elector" to "King in Prussia." His son was the no-nonsense, honest, and somewhat boorish Frederick William I (king, 1713–40), who restored the nation's prosperity and built up a fine army of 83,000 men—the fourth strongest in Europe. His delight was his "Giant Grenadiers"—men over six feet tall, recruited, kidnapped, or presented to him from all over Europe—who not only formed an impressive ceremonial guard but also served as an experimental unit for the king's testing of new weapons, tactics, and equipment.

Possibly Frederick William's major trial was his oldest son Frederick, a slovenly youngster who preferred literature and his flute to learning how to be a ruler and soldier and referred to his uniform as his "shroud." Worse, he disdained the German language and ways, preferring French—which Frederick William detested—and wrote poetry, all of it in French and most of it horrible. Growing increasingly crabbed with age and ill health, Frederick William feared his son would be another shiftless king like his own father. His corrections grew harsher and more public, especially after he discovered that Frederick was carrying on secret intrigues with the British government. In 1730 Frederick was caught trying to flee to England. His father treated him as a military deserter; Frederick had to watch the execution of a friend who had helped him and might have been ruthlessly punished himself had it not been for the shocked intervention of Holy Roman Emperor Charles VI, other rulers, and most of the Prussian generals. With the fear of death and his father upon him, Frederick submitted and took up his studies. Had it not been for a secret allowance from Charles VI, he would have had a pinched existence. Gradually he was given more responsibilities and authority, as well as a wife, whom he soon ignored except for occasional insults. He also corresponded with the famous French philosopher-historian Voltaire, to whom he pictured himself as a future philosopher-king.

On his father's death in 1740 Frederick certainly gave the appearance of one, lifting press censorship, reestablishing the Academy of Berlin, abolishing the use of torture in judicial proceedings, and allowing freedom of worship. After Frederick William's funeral, his giant guardsmen were disbanded. Frederick capped all this by publishing his *Antimachiavel (Against Machiavelli)* in which he denounced Machiavelli's concept of a successful "prince" (see page 40) and praised virtue, honesty, wisdom, and duty. Somehow, Frederick had grasped a basic truth of statecraft—get the fashionable intellectuals on your side and you can get away with murder! The educated classes of western Europe already were bemusing themselves with delightfully vague liberal notions—which would explode a half century later into the French Revolution. When Frederick presented himself as an enlightened and benevolent ruler, the European intellectual community gathered him to its collective bosom and ever afterwards found excuses for his most inexcusable behavior.

Frederick himself was a thorough autocrat. He might punish injustice and corruption, but there was no appealing any of his decisions. He was steadily increasing his army—the disbanded "giants" came parading back as his "Grenadier Guard Battalion." Neighboring states might have been warned when he occupied the then-independent principality of Liege because of a minor dispute and literally held it for ransom by its ruling prince-bishop. Voltaire, however, found this only a refreshing example of youthful resolution.

Four months after Frederick became king, Charles VI died, leaving his Hapsburg lands (Austria, Hungary, Silesia, and modern Czechoslovakia) to his twenty-three-year-old daughter, Maria Theresa. By years of patient diplomacy, he had persuaded Prussia and the major European powers to recognize this heritage, but he had not heeded his generals' warnings that a strong army was a better guarantee. As they had with Charles XII, the vultures gathered. Frederick at once demanded Silesia, a rich province north of Bohemia, promising military support in return. As Frederick himself wrote later, Austria's "finances were in disorder, the army run down . . . with all that a young princess without experience at the head of the government."[4] Maria Theresa, one of the truly great women of history, refused his demand. Frederick invaded Silesia, proclaiming himself the protector of its Protestant minority, and the War of the Austrian Succession (1740–48; in America, King George's War) was on. Frederick's infantry won his first battle at Mollwitz in 1741, though Frederick did not share the victory, having fled when his cavalry was ridden off the field early in the action. France, Bavaria, Saxony, Spain, and Savoy now moved in on Austria. England and Holland aided her. Frederick signed an alliance with France, but casually broke it when Maria Theresa ceded him Silesia in order to concentrate her scanty forces against France and Bavaria. In 1744, with Austria clearly winning, Frederick broke his treaty with Maria Theresa and reentered the war, but both Prussia and Austria were glad to make a separate peace again in 1745.

Thereafter, Frederick built up Prussia and his army. The peace obviously was only a breathing spell, but Frederick did himself no good by cheerfully insulting both the Czarina Elizabeth of Russia and Madame de Pompadour, who kept Louis XV and France in the folds of her nightgown. France and Austria formed an alliance; Sweden, Russia, Saxony, and—later—Spain joined them. England supported Frederick. The resulting Seven Years' War (1756–63; in America, the French and Indian War) strained Prussia to the utmost. The odds against Frederick seemed overwhelming, yet they were actually far less than they seemed. His enemies seldom were able to coordinate their offensives; when they did, they mismanaged them. Sweden had sunk to a minor nuisance, though the Swedes did gobble up Frederick's embryo navy in 1759. Russia had masses of tough troops, a mix of savagely disciplined regulars and Cossacks, Kalmucks, and other barbarians, notable mostly for the atrocities they inflicted on civilians. The Russian military still depended considerably on foreign officers (whom Russian enlisted men might shoot when things got sticky) and was stubborn, slow, and cumbersome. It had nothing much of a supply system and was always running out of food and having to go home. The Austrian army was valiant, its artillery and light troops the best in Europe, but it was impoverished and cursed with an unusally glimmer-witted and inert set of senior generals. Saxony had good troops but equally incompetent generals and a decaying royal family. The French were not what they had been. Maurice de Saxe could still win battles with them, but there was only one Maurice. Defeat or prolonged hardship would send droves of officers streaming back to the amenities of Paris and Versailles; the enlisted men might get out of hand and behave like Cossacks.

By contrast, Frederick had a big army at the peak of efficiency. Occupying a central position, he could strike at whatever enemy was the most threatening. Also, his western flank was covered by a strong force of English and north German troops in English pay, and England gave him financial support.

The fortunes of war varied over the years, but by early 1762 the drawn-out war had almost exhausted Prussia's limited resources; Frederick was as good as beaten. England suspended the subsidy he had been receiving. At this moment of despair, with even his crack cavalry guard unwilling to fight another battle, Czarina Elizabeth died. Her successor, the peculiar Peter III, admired Frederick and proposed to help him but was deposed by his wife Catherine (later to be called "the Great") and murdered by one of her lovers. Most of Germany was devastated, and all the combatants were exhausted. Under the terms of the peace treaty, Frederick kept Silesia. He spent the rest of his life working hard to rebuild Prussia. During 1778–79 he had one final spat with Austria in the War of the Bavarian Succession, which involved no battles but cost him 40,000 men from neglect, disease, desertion, and various small actions.

Frederick's true rank as a soldier is amazingly hard to determine. He has been the object of almost two centuries of largely uncritical praise from British and American historians. To contemporary Americans and Englishmen, he was "the Protestant hero" who single-handedly beat down Catholic France and Austria and savage Russia. To most Germans, however vividly they cursed him alive, he would become a legend of the great and perfect king and general.

The real Frederick was none of those. He had no religion and—so far as his actions show—no fear of God or the Judgment. He did maintain an excellent organization of *feldpredigers* (chaplains) to remind his troops of their duty to him as God's representative in Prussia, and he capitalized on the local religious feelings in Protestant areas. "If the people there are Catholic, do not speak about religion; if they are Protestant, make the people believe that a [pretended] ardor for religion attaches you to them."[5] There was little mercy and less gratitude in him; he talked much of honor but gave extremely few exhibitions of that quality. During the Seven Years' War officer casualties forced him to commission noncommissioned officers and young middle-class volunteers. Most of them served efficiently but, once the war was over, Frederick ejected all but a few of the most deserving, replacing them with foreign officers of noble birth when no suitable Prussians were available. There was an increasingly demonic quality about him that could terrify hardened veterans. His troops, especially the foreigners who made up at least half of his army, usually hated him, hoped to desert, but meanwhile fought like the devil.

As a general, Frederick paid his men little, kept them under the tightest discipline, but fed them well enough. To him, enlisted men were "lemons which you squeeze for the juice and then throw away."[6] He credited them with no sense of honor beyond feeling a pride in their regiment and played on that pride with an imaginative range of rewards and punishments. In the field, he happily treated his soldiers with rough Dutch-uncle humor and a lack of ceremony, occasionally punctuated by wallops with his cane, but taught them to fear their officers more than the enemy. Frequently, he treated his officers no better. His medical service was a disgrace, killing as many as four out of every five unfortunates who entered its hospitals. Worn-out and crippled privates might be given a license to beg if they were Prussian born, or simply expelled from Prussian territory if they were foreigners.

Like Alexander, Frederick inherited a first-class army of carefully trained, long-service professional soldiers. And, like Alexander, he had the supreme luck of never having to fight a first-class army or a first-class general on anything like even terms. His army was superior to its opponents in drill, discipline, and morale, but it lacked reliable light troops for screening and reconnaissance. It would normally win a battle, but it was at a grave disadvantage in prolonged campaigns of maneuver. Frederick therefore at first sought to bring on a major battle as soon as possible, hoping that he could achieve a decisive victory. If he won it with only minor casualties, as at Rossbach and Leuthen (1757), he could continue his campaign. But if he lost large numbers of his carefully trained men and horses, as he did at Prague (1757) and Zondorf (1758), it hurt him worse proportionally than it had the defeated enemy.

An excellent tactician when he put his mind to it, Frederick was frequently careless in his reconnaissance of enemy positions and stupidly disdainful of his enemies. He was quick and daring in maneuver and full of tricks; he never would acknowledge himself defeated, though he might have long spells of depression after a reversal. Several of his battles, especially the climactic fight at Torgau, were won by the pride and fighting spirit of his subordinates after Frederick had completely bungled matters. No other great captain has shown such an uncertain courage. Usually completely cool and in full control of the action, on two or three occasions he was little better than cowardly. He never admitted making a mistake, but he could learn from those he made. Facing superior Austrian artillery, he supplemented his field guns with heavy fortress artillery, which his men nicknamed *Brummers* (grumblers), and developed a small force of fast-moving horse artillery. At the same time, however, he treated his artillery officers as second-class citizens, and actually broke up the efficient military engineer organization Frederick William had left him. He meddled constantly in siege operations, which he did not understand, occasionally killing a few dozen of his soldiers out of sheer ignorance.

Frederick was an indifferent strategist. He kept the main part of his army under his personal command, shifting from one front to another as the situation required. Smaller forces, usually of his poorer troops, were detached to cover the other frontiers or his communications. Frederick left their commanders little freedom of action, often requiring them to take the offensive against much stronger enemies. Consequently, they were frequently defeated, and Frederick was merciless to them if they were.

In his first campaign Frederick had been able to seize Silesia because his blitzkrieg-type offensive took the small Austrian force there completely by surprise. During the next few campaigns he hoped to penetrate to Vienna and possibly to break up the Hapsburg realm. None of these, however, was successful; he won battles but was always forced out of Hapsburg territory in the end. His later campaigns were basically defensive, but he lost no opportunity to take advantage of enemy mistakes. In this he might be too overconfident. At Kunersdorf (1759), with 43,000 men, he encountered an entrenched Austro-Russian army 90,000 strong. Without reconnoitering their position, he attempted to envelop both of its flanks. His *Brummers* stuck in the sandy roads, and his army broke up. A quick pursuit would have ended the war, but the Russians dallied. There is a story that two of their officers sat down with a bottle to plan a pursuit—but the bottle won!

Frederick's best-known military work was *The Instruction of Frederick the Great for His Generals,* written in French in 1747 and revised in 1748 under the new title *General Principals of War*. Originally it was a ''Top Secret'' document; Frederick had only fifty copies made and entrusted them to a select group of officers under oath that they would not take them on campaign. A Major General von Czettrltz had the misfortune to be captured by the Austrians in 1760 with a copy in his possession. It was promptly translated into German and English and widely distributed. The book gives a splendid picture of the art of war as practiced in Prussia at that time (typically, the first two pages are devoted to the problem of desertion), but it depicts Frederick's views before the Seven Years' War, while he still favored shock action and the bayonet. The experience of having his troops massacred by enemy fire power during the early battles of that war changed his views. His *Elements de Castrametrie et de Tactique* (a revision of his *Instruction* written in 1771), *Testament Militaire* (1768), and *Testament Politique* (1752) increasingly stress the importance of fire superiority and the use of fire and maneuver. A final directive to his artillery in 1782 ordered that they were to concentrate on the enemy infantry and not squander ammunition on counterbattery fire against the enemy's artillery.

Napoleon thought Frederick had been fortunate in his enemies: "Frederick could not have maneuvered like that against me." He did praise Frederick's audacity, his speed of maneuver, and his determination: "it was not the Prussian army that defended Prussia for seven years against the three greatest powers in Europe, but Frederick the Great!"[7]

For the rest of the century, and beyond it, Frederick's system of drill and maneuvers was much imitated all over Europe, even in England, though under much protest from officers who had served in America. With it came that skimpy Prussian style of uniform that would so irk Lloyd. (Actually, it was a post-Seven Years' War style in Prussia. To save a few pennies, Frederick had reduced the amount of cloth allowed for each uniform, just as he had put his cavalry horses on a semistarvation allowance of forage.)

Contrary to common tradition, however, nothing much Prussian was adopted in America. Baron Friedrich von Steuben, drillmaster and Inspector General in our Continental Army, *had* served both on Frederick's staff and with one of the better free battalions (see page 101). He brought a certain Prussian sense of organization and order, but the drill and discipline he introduced were of another sort altogether, based on the realization that the American soldier was a reasoning individual and not one of Frederick's "walking muskets."

In France, the defeat of a French contingent at Rossbach had galled French pride as much as all of Marlborough's victories. The French Army began straightening itself out. And in Prussia Frederick kept increasing his army, but much of the spirit had gone out of it even before the 1778–79 "Potato War," so-called by his soldiers since most of the fighting was between hungry foraging parties seeking potatoes to ward off starvation. Frederick's stinginess; his emphasis on outward show, to the neglect of troop morale; his uncertain, searing temper; his abuse of deserving officers and promotion of brutal martinets, left it little better than an impressive facade. But so long as he lived—increasingly slovenly, increasingly misanthropic toward "this damned human race," fond of little besides his whippets—he kept Prussia half worshipful, half terrified.

Probably the greatest result of Frederick's career was its potent stimulation of German nationalism. After generations of seeing their country a battleground for French, Swedes, Austrians, Spaniards, Poles, and Danes, Germans now could boast a warrior king of their own, victorious over all comers. In 1945, when American troops entered Germany, pictures of Frederick, his generals, and his battles were everywhere. Probably they still are.

Frederick's cynical assault on Austria in 1740 unexpectedly released a new school of strategy that had been long maturing. By early 1742 the French had taken Prague, and Frederick's advance guard was gazing at Vienna with anticipation from the north bank of the Danube. Maria Theresa appealed to her customarily uncooperative Hungarian nobility, who gave her—with many chest-thumping protestations of loyalty—a mere four regiments of hussars. More profitably, she called up the half-wild borderers from her Turkish frontier: "the urgent necessity the Queen of Hungary found herself under . . . obliged her to gather together all she could find in her dominions, even the most barbarous nations, who till then, never had any intercourse but with the Turks, and which, in the first campaigns of Bohemia and Bavaria, treated us as such."[8]

Using what regular troops they could scrape together to fix and delay the French and Prussian advances, Austrian commanders loosed these irregulars across their enemies' communications and deep into their rear. There they did what came naturally, undoubtedly rejoicing over such easy pickings. As a French officer, Grandmaison, bluntly reported it:

> they overflowed Bohemia, Bavaria, and Alsace . . . incessantly harassed us and carried off our convoys, hospitals, baggage, foragers, detachments . . . in great numbers, by which the finest armies we ever sent beyond the Rhine were ruined without seeing or fighting against any other troops but Hungarians, Sclavonians, Woradins, Licanians, Croatians, Rusicans, Banalists, and Pandours. The Austrian hussars [carried] off generals and other officers from between two columns.[9]

Frederick likewise was harried home. The stubborn Moravian peasantry hid their food and bushwhacked isolated Prussians. He tried again in 1744 and soon found himself isolated. A whole regiment might be needed to escort a messenger from one Prussian column to another or to forage through a small village. Couriers, scouts, and reconnaissance parties vanished; Frederick could not maintain contact with Prussia. Meanwhile, Austrian regular forces concentrated against him, and Frederick could only make a miserable retreat, starving horses collapsing, troops mutinous and deserting in droves.

Early in the Seven Years' War these Austrian light troops heaped insult upon injury by raiding around Frederick's main army and seizing Berlin, departing with a contribution of 215,000 thalers and a dozen pairs of special fancy gloves for Maria Theresa. Possibly in retaliation, Frederick invaded Austria a third time in 1757–58 and got the usual treatment, losing all but 100 wagons out of a vital 4,000-wagon resupply train and an important forward magazine. Lucky to extricate his army without further serious loss, he thereafter kept out of Austrian territory until the Potato War. This invasion stalled just across the frontier, and once more the Austrian light troops played tag all through Frederick's rear area.

Austria's employment of such irregulars was nothing new. As early as the Thirty Years' War large numbers of irregular cavalrymen—usually termed Croats or Pandours—had been employed on the flanks of the main army for skirmishing and foraging. But the increasing use of such troops on large-scale strategic missions was something different; stiffly organized regular units found them almost impossible to catch and dangerous to handle if caught. These frontiersmen could operate over almost any sort of terrain and were excellent shots. Many irregulars received little or no pay but were allowed "plunder rights," meaning that they could keep whatever they chose to grab. (If paid, they plundered anyhow.) When they finished with a district, food, unbroken bottles, and virgins were rare indeed; they were often as dangerous to their own civilians as to the enemy. Lack of discipline and greed for plunder might divert them from their assigned missions, and only unusually tough and daring officers could handle them.

After suffering from these Austrian irregulars, France and Prussia began raising light troops of their own. Lacking the same raw materials that Austria had in plenty, they had to make do with Austrian deserters and adventurers of all nations, plus their own unemployed thugs and horse thieves. Frederick's Prussian hussar regiments had shown some promise for such service early in his wars, but after 1747 their training put greater emphasis on battlefield shock action than reconnaissance and counterreconnaissance. (Besides, Frederick usually had a fair proportion of them tied down patrolling the flanks and rear of his army to round up deserters.) He formed twenty-two "free" (volunteer) units, usually of battalion size, from foreigners and deserters for light infantry/light cavalry work, but only two or three of these were of any use whatever. Most were totally unreliable, surrendering easily or even deserting to the enemy. Frederick's one unit of effective Prussian light infantry—the green-coated "Foot Jaegers," armed with stubby rifles and recruited from foresters and gamekeepers—was enlarged but never properly handled; Frederick ended by replacing their rifles with muskets and bayonets. Beyond a doubt, Frederick never fully comprehended the strategic use of light troops; his concept of war was one fought under his tight personal control, and irregular batches of wild men loping through the hills far from his camps and battlefields did not fit into that. He *was* aware of how dangerous they could be but never really attempted to work out a counterstrategy.

By contrast, Prince Ferdinand of Brunswick, Frederick's brother-in-law, who commanded the Anglo-German forces in northwestern Germany during the Seven Years' War, made excellent use of his hussars, jaegers, and chasseurs. One of his hussar officers, Nikolaus, Graf von Luckner, became famous all over Europe. (In 1763 the French hired him as a lieutenant general; he would later support the French Revolution, become a marshal of France, and be guillotined, sneering at the mob around him.)

This "partisan" warfare, as it came to be called, naturally attracted a great deal of attention. Every aspiring officer felt the need to study it, and a large number of books on the subject appeared throughout Europe. Such service—hazardous, exciting, free from most of their armies' pipe clay and parading, with a chance of booty thrown in—attracted daring, quick-witted soldiers, and usually killed them off promptly if they weren't! Partisan units normally were enlisted only for the duration of the war; they fancied striking uniforms and went with a swing and a swagger. On campaign, they moved quickly and quietly, made night marches and fireless camps, followed little-known trails, and lived always on the alert. The one unforgivable fault in a partisan officer was to allow himself to be surprised. A favorite partisan organization was the legion—a combination of hussars or dragoons, light infantry or jaegers, sometimes with one or two light fieldpieces. The Queen's Rangers, an efficient

Loyalist unit during the American Revolution, had grenadiers, light infantry, riflemen, hussars, and Highlanders.

French free corps and those in England's pay were considerably less numerous than those available to Austria. Consequently, they were unable to devastate large areas on their own and so tended to work more closely with the main armies. In compensation, however, they were better sources of information, being able to report back rapidly, and under better control. A common employment for these partisan units was screening the movements of the main army by driving in the enemy's light troops and surprising his outposts over a broad front. Another typical mission was a deep penetration of the enemy's position, directed at vital supply dumps, field bakeries, or supply trains while at the same time collecting information on enemy strength and dispositions.

European partisan warfare came to North America during the French and Indian War with experienced officers such as the Swiss Henry Bouquet of the 60th (Royal American) Regiment of Foot. It blended easily with the similar teachings of American ranger captains like Robert Rogers and Joseph Gorham. The American Revolution saw some British commanders—especially "Gentleman Johnny" Burgoyne in 1777—attempt to employ Indian war parties in the same fashion that Maria Theresa had used her light troops. Indians, however, though happy to kill, loot, and burn, were not interested in drawn-out campaigning. Light troops of both armies waged a constant struggle in the lower Hudson Valley, supplemented by pro-British "Cowboys" and pro-American "Skinners," both at least half bandits. In the closing year of the war, Nathanael Greene made masterly use of strategic partisan warfare across the southern states. Fixing the British regular forces there with his own small army, he utilized existing partisan bands—in particular, that of Francis Marion—to destroy the British communications system and to put down any Tory groups favoring the British. To stiffen Marion's militiamen, Greene reinforced him with Lee's Legion, a highly trained, smartly uniformed vest-pocket task force of Continental light infantry and light dragoons, commanded by Henry ("Light Horse Harry") Lee, who later would have a son named Robert E. Lee. This strategy was most effective; their supply lines cut, the British forces in the interior of the Carolinas had to fall back into the Charleston and Savannah areas, where they were blockaded for the rest of the war.

Partisan strategy had several odd aspects. It did not originate with any particular great captain. It had many able practitioners but no one admitted master. It employed guerrilla-type tactics—surprise attacks, ambuscades, heckling and sniping, night raids—but, unlike the average guerrilla, the partisan operated largely in hostile territory where he could expect neither help nor information from the inhabitants and went in constant danger from them. Finally, the Austrian light troops had no appreciable effect on the armies of the French Revolution and Napoleon, probably because these armies were far more flexible than those of Louis XV.

One American parallel with large-scale European partisan operations would be the Civil War cavalry raids by Nathan B. Forrest, John H. Morgan, and Earl Van Dorn, which hamstrung Federal offensives by destroying their supporting railroads and supply bases in 1862. These raiders, however, remained in friendly territory. When Morgan ventured north of the Ohio River in 1863, he was promptly whipped and captured.

Toward the close of the eighteenth century another Englishman gave Europe lessons in strategy—and an enduring puzzle. To be exact, he was Welsh: Henry Humphry Evans Lloyd (1720–83). He left "no history either of himself, or of the place of his nativity."[10] John Drummond, a Scots officer in the French service who met him in Paris in 1744, remembered that Lloyd was said to be of a respectable family (another account makes him a clergyman's son), had received a liberal education, and had studied with a lawyer. For some reason he had decided to come to Paris in hope of getting a commission in the French army, which—since he lacked any influential friends there—proved impossible. Possibly out of hunger, he accepted the offer of some English priests to become a lay brother in their religious house. He was, however, giving lessons in geography and field engineering to Scots and Irish officers in the French service, and Drummond became one of his pupils. He was with Drummond at the battle of Fontenoy in 1745; there his skill at mapping gained him entry to the French corps of engineers where he served under Drummond "on horseback, as an assistant draftsman, with the pay of a sub-ensign."[11]

Lloyd next joined Prince Charles Edward ("Bonnie Prince Charlie") the "Young Pretender," grandson of James II, in his famous invasion of the British Isles that same year. After reaching Scotland he was sent, disguised as a clergyman, on a lone reconnaissance of the English coastline from Milford Haven in southern Wales around to London, apparently to pick out good landing places for a French fleet that was to bring reinforcements for Prince Charles. Stormy weather held up the fleet; when Drummond—captured in late 1746 at Prince Charles's last, lost battle of Culloden—was brought to an apparently comfortable London imprisonment, he found Lloyd already there. Lloyd had been arrested for nothing more than suspicious conduct—or so he said. Pretending to meet him for the first time, Drummond hired him as a tutor; in 1747, after a little backstairs wire-pulling, they were allowed to return to France. There, at the siege of Bergen-op-Zoom, Lloyd won promotion to major for "infinite service in mounting batteries, in choosing ground, and exploring mines, as well as in opening sluices."[12] Drummond then went to Spain, Lloyd into the Prussian service, but in 1754 he was back in France where he was given a secret mission to complete his reconnaissance of the English coast, the French being ready to attempt an invasion. Having no military air whatever, he was able to pose as a middle-class man of business. He certainly made a thorough survey, after which he reported the south coast of England too rough and broken for large-scale military operations.

The Seven Years' War found Lloyd in the Austrian Army. Initially, he was a staff officer, but in 1760 (so he later claimed) he was "intrusted with the command of a considerable detachment of infantry and cavalry, with orders never to lose sight of the Prussian Army, which he punctually complied with and was never unfortunate."[13] Nevertheless, Lloyd then switched to the Prussian service under Prince Ferdinand of Brunswick. The Seven Years' War done, he signed up with the Russian Army against the Turks, attaining, he stated, a grade equivalent to that of major general in the British service. Unfortunately, he was denied the prestigious Order of Saint Anne because of his "plebeian" birth and so departed, more in anger than in grief. Thereafter, he was in Italy and Spain for unclear reasons. Drummond met him again in London in 1776. Lloyd styled himself "general" and had "made his peace" with the British government, which was paying him a very respectable pension. He had married a Scottish lady out of a family fervently loyal to Bonnie Prince Charlie; their son would be a clerk in the British Foreign Office and a German-language expert.

It was a strange, whirligig career. Other accounts have Lloyd being trained for holy orders in the Jesuit college in Rome, and serving as military engineer in the Spanish Army. Lloyd may have been a double agent, spying for both France and England and possibly for others. But it is equally likely that he was something tougher—an eighteenth-century James Bond, hiding a career as a British secret agent inside one as a soldier of fortune. It could be that Lloyd used Drummond as a stalking horse. (That would explain how Drummond happened to find him waiting in comfortable captivity in London, ready to pick up their former association.) Another possibility is that Lloyd *was* caught spying in 1746 and that British intelligence "turned him around" to work for them. On the other hand, when Drummond first met him in Paris in 1744, Lloyd was already a capable military engineer and cartographer. Such specialized skills demanded a considerable knowledge of mathematics and were not picked up in a religious house or a lawyer's office, and seldom in the British service, for that matter. Also, French military engineers of this period were considered the world's finest. If Lloyd could achieve promotion among them—as at Fontenoy and Bergen-op-Zoom—he must have been unusually capable and courageous. Yet where *did* he learn that profession?

In short, the man himself remains a mystery.

Lloyd's writings, however, were mostly blunt and useful. The principal one was his three-volume *History of the Late War in Germany Between the King of Prussia and the Empress of Germany and Her Allies,* published in London during 1766–81. Lloyd's unusual technique in describing a battle was to compare the versions of the two opposing armies, with such comments as he considered necessary. The result was a somewhat ponderous work but a very useful one, probably the first serious work on strategy written in Europe. In 1779, with the best of the British Army pinned down in America and England faced with a very real risk of invasion, Lloyd published *A Political and Military Rhapsody on the Invasion and Defense of Great Britian and Ireland.* This certainly embodied the reconnaissance he had made years earlier for the French. Its general sense was that an invasion of England could only end in disaster for the invader, through he might have better luck in Ireland, but it gave a thorough review of the topography of southern England, providing a handy reference work for either invader or defender.

Whatever service Lloyd actually saw, he had looked at history, men, and war with clear eyes. His books resemble ancient mansions in which one suddenly finds odd and interesting little side rooms in unexpected corners. He interrupts his *Rhapsody* to lecture that naval superiority is *the* essential for England; also, that England should replace its army with "twenty or thirty thousand marines" and confine its actions on the European continent to "hiring" allies. With a "fleet having on board twelve or fifteen thousand marines . . . England could seize its enemies' colonies and keep them . . . in continual anxiety in every part of the world."[14] (Napoleon would later express puzzlement and relief that England had not used this strategy against him.) Glancing at America, he observed that republics generally made a mess of their military affairs except in times of absolute crisis; then, especially if inspired by "civil or religious principles, they generally become invincible," but once the danger was past, they went back to their former inefficiency. Therefore, he wrote,

> I am so convinced of the truth of this reasoning that I have not the least doubt, if we could hold New-York, Long Island, Rhode Island, and Philadelphia, and cease to make those fruitless and unmeaning excursions into the American woods, that the Congress and the rebel people, no longer united by their sense of fear, would soon dissolve their confederacy, and a more favorable oppportunity would offer of restoring peace and union between them and the mother country.[15]

Lloyd's inclusion of Philadelphia in his list of places to be held suggests a relative ignorance of North American geography, but the psychological effect of his proposed strategy on the largely indifferent and war-weary American public is worthy of contemplation. As for the British troops in North America, Lloyd would have used them to "strike some capital stroke in the West Indies" against the French and Spanish colonies there. (Today, it is hard to realize that those overpopulated, poverty-stricken little places were once the fabulously valuable Sugar Islands that many Europeans considered far more important than all the wood lots in North America.)

Lloyd was a strong advocate of the Strategy of Evasion. If on the defensive, he would avoid battle, delaying the enemy's advance with part of his army by gradually falling back from one strong position to another while the rest of his forces attacked the enemy's communciations. Lloyd's model general would have an exact knowledge of the topography of the area in which he is operating, and therefore "may reduce military operation to a geometric preciseness and may forever make war without being obliged to fight [a major battle]."[16] If he were the attacking general, however, he would advance as swiftly and as vigorously as possible, since anything less than a complete victory would eventually mean failure. (Douglas MacArthur's "There is no substitute for victory" only puts the same thought more dramatically.) Lloyd felt that it was almost impossible to achieve such decisive "celerity" with the armies of his day because of their rigid organization. He therefore proposed to divide "his" army into five "corps" which could move more or less independently, yet maneuver in concert—obviously a forecast of the Napoleonic army corps. Finally, he wanted the disorderly hordes of light troops replaced by smaller, better disciplined forces that could take a hand in pitched battles.

Much of Lloyd's thought agrees with Saxe's: Soldiers' uniforms should be practical rather than ornamental; pikes should be retained for shock action and defense against cavalry. (This constant desire for pikes rose from the fact that eighteenth-century infantrymen tended to be poor shots; cavalry frequently could ride right over them.) The system of feeding troops in the field should be improved, and a general must concern himself with the "philosophy of war," which involved an understanding of the passions that governed soldiers' behavior.

Frederick the Great was not Lloyd's special hero; he was, in Lloyd's informed opinion, too reckless and too ready to risk everything on an unnecessary battle. Lloyd was especially distressed by the slavish imitation of Prussian drill and dress throughout Europe after the Seven Years' War: "short cloaths, little hats, tight breeches, high-heeled shoes, and an infinite number of useless motions."[17]

Lloyd's views on warfare had a definite effect on nineteenth-century soldiers. Soon translated into French and German, his *History of the Late War* infuriated many loyal Prussians. Napoleon thought Lloyd old-fashioned—that emphasis on pikes seems to have particularly galled the Emperor. Jomini (see chapter VII) made greater use of Lloyd in developing his theories, though he found some of the

Welshman's ideas unusually bizarre. But Lloyd was probably the first European military writer to isolate and state definite principles of war: "An army superior in activity can always anticipate the motions of a less rapid enemy, and bring more men into action than they can at any given point, though inferior in numbers. This advantage must generally prove decisive and ensure success."[18]

Very appropriately, Lloyd's last years furnished the same puzzles as his youth. No sooner had his *Rhapsody* been published than he was put under great pressure, from an undisclosed source, to suppress it and give up all copies in return for a considerable payment. He agreed, and the book at once became a rarity. The identity of the pressuring party is uncertain. By one account, it was the British government that wanted to quietly get rid of a topographical study that would have been immensely helpful to an invading army. Another version is that the French wanted to expunge evidence of their plans to launch such an invasion. Whoever it was, Lloyd became the fastest-selling military historian on record.

Lloyd died in Brussels in 1783. The British government—or at least someone identified as a British agent—seized all of his private papers.

VI

YANKEE DOODLE

> During these two years the Americans have trained a great many excellent officers who very often shame and excell our experienced officers, who consider it sinful to read a book or think of learning anything during the war . . . I must admit that when we examined a haversack of the enemy, which contained only two shirts, we also found the most excellent military books translated into their language. For example, Turpin, Jenny, Grandmaison, La Croix, Tielke's *Field Engineer,* and the *Instructions* of the great Frederick to his generals I have found more than one hundred times.
>
> JOHANN EWALD, DECEMBER 1777[1]

A captain of Hesse-Cassel jaegers, acknowledged as one of the best outpost officers with the British forces in America during the Revolutionary War, Ewald studied his enemies with a coldly professional mind. He wasn't certain how much American officers actually learned from their books, but he respected them for their efforts.

American officers had to train themselves. A few had seen service during the French and Indian War but only as company or regimental officers. George Washington had been a brigadier briefly in 1758, commanding a brigade of provincial troops. So far as we know, his military reading was never extensive. In 1756, as colonel of Virginia provincial troops, he secured a copy of Humphry Bland's *Treatise of Military Discipline,* which dealt with tactics and discipline.[2] During the final campaign against Fort Duquesne in 1758, Gen. John Forbes, the commander of the expedition, introduced him to Capt. Joseph Otway's translation of Montecuccoli's *Commentarii Bellici* as a handy book on irregular warfare. As commander in chief of the Continental Army besieging Boston in 1775–76, he had to give himself a concentrated course of gunnery in order to use his artillery effectively, since none of his officers had yet accumulated the necessary knowledge. But he was otherwise too desperately engaged in being both the commander and most of the general staff of his improvised army to concern himself with further theoretical self-education.

Washington, however, came to his final victory by courage, prudence, hard work, patriotism, and character rather than by any native knack for strategy. He learned how to use the rugged American terrain to make up for his usual lack of numbers and could strike like an irked rattlesnake at unexpected moments. Unfortunately, his two surprising skills—the creation of an efficient espionage network and an ability at deception that frequently left everyone puzzled as to his exact strength and situation—were rarely appreciated by later American commanders.

The students among Washington's subordinates included Nathanael Greene, Henry Knox, and Wayne, and there can be little doubt that they profited by their reading of the great captains. Nathanael Greene, the limping, clear-minded, reformed Quaker who had studied even as he worked at his forge, waged a campaign much after the style of Turenne across the southern states through 1781–82. However, a little learning could be a dangerous thing. At Germantown, October 1777, Washington caught the British army outside Philadelphia by surprise. Though hampered by fog and incompetent guides, his attack put the overmatched British on the run rearward. In the confusion, some hundred English infantrymen barricaded themselves in a large stone house from which they opened fire on the American reserve division as it appeared. Confronted with this unexpected obstacle, most of Washington's staff wanted to leave a regiment to contain the house and get the rest of the reserve forward. (In World War II's more vulgar language, ''Bypass and haul ass.'') Unfortunately, Washington's Chief of Artillery, Brig. Gen. Henry

Knox, remembered a phrase from his omnivorous military studies during his years as a Boston bookseller: "While penetrating an enemy's country, you must not leave an occupied castle in your rear."[3] Knox was a large, robust man with a very loud voice; Washington took his advice, and most of the reserve got involved in a futile attempt to take the house. Knox further bollixed the business by inexpert use of his field guns. Meanwhile, the drive of the American assault ran down as British reserves arrived; worse, the sound of heavy firing behind them unsettled the leading American divisions. And so things promptly came apart.

As general and president, George Washington made the United States independent "with a rank due to [it] among nations"[4] But England, increasingly more than distracted by her quarrel with France, had certain implicit reservations. In 1808, as a climax of continuing provocation, the British ship H.M.S. *Leopard* seized four seamen out of the American frigate U.S.S. *Chesapeake*. Even pacific Pres. Thomas Jefferson was moved to add new regiments to his tiny army. Among the deserving members of his Republican party, "swaggerers, dependents, decayed gentlemen and others . . ." unfit for any military purpose whatever[5] whom he made officers, was Winfield Scott, a colossal young Virginia gentleman attorney.

Commissioned a captain of light artillery, Scott had everything to learn. He had trouble with his company records; also, he proclaimed his commanding officer, Brig. Gen. James Wilkinson, a finished scoundrel. This sentiment was eminently accurate: Wilkinson had been in the pay of the Spanish government for years and was currently swindling his men out of their proper rations. It was not, however, sensible for a mere captain to say so repeatedly in public. Scott abruptly found himself court-martialed and sentenced to twelve months' suspension from the service, without pay.

Wilkinson unintentionally had done Scott a favor. On the advice of a friend with an extensive personal library, Scott settled down to a careful study of his profession, with emphasis on the British Army, including such works as William Armstrong's *Practical Considerations on the Errors Committed by Generals and Field Officers, Commanding Armies & Detachments, from the Year 1784 to the Present Time*. Thereafter, he built up a portable military library of his own, including such up-to-date works as Paul Charles Thiebault's *Manuel général de service des etats-majors generaux et di-*

visionnaire des armées, the basic staff manual of Napoleon's armies;[6] Jomini's *Traité des grande tactique;* and French infantry drill regulations.[7] Skilled and tireless as drillmaster, staff officer, and combat leader, he ended the War of 1812 as a much-lauded brigadier general.

On an extended visit to Europe in 1815, he studied the Allied armies occupying France after Waterloo, visited leading European commanders and scientists, and collected "every work (not obsolete) on the service, police, discipline, instruction, and administration of an army."[8] He also studied French cooking and English military tailoring, gourmet food and splendid uniforms being serious matters to him.

Scott's subsequent work as instructor and administrator to the U.S. Army was thus informed and thorough, his knowledge of Napoleonic strategy and tactics that of a studious contemporary who had no need of Jomini's interpretations. Between 1821 and 1829 he either wrote or helped prepare a series of regulations governing every aspect of the army's existence and stressing swift movement, individual marksmanship, maximum fire power, military sanitation, and the evils of freshly baked bread and fried food. He was "Old Fuss and Feathers" to the army—very much the military patrician, a difficult subordinate of hair-trigger temper and too-ready pen, but always responsible and a gentleman. He built the small, expert Regular Army that remains the basis of our military tradition.

Too junior to have had much interest in strategy during the War of 1812, Scott began the Mexican War in a chill of political disfavor. Pres. James Polk, who sometimes gave the impression that there was no God except the Democratic party with himself as its sole prophet, promptly decided that Scott was "rather scientific and visionary,"[9] because—instead of dashing off at once for the Texas frontier—he proposed to remain in Washington for some weeks to organize the mobilization of the twelve-month volunteers with which Polk proposed to fight his war, and a supply system that could support armies operating in Mexico. Therefore Polk nailed Scott's uniform coattails to his office chair in Washington and let Brig. Gen. Zachary Taylor command the American forces in northern Mexico. Taylor won small victories but none sufficient to make Mexico ask for peace. Increasingly thwarted in his desire for a short, profitable war, Polk decided on the capture of the Mexican seaport of Vera Cruz, followed if necessary by an advance inland against Mexico City. Unable to find a commander who qualified as a loyal Democrat, Polk reluctantly put Scott in command of the expedition.[10] With keen help from the Navy, Scott quickly took Vera Cruz, moved inland, and routed the Mexican army at Cerro Gordo. Then the enlistments of his Volunteer regiments ran out. He sent them home, almost two months early, to get them safely out of Mexico before the yellow-fever season began its annual ravage of the coastal regions. Thereafter, with barely 6,000 men, he halted for three months at Puebla, in the center of Mexico, awaiting reinforcements while another Mexican army gathered. Reinforcements finally came, but they were green troops and far fewer than promised.

Scott's situation was unenviable. He was almost two hundred guerrilla-haunted miles from Vera Cruz; Mexico City was seventy-odd miles farther on across the mountains. He had neither wagons enough to maintain a supply line to Vera Cruz nor troops enough to guard it. Scott's solution was to call forward all his security detachments between Vera Cruz and Puebla, leave his sick and wounded in the latter city under a small guard, muster 10,738 officers and men and a large supply train, and march on Mexico City. For an army to thus abandon its line of communications in the midst of a hostile country was a thing unheard of. Napoleon had remarked that it was something he himself never dared attempt.[11] Now, the Duke of Wellington could only comment, "Scott is lost!"[12]

It was August 7, 1847 when the American advance guard sent a "Cerro Gordo shout" roaring through the morning as it left Puebla. By September 17 Mexico City was quiet and submissive. Scott had shattered an army of 30,000 Mexicans, strongly fortified in and around their own capital. As more reinforcements reached Mexico, the Vera Cruz road was reopened. After some months of dickering, Mexico made peace. Meanwhile, since supplies from home were still insufficient, Scott fed, reclothed, and reequipped his army from Mexican sources and nevertheless maintained excellent relations with most of the conquered.

Wellington dubbed him the "world's greatest living soldier" and his campaign "unsurpassed in military annals."[13] The plain daring of Scott's strategic concept still has no equal in American history; the skill with which he executed it, always seeking to outflank and surprise his enemy, is outstanding. His example undoubtedly influenced Grant's final campaign against Vicksburg and probably Sherman's promenade across Georgia. MacArthur's Inchon landing and Westmoreland's "air cavalry" surprise stroke at Ia Drang had some of its audacity.

In late 1860, as Southern states began seceding from the Union, Scott was seventy-four, chronically ill, racked by "dry scars that ache of winter nights"[14] from three wounds taken in Canada and a fourth in Mexico, and still General in Chief of the Army. Through 1853–57 he had brawled with Jefferson Davis, Secretary of War in Pres. Franklin Pierce's cabinet. Davis—"proud as Lucifer and cold as a lizard"[15]—had deliberately set about stripping the general in chief of effective authority. Enraged, Scott had moved his headquarters from Washington to New York. Now, summoned by Pres. James Buchanan, he left his sick bed to return. It was desperate work again for the feeble, corpulent old soldier, with Buchanan fluttering, Secretary of War John B. Floyd an active traitor, trusted Army officers such as Robert E. Lee resigning to "go South," and the whole nation in high confusion. Driving himself through up to seventeen working hours a day—always relieved by an appropriate dinner—he was the sternly professional pivot on which the North's military preparations turned. But, unable to mount a horse or walk more than a few steps unaided, he could not long endure the strain. He hoped that Henry W. Halleck, whom he had recalled to active duty as a major general, might take his place. However, another recalled West Pointer, George B. McClellan, came to Washington to take over the Army of the Potomac and at once launched a deliberate campaign to force Scott into retirement. One of his major complaints was that Scott would not join him in looking for "boogers" under the bed—McClellan's firmest conviction being that slavering hordes of Confederates were about to pounce on Washington. Unfortunately, McClellan was also a personality boy, radiating an impression of military competence that charmed cabinet members and raw recruits alike. Elbowed into the background and neglected, Scott retired in October 1861.

Sometime before going, however, he had roughed out a strategic plan for the suppression of the South's rebellion. He had few illusions; he expected the war to last for three years (with the "fury of the noncombattants" causing great trouble thereafter)[16]; he had no confidence in the steadiness of three-month volunteers. He wanted to expand the Regular Army and raise strong forces of three-year volunteers, trained and disciplined for four and one-half months; meanwhile, the Navy would clamp a tight blockade on all Confederate-held ports. Then, his land forces ready, a fleet of powerful river gunboats built, he would send a major expedition down the Mississippi, splitting the Confederacy and completing its isolation. This, Scott hoped, would cause a swell of pro-Union sentiment in the seceded states and end the war with comparatively little bloodshed. If it did not, the South would be invaded and conquered. His great worry was that the North would not pause for careful preparations—which it did not. Political pressure and popular excitement sent the three-month soldiers ''On to Richmond''—and defeat. Since it proposed to squeeze the Confederacy to death, Scott's strategy was dubbed the ''anaconda plan'' and made the subject of much newspaper hilarity. But it remained the basic strategy by which the Civil War was finally won.

Dennis Hart Mahan, the Army's best-known theoretician prior to the Civil War, was an utter contrast to Winfield Scott. The son of an Irish immigrant workman, shy and frail, he was graduated from the U.S. Military Academy in 1824, ranking first in a class of thirty-one. While a cadet, he had served as ''acting assistant professor of mathematics''; on graduation, Superintendent Sylvanus Thayer kept him at West Point as an instructor in mathematics and engineering. His work was outstanding, but his health continued uncertain. Thayer, therefore, arranged for him to spend four years in France attending the famous School of Application for Engineers and Artillery at Metz. (American doctors of that period somehow thought the French climate mild and restorative!) There he studied the construction of field and permanent fortifications, the attack and defense of fortified places, and the theory and practice of artillery. Only a small part—possibly a tenth—of the course dealt with the art of warfare. Instruction was by lecture and demonstration; there were no texts. Students were expected to take notes and supplement those by outside reading.

Two years after returning to West Point in 1830, Mahan became a full Professor and head of the Department of Civil and Military Engineering. (He modified this title on several occasions, one change being to add ''and the Art of War.'') He greatly enriched (cadets had other terms for it) the content of its course. Since there were no good English-language engineering texts, Mahan prepared his own, printing them on a lithograph press he had brought back from France. As an instructor, he earned the cadet nickname "Deadly Mahan" for his meticulous perfectionism, his ability to detect those cadets who had not mastered their lessons and to really make them regret it. His thin, high-pitched voice could lash like a wire whip. Twenty years later, generals still shuddered over the memory of being so caught—"most particular, crabbed, exacting . . . a little slim skeleton of a man and always nervous and cross . . . But I would give my fortune to know what he has in that little head of his."[17] He heaped extra work on cadets; if not pleased with their drawings for various engineering problems, he might confine them to a classroom for a Saturday afternoon, with a sentry at the door, until the work was redone to his satisfaction.

He loved West Point and, in Army speech, became a "homesteader" there. He never commanded so much as a squad, never heard a shot fired in anger, never displayed the least interest in being so educated. He very seldom wore a uniform and always carried an umbrella when outdoors. His son, Alfred Thayer Mahan, joined the Navy.

Mahan did not initiate instruction in military art and strategy at West Point. Earlier professors of Engineering had mixed it in with their instruction on fortification and siegecraft. Claude Crozet, formerly an artillery officer under Napoleon, had used S. F. Gay de Vernon's *Treatise on the Science of War and Fortification,* the official text of the École Polytechnique, as translated by a Capt. John M. O'Connor, who added a few pages of extracts from Jomini and Henry Lloyd.[18] The book cost the horrifying sum of twenty dollars—on a rough guess, the equivalent of a hundred dollars today.

Mahan's contribution was double: He separated military art from the engineering course and produced his own textbook. Normally, he taught six lessons a year on the "science of war," sometimes with a few extra "review" lessons—not enough to make much of a dent in the average cadet's consciousness but all he could find time for. Like his predecessors, he also used military history to illustrate his lessons in military engineering.

His own text, *An Elementary Treatise on Advanced-Guard, Out-Post, and Detachment Service of Troops, and the Manner of Posting and Handling them in the Presence of the Enemy, With a Historical*

Sketch of the Rise and Progress of Tactics, etc. etc. (shortened in Army talk to *Out-Post*), first published in 1847, was a slim book—actually, a field manual—that seemed rather overwhelmed by its title. In part, it was an adaptation and condensation of Gay de Vernon's text, but its major source was Jomini. Mahan considered Napoleon an almost perfect general, lamenting only his neglect of fortifications. (This attitude would be natural for an engineer but indicates that Mahan had little direct knowledge of Napoleon's campaigns; Napoleon's correspondence and *Maxims* are filled with reference to both field and permanent fortifications.) Parts of the book were not adjusted to American warfare, especially the portions dealing with cavalry wherein Mahan expounded on hussars and other exotic European varieties.

All that aside, *Out-Posts* contained a number of new ideas and considerable terse common sense. Much of this already had appeared in Halleck's *Elements of Military Art and Science,* published in 1846, but such cross-fertilization would be natural. Mahan considered the waging of war an art. There were definite principles of strategy, and these could be learned from the study of military history, which was the only substitute for actual experience. However, the properly educated, experienced soldier would realize that "in war, as in every other art based on settled principles, there are exceptions to all general rules. It is in discerning these cases that the talent of the general is shown."[19]

Among his principles Mahan stressed "celerity" (mobility and surprise), "concentration" (*mass*), *maneuver,* and the use of interior lines. Slow and overprudent generals could expect misfortune. The "true end of victory" was to "do the greatest damage to our enemy with the least expense to ourselves."[20] Since American militia did their best fighting, as at New Orleans, from behind entrenchments, he urged that they be taught how to construct them. More important, he recommended the creation of additional Army schools where officers could receive advanced instruction in strategy, tactics, and administration. He read correctly the lessons of the Crimean War, 1854–56, on the killing power of the new rifled muskets and the value of field fortifications, but a monograph he produced on Indian warfare had little application to the nomadic horse-Indian tribes the Army was encountering beyond the Mississippi.

Mahan's contribution to West Point's extracurricular activities was the Napoleon Club, of which he was the president. Its membership of professors, assistant professors, and instructors studied Napoleon's campaigns, apparently in some detail. Supt. Robert E. Lee gave them the use of a classroom, which members decorated with large maps of Central Europe, Italy, and Spain. One of the club's most energetic members was Capt. George B. McClellan, who obviously did not profit from it.

During the Civil War most of the senior generals in both armies had been Mahan's students, and *Out-Posts* was a common reference work on both sides—to Mahan's intense disgust, southern publishers pirated his book. Possibly this inspired him to bring out an 1861 revision, with additional material (not too well researched) on the evolution of the art of war from Greek and Roman times. In the war itself he had little, if any, part; possibly he once was called on for advice concerning the fortification of Washington. His classes continued through war and into peace. One cadet, who also was a Civil War veteran, caught it when he rashly used his combat experience instead of "the book" to diagram an outpost system.[21] Mahan had put his whole life into his teaching. When he fell overboard from a Hudson River steamboat and drowned in 1871, after learning that he must retire because of age, there were potent suspicions of suicide.

Among Mahan's pupils was Henry Wagner Halleck, eldest son of a hardscrabble Mohawk Valley farmer. At sixteen he had run away from the constant physical labor and mental starvation; his maternal grandfather had found him, sent him to school, gotten him an appointment to West Point. Graduating in 1839, third in a class of thirty-one cadets, he naturally went into the Engineers, but remained at the Military Academy for a year as a French-language instructor. Thereafter, while working on the fortifications of New York harbor, he was noticed by Winfield Scott, who wanted a study on coastal defenses. Halleck's report earned him six months of advanced study in France. In 1846 he published *Elements of Military Art and Science,* a deeply researched review of the whole art of war as it might be applicable to the United States. His sources were mostly French and Napoleonic, with special reliance on Jomini whom Halleck considered "the best of military judges,"[22] but his bibliography included Napier, Guibert, and Scharnhorst and even the French translation of one Russian work. Saxe influenced his ideas on leadership and the need of keeping soldiers' "legs" exercised, but his

1862 demand for improvements in Army cooking came from sampling it in the field.

Halleck's intent was to develop a coherent military policy for the United States. He wanted a small, balanced, highly trained Army that would serve as a cadre for recruits called up to meet national emergencies. Only professional, properly educated officers were qualified for high command. (Had many politicians read his book, his career might have suffered an early blight—Halleck's comments on their current policy of paying off political debts by giving unqualified civilians commissions in Regular Army regiments were acrid.)

Halleck gave much attention to logistics. His study of Napoleonic warfare convinced him that an advancing army frequently could "live off the country." The operation should be carefully managed, civilians should not be abused, but the army should take whatever food, forage, and transportation it required as it moved forward. An army operating in this fashion, with only enough wagons to carry its most essential supplies, would be able to march and maneuver far more rapidly than one clogged by long baggage trains. Halleck preached this "hard war" doctrine from the beginning of the Civil War, though most Union commanders rejected it as inhumane, impractical, or both. Grant was brought to appreciate it only in December 1862, after Confederate cavalry destroyed his supply base; Sherman, a year later. But it was this roughneck system that made possible the sweeping campaigns and raids that shattered the Confederacy.

Halleck seemingly had fulfilled his Army nickname "Old Brains." Next, the Mexican War carried him off to California. On the pitching, wallowing transport that carried them around the Horn, while William T. Sherman and other officers played cards, Halleck translated Jomini's *Vie Politique et Militaire de Napoleon.*[23] His exposure to combat was brief, though it brought him a brevet promotion for valor, but he made himself indispensable in the local military government in California and the subsequent organization of the state government. Leaving the Army as a captain in 1854, he soon became a sort of universal genius to California, successful in law, mining, railroading, banking, and architecture and influential in politics. He wrote two books on mining law and another (published in 1861) on international law that remained a standard text for years.

With all this, he found time to marry a granddaughter of Alexander Hamilton and to assist in the organization of the state militia.

When the war burst, Scott remembered Halleck and got him recalled to active duty as a Regular major general. He was sent out to St. Louis where Maj. Gen. John. C. Fremont had created an ineffable mess out of the whole military situation. It was making bricks without straw, amid confusion and corruption, at the far end of a supply line that delivered weapons, equipment, and clothing in grudging dribbles. Missouri was lousy with Confederate guerrillas and equally barbarian Unionist irregulars. By working endless hours with his scanty staff (and enduring an attack of measles), Halleck organized and armed a sufficiency of green regiments and pitched them forward under Samuel Curtis, Ulysses S. Grant, and John Pope. He gave those officers general missions—told them what to do and let them decide how they would do it—and fed in supplies and reinforcements to keep their offensives moving. The result was a string of victories—Fort Henry, Fort Donelson, Pea Ridge, New Madrid, Shiloh, and Island No. 10.

Halleck's subordinates controlled the actual fighting and got the glory of battles won, but the basic responsibility was Halleck's. However, when he took personal command of his field forces after "bloody Shiloh," he proved extremely cautious. His capture of Corinth, Mississippi, broke the Confederacy's best east-west railroad, but he made no real effort to destroy the battered Confederate army opposing him.

Abraham Lincoln (who had begun his study of the art of war with Halleck's *Elements*) then brought Halleck to Washington as General in Chief of the U.S. Army. Halleck did not want the honor; he preferred the West, where he had a relatively free hand and had decided what needed to be done next. However, being a good soldier, he went into what he would soon consider a political hell. In Washington, everyone's thumb was in the soup: Lincoln, who was becoming a knowledgeable strategist, was always ready to intervene; Secretary of War Stanton was able but erratic and irascible; the Secretary of State and Secretary of the Treasury meddled incessantly with military affairs; Congress had its Joint Committee on the Conduct of the War, a star chamber of extreme Republicans. In Halleck's view, "There are so many cooks. They destroy the broth."[24]

He arrived in Washington in a swelling crisis reaching from the outskirts of Richmond to Sioux villages in Minnesota. His concept of generalship would not work there because too many of his new subordinates were incompetent and unwilling. Gen. George B. McClellan had formed the Army of the Potomac in his own image and considered it his personal property; Gen. Don Carlos Buell of the Army of the Ohio simply would not move. Halleck, therefore, chose to consider himself simply a military adviser of the Secretary of War and the President. Knowing the trade languages and thought patterns of both politicians and soldiers, he could be a buffer between the two. He attempted, often unsuccessfully, to deter Lincoln from entrusting major commands to unqualified politician-generals, and managed, together with Lincoln, to shape a winning strategy, especially in the west. Here he directed the Union armies in true Jomini fashion against strategic points, such as Vicksburg and Chattanooga, on the enemy's supply lines. If the Confederates stood to fight for them, their armies could be defeated and possibly destroyed, as at Vicksburg. In the east, with no real strategic point except the Confederate capital at Richmond, it sufficed to operate against Lee's army, with the hope of catching it at a disadvantage and crippling it. By February 1864 the Confederacy had been so compressed that Halleck could write Grant that the objectives of the next campaign must be the two remaining Confederate armies. The next month Grant replaced him as General in Chief. Halleck became Chief of Staff with little change in functions. Grant, moving into the field with the Army of the Potomac, made major strategic decisions for the seventeen different Federal commands; Halleck handled the massive operational and administrative details and served as a link between Grant and Lincoln. His task was vital to the final Federal victory, and Halleck was probably the only officer capable of handling it.

Halleck had frankly hoped for high command and military renown. Instead, he quickly passed into the Civil War's accepted tradition as a petty meddler, indecisive and stupid—a mere clerk. Civil War generals came in all sizes, but most of them sought to be dramatic battle captains; even plain and quiet Grant could manage in a crisis, if only by being apparently imperturbable. Halleck was a book soldier, a headquarters operator, a planner, administrator, and manager. Plump, double-chinned, balding in front, slow in movement and speech, unimpressive on horseback, he did not *look* like a general. His official manners were impersonal, blunt, sometimes secretive, with no tincture of graciousness. He did not suffer fools gladly or otherwise and could be equally short-spoken with irate congressmen, bumbling generals, and delinquent shavetails. Most of his work went unnoticed by the men who profited by it; when it was noticed—as in his efforts to purge the Army of unqualified officers—it made him more enemies than admirers. He was the protector of competent officers—among others, Sherman, whom he nursed through a nervous breakdown and George H. Thomas, whom he sheltered from Grant's enmity—but even Sherman would turn on him. But Lincoln, however, though he might use Halleck as a combination stalking horse and whipping boy, nevertheless found Old Brains the fitting accolade: "Halleck is wholly for the service. He does not care who succeeds or who fails [just] so the service is benefited."[25]

In short, Halleck was a general out of the twentieth century—the manager-type commander, like Dwight D. Eisenhower, who never sees the battlefield. Unfortunately, he considered it sufficient to do his duty without sheltering himself behind a public relations persona, as exemplified in Ike's famous grin. He finally would find justification from military historians after World War II—writers who were both more scrupulous in their research of Civil War records and more appreciative of the problems of command in major wars.[26]

Of the Civil War's successful commanders, four—Grant, Robert E. Lee, Sherman, and Thomas J. Jackson—were internationally famous before the war was done, Lee foremost among them.[27] Yet, in the complete meaning of the word, Lee was never a strategist in the sense that Scott, Halleck, and Grant were. A commander of rare character and devotion, a skilled tactician, he fought for his beloved Virginia and seldom concerned himself with the rest of the Confederacy. He had few equals as a defensive fighter and counterpuncher, always balancing his limited resources against his opportunities, but always preferring the possible loss from aggressive action to the probable loss from not acting. However his two invasions of Northern territory were scrambling, improvised affairs—little more than large-scale raids. Both times he was lucky to get his chewed-up army back into Virginia. Withal, he was complete within himself—the "Marble Model" his fellow cadets at West Point had dubbed him—a courteous, fearless soldier who won the reverence of his most irreverent soldiers. He made few phrases and left little counsel, except his deeds, for future generals.

Nonetheless, Lee's campaigns were widely studied in Europe. His popularity in England was enormous; he and Stonewall Jackson were accepted as paragons, unequaled by any Union commander; Lee's final defeat was explained away as the result of the overwhelming odds of men and guns massed against him. The same belief was sedulously nursed in the United States by novelists and semihistorians of the moonlight-and-magnolia-blossoms school. Only recently has the cold fact emerged that—though Lee won battles on the front porch of the Confederacy—the relentless Scott-Halleck-Grant strategy of the North tore the rest of the house apart behind him, to bring him down in the final collapse.

Like Lee, whose deadly right arm he became, Jackson fought for Virginia, but with the difference that he sometimes spoke his mind as to how wars might be won. A religious zealot who might have ridden with Cromwell, odd, crabbed, secretive, dedicated, apt to behave strangely when exhausted, he was also a fond husband who addressed his wife with little Spanish terms of endearment, remembered from less godly days in Mexico when his only worry was that the war might end before he had the chance to really distinguish himself. Mahan had taught him and may have inspired him to some study on his own. Though the copy of Napoleon's *Maxims* (a gift from Jeb Stuart) that traveled with his Bible and some lemons in his saddlebags apparently remained unread,[28] his recorded thoughts have a Napoleonic ring and clarity, sometimes practically echoing the emperor's.

> Only thus can a weaker country cope with a stronger; it must make up in activity what it lacks in strength. A defensive campaign can only be made successful by taking the aggressive at the proper time. Napoleon never waited for his enemy to become fully prepared, but struck him the first blow. . . . And when you strike and overcome him, never give up the pursuit as long as your men have strength to follow. . . . Never fight against heavy odds if, by any possible maneuvering, you can hurl your whole force on a part, and that the weakest part, of your enemy. . . . To move swiftly, strike vigorously, and secure all the fruits of victory is the secret of successful war.

Jackson also has one principle especially his own: "Always mystify, mislead, and surprise the enemy."[29] No commander ever held his plans more tightly within his own mind or released them so grudgingly. In fact, Jackson occasionally overdid it, confusing his own subordinates as thoroughly as he might the Yankees. From the first battle of Bull Run he was "Stonewall" Jackson—by tradition for the stubbornness of his defensive stand, but possibly through a misunderstanding of an epithet directed at him by a dying Confederate officer who felt that Jackson had held back instead of advancing to support his shattered command.

Jackson was more fortunate than most soldiers in his biographer. George F. R. Henderson, a young British officer who had studied history at Oxford and won distinction in Egypt, had become fascinated with our Civil War and had visited many of the East Coast battlefields. An efficient scholar, gifted teacher, and effective writer, he concluded that the Civil War held far more useful lessons for soldiers than the more recent Franco-Prussian War could offer. His *Stonewall Jackson and the American Civil War,* published in 1906, was based on a review of the *Official Records* of the Civil War and detailed correspondence with Jackson's former associates. An outstanding work, it ended by unintentionally making Jackson somewhat larger than life. Undoubtedly, those "unreconstructed" Confederate veterans who helped Henderson remembered Jackson's feats, in Shakespeare's phrase, "with advantages." Also, like many contemporary American military writers, Henderson did not entirely understand the American command system. But his book was deservedly popular; British officers facing promotion examinations studied it; and its teachings were credited with helping to bring the Boer War to a successful ending. To Henderson, Jackson's greatness was in his strength of character and his ability to apply the principles of war to changing situations.

Ulysses S. Grant remains something of an enigma. William T. Sherman, his good friend and favorite lieutenant, once remarked that Grant was a mystery to him—and probably to himself. Most of the officers who served with him found him reliable and considerate, a good comrade, without show or pretense, modest but fully conscious of his own worth. Behind that was a man who could cherish grudges for years and probably never forgave a personal enemy. His temper seldom flared except when he saw animals being abused. Mostly Grant was a plain, friendly person, with a painful sensitivity buried behind his impassive public face. Even as a lieutenant general and general in chief he was easily overlooked; amid a crowd he could give the impression of being off somewhere else. He was a devoted husband and father who wanted his family with him whenever possible. Without them, out on the West Coast after the Mexican War, he had turned to whiskey for comfort, gotten crossways with a martinet post commander, and resigned his commission. The affair remains mostly unexplained; probably homesickness had much to do with Grant's decision. But the enduring result was that Grant—normally a very temperate man in an army of two-fisted drinkers—would be regarded for years thereafter as an incipient dipsomaniac. After he left the service, all his luck was bad; when the war came, he had difficulty getting back on active duty. Once back, he made himself noted as a competent, hardworking officer who was eager to get the war moving.

Undoubtedly, it was fortunate that he came under Halleck's command, though they suffered mutually from it. Grant's more motherhen biographers also have suffered from the injustices he supposedly experienced at Halleck's hands, but most of these either never happened or have been much exaggerated. Grant had a sloven, careless streak, guaranteed to stimulate Halleck's talent for being an officious fusspot. But the two men complemented one another and slowly came to be an effective strategic team.

The envelopment and capture of Vicksburg in 1863—a small-scale repeat of Scott's campaign against Mexico City—was Grant's finest feat. Promoted to the newly revived grade of lieutenant general (the first man to hold it since George Washington) in early March 1864, and so replacing Halleck as general in chief, Grant sought to bring the full power of the United States into several coordinated offensives around the whole perimeter of the battered Confederacy. Unfortunately, 1864 was an election year; Lincoln felt it essential that several potent political generals be given command of important minor offensives, the most important case being that of cockeyed and ambitious Maj. Gen. Benjamin F. Butler. While the Army of the Potomac (accompanied by Grant) moved against Lee's army north of Richmond, Butler was to bring a strong corps up the James River and snatch Richmond behind Lee's back. Butler discharged his mission with all the grace of an idiot child entrusted with a bag of eggs and a big stick, managing to get himself defeated by a much smaller Confederate force.

Maj. Gen. Franz Sigel, ordered to move up the Shenandoah Valley and cut Lee's communications to the west, once again displayed his talent for disorderly retreats. These reverses left the Army of the Potomac little choice but a grinding battle of attrition against Lee, a bloody business that got Grant the reputation of "Butcher." Until the end of the war, Grant's major concern would be to coordinate the various Federal offensives, great and minor, most of which went too slowly to please him. In urging them on, he had one constant theme—*push* the enemy, break up his communications, take what you need from the country, "much is expected." In the end, his pursuit of Lee to Appomattox was a model operation; his management of Lee's surrender one of America's honorable traditions. Thereafter, it was a downhill slope. For two terms he would prove an increasingly incompetent president; he would vainly seek a third term, fail in business, and—to provide for his family—heroically write his memoirs while dying of throat cancer.

Grant's military skills seem to have been innate, sharpened mostly by experience. He was no student of the military art, though his early taste in novels did run to Charles Lever's *Confessions of Harry Lorriquer* and *Charles O'Malley,* with their colorful (and inaccurate) background of Wellington's campaigns. As a new colonel in 1861, he rather ignored the changes in infantry drill regulations since he had left the Army seven years before. He had the advantage of relatively slow promotion (by Civil War standards) through 1861, giving him the opportunity to relearn his trade. Probably his strongest characteristics were common sense and combativeness. He was aggressive, and his aggressiveness was not sated by a battle won so long as the enemy force was not destroyed. He could grasp the whole war across all its shifting fronts, and usually he saw it clearly. Implacable and tireless, he gripped control of the fighting far more tightly than Halleck had done, breaking or putting aside any subordinate who would not do his will.

Grant's *Personal Memoirs* are definitely worth reading, but they are not accurate military history. Aging, dying, half remembering old disappointments, imagining others, he left a twisted tale, lit by amazingly deft cat's-paws of unexpected wit. His example of what an average-seeming officer could do, given opportunity, abides. And he put his sense of strategy into one sentence that perfectly expresses the traditional American school of war: "Find out where your enemy is; get to him as soon as you can; hit him as hard as you can, and keep moving on."[30]

Redheaded, red-bearded William Tecumseh Sherman had been

christened "Tecumseh," but his Catholic foster mother would have no such heathen name and had her priest do the job over; it was St. William's Day, and so Sherman became "William T." He was intelligent, lively, and endlessly talkative, restless in mind and body. At West Point he was noted for his skill in smuggling food out of the mess hall and recombining it as "hash" over the fireplace in his room. Fortunately, he spent his plebe (freshman) year with big George H. Thomas of Virginia as a roommate. Thomas, a model of military decorum otherwise, was known to ask an intruding upperclassman bent on hazing them whether he preferred to leave by the door or window. Sherman graduated in 1840, sixth in a class of fifty-two cadets, but his sloppy, careless dress and behavior kept him a cadet private until graduation. Sidetracked in California during the Mexican War, he left the Army in 1853 to try his fortune variously as banker, real estate dealer, and lawyer. Despite honesty and hard work, he was never successful; he finally found a satisfying career as head of the Louisiana State Seminary, but it was 1859, and the Civil War ended that. He had married a foster sister, who mothered him with more affection than respect and gave him no say in the education of their children.

Years of failure, a regretful fondness for the South and southern friends, an intense impatience with the North's inexpert mobilization, left him in mental turmoil; despite the urgings of potent political backers, he wanted only a colonel's commission. He fought stoutly at Bull Run, was promoted to brigadier general and sent to Kentucky, then in the confusion of deciding which side it would join. Overwork, the difficulty of getting weapons and equipment for his raw troops, the greater difficulty of getting reliable information on the location and strength of the enemy, twisted his nerves. He began imagining vast hordes of Confederates moving against him and compounded this by declaring war on all visible war correspondents. Generally a top-lofty, scurrilous lot, they responded by hints that he was insane, treasonous, or both. Transferred to Halleck at St. Louis, his delusions worsened. Sent on an inspection tour, he "stampeded" and ordered a general withdrawal to escape a Confederate offensive that existed only in his imagination. Halleck put him on leave; when Sherman returned, he gave him gradually increasing responsibilities that utilized Sherman's adminstrative ability until he could be returned to troop duty.

During this period of recuperation Sherman was busied forwarding supplies and reinforcements to Grant, then engaged in his Fort Henry-Fort Donelson campaign. A rare friendship developed—Grant silent, cogitative, deliberate; Sherman garrulous, quick-witted, enthusiastic—battle-ax and rapier. They shared a bitterness of barren years, the urge to press the war. Sherman became Grant's right arm, just as Jackson had been Lee's, but on a far greater scale.

Unlike Grant, Sherman had studied the art of war; until the Vicksburg campaign, he made war by the book. He considered Grant's final campaign against that city a dangerous gamble and advised against it. Thereafter, Sherman's methods changed. He lived off the country as much as possible and relaxed his discipline and his efforts to protect Southern civilians from his marauding troops. From his earlier despair, his later hard-won balance, he began to visualize himself the embodiment of the avenging wrath of an outraged republic. Always, he talked and wrote far more ferociously than he acted, harsh as his actions sometimes were: "the present class of men who rule the South must be killed outright" was a typical effusion.[31] He developed into an able army group commander, capable of handling 100,000 men with efficiency. (One forgotten Confederate claimed to have heard Sherman order: "Creation, attention! By Kingdoms, left wheel! March!") Normally he shrank from set battles, preferring to maneuver his enemy out of position. When he did attack an enemy confronting him, as at Chickasaw Bluffs (1862) or Kennesaw Mountain (1864), his tactics were uninspired. First to last, he never exactly won a battle. He had, however, taken Halleck's theory of living off the country a long ways beyond even Grant's application. To him, the objective became neither the conquest of enemy territory nor even the defeat in battle of the enemy's armies; rather, it should be the undermining of the morale of the enemy civilians and of the economy that supported the enemy's war effort. (He had shown something of this tendency from the start: His failure at Chickasaw Bluffs probably resulted from his halting for a day to demolish a jerkwater railroad, giving Confederate reinforcements time to get into position.) He would imitate Scott in Mexico, Grant at Vicksburg, and abandon his supply line and strike off through enemy territory. His march across Georgia for 200 miles from Atlanta to Savannah and a second march from Savannah north some 425 miles to Raleigh, North Carolina, were really gigantic raids to destroy railroads, foodstuffs, and industries and convince all Confederates that, since their government could not protect them anywhere, they had lost their war. Sherman considered this "statesmanship." These raids gutted Confederate morale, both civilian and military; the wrecked railroads made it impossible to keep up the supply of Lee's army, which still clung to Richmond. At Raleigh, Sherman was in fine feather indeed. He had conceived these raids and persuaded Grant and Lincoln to allow them; he had carried them through with skill and swagger. When it came to negotiating the surrender of the Confederate forces facing him, he happily expanded the process into a veritable treaty of peace that he felt the President should sign at once without alteration. He was quite hurt by the reception that the government—in a turmoil after Lincoln's assassination—gave his attempt at statesmanship.

Sherman has been called the first advocate of total war and a totalitarian general. Such titles probably would have shocked him; for all the overloose discipline he allowed, for all the bloodcurdling phrases he delighted in, he was a sensitive, merciful individual. He haunts Southern mythology as a degenerate version of Attila the Hun, but by the contemporary European standards his frolic across Georgia was a Sunday school processional. He also has been called the "first modern general." Liddell Hart (see page 215) was particularly fascinated by Sherman's campaigns; his study of them stimulated the development of Hart's pet theory of the "indirect approach"; he also applauded Sherman's insistence on reducing his supply trains to the absolute minimum to increase his army's mobility. Thrown in with similar studies on Genghis Khan and Nathan B. Forrest, Sherman furnished justification for Hart's pre-World War II theories on deep strategic penetrations by armored and airborne troops.

In all this there is merit, but Hart (and others) sluffed off some awkward facts. Sherman began his march through Georgia with the pick of his troops, turning his back on a strong Confederate army under Lt. Gen. John B. Hood. Grant and Lincoln had wanted him to deal with Hood first, but Sherman found this frustrating and preferred to "make Georgia howl." He left Thomas to handle Hood with a scattered force that included many green recruits, dismounted cavalrymen, blacks, and armed quartermaster employees. Fortunately, Thomas managed splendidly despite much uninformed heckling from Grant. Sherman's raiding met no noticeable resistance in Georgia and no effective opposition in the Carolinas simply because, after years of grinding campaigns and battles, the South had no appreciable forces between Hood's army and Lee's—while the North had available naval and military strength to resupply and reinforce Sherman along the coast from Savannah northward. In plain words, Sherman's idea worked only because the South was already a hollow shell and Sherman had the wit to see that it was such. His raids undoubtedly shortened the war; they certainly did not win it.

Sherman's contributions to the art of strategy were unusual. His *Memoirs,* published in 1875, were written, as he said, "not as history but as a recollection of events."[32] Chatty, lively, and interesting, it produced a gale of objections from men who felt—often with reason—that Sherman never acknowledged his own errors and was liberal in assigning blame to other officers and to any army except his own pet Army of the Tennessee. A popular and pungent public speaker in his older years, he might define military fame as "to be killed on the field of battle and have our names spelled wrong in the newspapers" and war as "all hell." But once he defined strategy: "Common sense applied to the art of war. You have got to do something. . . . You can't go around asking corporals and sergeants. You must make it out of your own mind."[33]

To better stock and sharpen the minds of American Army officers, in 1881 Sherman (then Commanding General of the Army) created the School of Application for Infantry and Cavalry at Fort Leavenworth. This taught mostly practical subjects but included the "science and practice of war." It would gradually develop into the Command and General Staff School. Sherman also supported the existing Artillery School and developed an Engineer School of Application. He could do little to stimulate better instruction at West Point, where the faculty would remain complacently inert for years, convinced there was no need for change since their graduates had just won a mighty war, but he could encourage studious young officers such as Emory Upton.

Upton—savagely devoted, heroic, often wounded, something of a northern Stonewall Jackson—was graduated from West Point in 1861; by 1865 he was a brigadier general commanding a crack cavalry division rampaging through Nathan B. Forrest's last despairing stand in Alabama. Many considered him the best tactician the war had produced in either army. Unlike Sherman, he was a strict disciplinarian and a demon drillmaster, yet popular with his men. He kept a portable military library in his tent, studied it regularly, and said his prayers every night. Upton's interests were two: Develop a coherent national military policy for the United States and work out new tactics to fit the breech-loading, repeating weapons then coming into general use. But he also was interested in encouraging the study of military history and the art of war among Army officers, especially those in isolated western posts who might otherwise squander their lonely evenings as Capt. U.S. Grant had done. Sherman sent him on a world study tour that resulted in *The Armies of Asia and Europe*,[34] giving an exact picture of the world's armies in the late nineteenth century, with special emphasis on the virtues of the German general staff and military organization. Besides his drill regulations, Upton wrote frequently for publications such as *The Army and Navy Journal* and performed as the Army's intellectual handyman. Though his unfinished *Military Policy of the United States* was not published until 1904 (tortured by a probable brain tumor, Upton had committed suicide in 1881), its controversial contents had long been known to a good many fellow officers.[35]

His urgency in increasing the Army officers' knowledge of strategy, tactics, and the whole art of war was ably supplemented by officers such as John Bigelow, whose *The Campaign of Chancellorsville: A Strategic and Tactical Study* is still one of the finest battle studies ever published, and Arthur L. Wagner, who had a major share in the growth of the Fort Leavenworth schools and the organization of the advanced Army War College, authorized in 1900. One of Wagner's works, *The Service of Security and Information*, was the first major study on military intelligence published in this country. Studying the works of such soldier-scholars, as the nineteenth century ebbed into the twentieth, attending the schools that Sherman had launched, the American Army officer came to know himself a member of a learned profession with worldwide interests.

The Spanish-American War and World War I offered no real opportunity for strategic achievement. During the latter, Gen. John J. Pershing, commander of the American Expeditionary Force (AEF) in France, could take pride in having taught his rawly raised soldiers the rudiments of mobile warfare that carried them through Belleau

Wood and the Meuse-Argonne, but his major triumph had to be the defeat of British and French attempts to reduce the AEF to a replacement pool for their own riddled armies.

Among the colonels of the AEF were brilliant men—driving young George S. Patton, Jr., and George C. Marshall, whose skills as a staff officer and administrator would not only make him army Chief of Staff during World War II but also too indispensable in that office to be spared for the troop command he wanted. There was keen-minded Fox Connor, who had his junior officers study Clausewitz and foresaw the shape of World War II but would be too old for service when it came. (Connor is remembered for the enunciation of one overlooked strategic truth—that dealing with an enemy is far simpler than dealing with an ally.) And there was dramatic Douglas MacArthur, who would be the one outstanding American strategist of World War II.

It once was told in Scotland that the only things more ancient than the MacArthurs were the hills, the sea, and the devil. One branch of them were hereditary pipers—a post of pride and repute—to the bloody-minded MacDonalds of Skye, who considered themselves lords of the western isles. And Douglas MacArthur could be as remote and chill as the outermost Hebrides in winter mist and as clamorous as any drunken piper.

His father Arthur MacArthur, had been made a first lieutenant and adjutant of the 24th Wisconsin Volunteer Infantry Regiment in 1862 at the age of seventeen. It is ancient Army tradition that his voice still was apt to crack into an adolescent treble while reading orders to the assembled regiment, but he swiftly made himself remarkable for courage and efficiency. Passing into the Regular Army in 1866, he went through Indian campaigns and the conquest of the Philippines, where he did excellent work as military governor, to retire in 1909 as a lieutenant general. An officer who served with him left this edged description: "I thought that Arthur MacArthur was the most flamboyantly egotistic man I had ever seen—until I met his son."[36]

Young Douglas MacArthur grew up an Army brat on frontier posts through years when the Army was very much a Spartan world unto itself. (His older brother, Arthur MacArthur II, became a Navy officer and is long forgotten.) When he entered the U.S. Military Academy in 1899, his mother, a gentlewoman of Virginia, established herself at the Academy hotel. Douglas visited her daily. Being a general's son, he was hazed unmercifully but rose to Cadet First Captain and graduated first in a class of ninety-three. Eleven years later he was a major of engineers. With America's entry into World War I he took part in the organization of the 42nd ("Rainbow") Infantry Division, going overseas as a colonel and its chief of staff; later, after promotion to brigadier general, he commanded one of its brigades. He was a courageous, skillful, and dashing combat soldier who wore neither steel helmet nor gas mask with his nonregulation uniform, carried no weapon except a riding crop, and did not spare himself. He was twice wounded and received twelve decorations, including two awards of the Distinguished Service Cross.

In June 1919 he was appointed Superintendent of the U.S. Military Academy, which—its course of instruction having foolishly been cut back to one year during the war—was in near chaos. MacArthur introduced a more commonsense discipline and a cadet honor system, emphasized intramural athletics, and attempted to modernize the curriculum, his reaction to the existing instruction being, "How long are we going on preparing for the War of 1812?"[37]

Most of his proposed changes were justified, if not long overdue, but he attempted, as he would later half admit, to accomplish too much at once. The Academy's permanent professors were thoroughly set in their ways. Five of them had been professors when MacArthur entered West Point and were not about to be hustled by any young whippersnapper, no matter how bemedaled. MacArthur's aloof ways did not help matters. He ignored (when he did not snub) West Point's social life, wore a peculiar uniform, and returned salutes with a flip of his riding crop. Cadets were as unimpressed as their professors, remarking that he always seemed to be "gazing at distant horizons" as though posing for a publicity photo.[38] On his departure, his successor canceled many of his most pertinent reforms.

After serving twice in the Philippines and being promoted to major general, MacArthur was named Army Chief of Staff in 1930. Because this assignment coincided with the Great Depression, his efforts to modernize the Army had only limited success, there being no money for improved aircraft and armored vehicles, or trucks to replace mule teams. He also had to handle the difficult problem of the Bonus Army in 1932, which brought him much criticism. On completion of his tour as chief of staff, he was appointed military adviser to the newly independent Philippine Republic. During these years he acquired a hardworking young staff officer named Dwight D. Eisenhower, whose friendly grin concealed an incinerating temper. Old Army gossip had Eisenhower later describing this service as studying dramatics under MacArthur—and MacArthur remarking that Eisenhower (by then a general) was the best file clerk he had ever had.

Though he retired from the U.S. Army in 1937, MacArthur continued in the service of the Philippines as a field marshal, replacing his riding crop with the corncob pipe, dark glasses, and gold-braided cap that became his famous personal insignia. His far-reaching plans for the defense of the Philippine archipelago against foreign enemies and domestic brigands would require ten years and much money to implement, and neither was given him. Also, it may be that he concentrated too much on the "big picture" of his program and neglected the gritty details of troop training and equipment. He did marry happily in 1937, giving his life a domestic anchor. (A first marriage, to a socially active woman, had ended by mutual consent in 1929: In Army folklore grim old General of the Armies John J. Pershing had predicted the divorce because there was only one full-length mirror in the MacArthur's quarters.)

With Japanese aggression through the Far East building rapidly into war, MacArthur was recalled to active duty in 1941 as a lieutenant general commanding U.S. Army forces in the Far East. He made a determined but fruitless defense of the Philippines against the Japanese invasion, finally escaping—on orders from Pres. Franklin D. Roosevelt—to Australia before the final surrender, to take command of the Southwest Pacific area. Here, beginning with a shoestring force, he first checked the Japanese, then launched a swelling counteroffensive across the island chains northward to Luzon. In this he brilliantly combined amphibious, naval, air, and airborne operations to strike deep and unexpectedly between Japanese strongholds, slashing their communications and leaving bypassed Japanese forces to wither on the vine from starvation and disease. At the same time he had energy to spare for a private war with the U.S. Navy high command.

Older and more experienced than other American commanders, MacArthur was a thoroughly difficult subordinate. He had been Army Chief of Staff when Gen. George C. Marshall, who now held that post, had been only a colonel, and he did not forget it. He disdained the ''Germany first'' priority the United States and its allies had established, and could regard the vital Anglo-American liberation of North Africa as a Marshall-Eisenhower plot to divert troops and shipping that should have been sent to him in the Pacific. He was expert at presenting himself to the American public as a devoted martyr, abandoned to fight his desperate battle against great odds, and was willing to be considered as a possible Republican candidate for President in the 1944 elections. But he won battles, efficiently and usually without excessive casualties. He did it with ruffles and flourishes; endless photographs showed him striding through water, jaw set, onto yet another barely established beachhead. It was a hard act for more conventional commanders to match, and none of them did.

In December 1944 MacArthur was promoted to the five-star grade of General of the Army, but his plans for the invasion of the Japanese home islands were cut short when the nuclear bombing of Hiroshima and Nagasaki forced that nation's surrender. In August 1945 he became Supreme Allied Commander in Chief, Allied Forces in Japan, and as such presided over the demilitarization, reform, and reconstruction of Japan—probably his greatest achievement. From the start, he dealt bluntly with Russians, Japanese Communists, and—on occasion—with the U.S. State Department. More and more he became the reincarnation of a proud proconsul ruling a frontier province of the Late Roman Empire—the sort of proconsul who gave his emperor back in Rome troubled days and restless nights. His aloof reign (for such it was) was interrupted by the Communist invasion of South Korea in June 1950. On orders from Pres. Harry S. Truman, MacArthur immediately aided South Korea and in July was named commander of the UN forces mustered for that purpose. Unfortunately, he had allowed the American forces in Japan to become slovenly and ill-trained, with substandard equipment. The American intervention came close to disaster before MacArthur could cut the North Koreans' communications by an amazing amphibious envelopment at Inchon. Thereafter, MacArthur's forces pushed into North Korea but were badly defeated by an unexpected Chinese Communist counteroffensive. Deeply angered (he had been certain that the Chinese would not risk intervention: If they did, he was equally certain that his air force would detect and destroy any Chinese troops entering Korea), MacArthur demanded that the full military strength of the United States be put at his disposal and that he be authorized to carry the war into Communist China. Fearful that such action would precipitate World War III, deeply concerned over the Russian threat to Europe, Truman refused—and MacArthur deliberately thrust his opinions before Congress and the America public, insisting that there was no substitute for victory. In years to come, history may yet prove that he was right, but his actions were rank insubordination; Harry Truman relieved him. In Korea a mighty storm suddenly flung snow and hail across the front, until one soldier marveled, "Say, do you suppose MacArthur *was* God, after all"[39] MacArthur took his punishment like a soldier; he ws given a hero's welcome in the United States and made a stirring defense of his policies before Congress. And then, as old soldiers do, he faded away.

As a soldier and individual, MacArthur was unique and probably beyond complete explanation. Of unusual intelligence, highly intuitive, he could sense a subordinate's potential or an enemy's future course of action. His self-confidence and high courage blended into a sense of destiny; his strong religiousness could verge on mysticism. He expressed himself grandly in a thoroughly personal style of speech and prose. He could win the unquestioning loyalty of most men, making them feel themselves valued assistants and thus eliciting their best efforts, without encouraging the least familiarity. A minority disliked him on first sight, considering him arrogant and too obviously law unto himself. Dramatic in his public actions, practical in essential matters, he remained habitually solitary and remote from many human activities and interests. Sports fascinated him—the final score of the annual Army-Navy football game seemed as important as the result of a major battle. Not surprisingly, age and many victories made him more insensitive and self-centered. Except for Inchon, his conduct of the Korean War lacked his former fine feel for the battle's ebb and flow.

Though he wrote and spoke frequently and on many subjects, he left few specific observations on strategy. He was a convinced and constant student of military history, especially of Napoleon, the Civil War, and Genghis Khan. From them he drew his insistence on mobility and rapid *offensive* maneuver, evading the enemy's main forces to strike at their communications. "We hit them," he said of the Japanese in World War II, "where they weren't."[40] From his own experience he stressed the smooth synchronization of ground, air, and naval forces and the need for unquestioned *unity of command*. Always he emphasized moral factors—the will to victory; devotion to "Duty, Honor, Country"; and the unique mystique of military service.

> The letters blaze on history's page,
> And ever the writing runs,
> God, and honor, and native land,
> And horse and foot and guns.[41]

VII

THE GUNNER AND THE GUNNER'S DISCIPLES

Read over and over again the campaigns of Alexander, Hannibal, Caesar, Gustavus, Turenne, Eugene and Frederick. Make them your models. That is the only way to become a great general and to master the secrets of the art of war.

The Art of War is the giant among the branches of learning, for it embraces them all.

NAPOLEON[1]

Napoleon I, Emperor of the French, is an epitome of the Great Captain. He fought as many battles as Alexander, Caesar, and Frederick together, under the most varied circumstances of weather, terrain, climate, and type of enemy. His mastery of mass warfare and his skill in raising, organizing, and equipping mass armies revolutionized the art of war and marked the beginning of modern warfare. The extent of his campaigns forced the development of logistics as an essential partner of strategy.[2] He continues to influence strategic planners today both through his own words and deeds and through the writings of his utterly dissimilar, disputatious disciples, Antoine Henri Jomini and Carl von Clausewitz.

Though his career ended in defeat and exile, this hardly affected the esteem that even his enemies felt for his military skill. He remains acknowledged as the ablest soldier of his age and in all military history. As J.F.C. Fuller wrote of another general: "In order to understand [generalship] we must at once strike from our minds the popular illusion that generalship can be measured by victory or defeat; for there is something more subtle than this in the drama of war, namely the art of those who wage it."[3]

By an odd twist of fate, it is only recently that we have been able to actually "know" the living Napoleon. His personal life and character, his political aims and methods, even aspects of his military career and strategy, have been mishandled by most historians—often intentionally, frequently from the difficulty of properly evaluating the available source material, sometimes out of built-in national bias. (A few English semihistorians still insist on retelling how a squat little Corsican bounder named Bonaparte misbehaved until he finally met his overdue comeuppance at the immaculate hands of that true-blue English gentleman, the Duke of Wellington!)

Even fair-minded historians found their available sources full of booby traps. While he lived, enemy propaganda presented Napoleon as a monster who relished murder, treachery, theft, incest, blasphemy, and any other possible evil. The counterblasts of his supporters sometimes went to almost equal extremes in lauding him. The most misleading truth twisting, however, came from people who had served him to their profit, but—in hopes of making an equally profitable peace with the Bourbons who supplanted him after Waterloo—turned to defaming him. Prominent among them were former close associates of Napoleon such as Louis Antoine de Bourrienne, the Duchess of Abrantes, Claire de Remusat, and Marshal Auguste Marmont. The memoirs such people wrote, or had ghostwritten, were accepted as indispensable reference works by too many writers, though most of them are worthless and even the better ones contain much untrustworthy material. Only during the last few decades have English-language historians really managed an accurate recreation of Napoleon as an individual human being, as well as a ruler and statesman.[4]

There has never been any certainty as to whether Napoleon's eyes were gray or blue; his hair was a reddish-brown chestnut. Of roughly average height for that period (a little over five feet six), he was broad shouldered and deep chested, not particularly muscular but full of health and activity. His face usually impressed observers with its combination of strength and male beauty and its clear pallor. Impatient and restless to be at his next task, he was sensitive to cold, unpleasant odors, disorder, and dirt. He loved open fires, leisurely hot baths, books, music, fine horses, black cherries, and ghost stories. When he played games he often cheated, but always paid back his winnings. Usually held under tight control, his temper could explode in annihilating storms of abuse, sometimes accompanied by a blow from the imperial hand or riding whip. His friends (he had friends and valued them greatly) would ignore these displays, knowing they would pass quickly and that his cold good sense would reassert itself. He was generous with friend and foe, humane, and always grateful for favors done him when he was young, poor, and lonely. By contrast, he ran his imperial household and wardrobe on a tight budget, favored a mediocre red wine, and drank it mixed with water. Except for state functions, his meals were simple and soon over—wise guests dined comfortably at home before eating with him. And, contrary to all tradition, he was a hero to his valets; his second one went willingly to St. Helena with him.

His intelligence—swift, mathematically precise, analytic—was teamed with equally formidable powers of concentration and industry. He spared neither himself nor his assistants, often working into the small hours of the morning or getting up in the middle of the night to complete unfinished business. He compared his mind to a filing cabinet: Each subject had its separate drawer; when he finished with one, he closed it and opened another; the contents of the various drawers never got mixed. When he was weary he closed all the drawers and was asleep, whether for the night or just a snatched catnap during a lull in a battle. His near-total memory was a terror to the careless and the procrastinating. In sum, he was a very human person, endowed with certain extraordinary qualities that sometimes made him seem a little more than, a little different from, human.

Napoleon Bonaparte (1769–1821) was born in Ajaccio, Corsica, shortly after French rule had been established over that unruly island. His beautiful mother, though pregnant with Napoleon, had accompanied her husband through months of guerrilla battling. Both parents were descended from ancient noble families of central and northern Italy. They were well enough off by Corsican standards, since their small estate kept them supplied with olive oil, bread, and wine. But Corsica was a poor backwoods place where one Corsican was about as good as another; its nobility had no special privileges and were leaders only as they proved themselves worthy. Napoleon's mother brought him up strictly as a nobleman's son, but he played with neighborhood children and knew how commoners lived. Corsican family feeling was tight and strong. Napoleon was always an affectionate, dutiful son and the protector of four brothers and three sisters—a squabbling, demanding, scratchy lot that gave him little but more trouble in return.

Napoleon's father soon became a leading supporter of French rule. As a reward, several of his children were admitted to French state schools; at the age of ten, Napoleon entered the recently established military academy at Brienne. His Corsican accent and small-town ways made him an oddity at first, but he did well enough to be accepted into the advanced École Royale Militaire (Royal Military School) in Paris in 1784. Completing the nominal two-year course in one year, he was commissioned, at the age of sixteen, a second lieutenant of artillery. He was poor but apparently happy enough, working hard at his profession and reading an amazing variety of books. His father had died in 1785; to help his mother, Napoleon took over the raising of his younger brother Louis. He spent much time on leave in Corsica, for there were family problems to handle and his older brother, Joseph, was a wet dishrag in any emergency. Like the majority of French artillery officers (mostly from the middle class or poor minor nobility), Napoleon welcomed the French Revolution but soon was disgusted by its mob violence. His attempt to exploit it to his own advantage in Corsican politics, however, ended in 1793 with the entire Bonaparte family exiled and outlawed by the pro-royalist, pro-English faction.

Back in France, with his family to support, Napoleon returned to his duties as a captain of artillery. His skill and bravery (he was twice wounded) at the siege of Toulon won him promotion to brigadier general and subsequent assignment as Chief of Artillery to the French Army of Italy. There, in addition to his normal duties, he functioned as the commanding general's brain, achieving considerable distinction. His next assignment, however, was the command of a brigade of infantry in the Vendée (western France) where a brutal civil war went on and on. Not wanting such service, Napoleon took the risk of remaining in Paris, enduring official threats, dirty lodgings, and slim meals. (Most of his pay went to support his family.) He was deeply interested in completing the French conquest of Italy. His continuous memoranda on that subject finally got him an assignment on the war ministry's central planning staff, but that did not promise a brilliant future. A conservative counter-revolution in Paris itself rescued him from this desk job. The one capable Regular general in the capital, Napoleon was placed in command of the few troops available, quickly squelched the uprising by a liberal application of canister, and was promoted to major general.

Proud, poor, always working, Napoleon had never been at ease with women. Now a tiny, graciously fading widow, Josephine de Beauharnais, caught him. She was six years older than Napoleon, with two children and poor teeth—extravagant, gentle, elegant, and completely feminine. Napoleon fell in love as thoroughly as he did everything else. They were married in March 1796; on their wedding night Josephine's poodle, Fortune, objected to Napoleon's presence: "I have the marks on my leg to show what he thought!"[5] Two days after their wedding he rode south to take command of the Army of Italy. A devout believer in wringing pleasure from every moment, Josephine soon was cuckolding him. She seldom answered his fervent letters. Later, there were stormy scenes; he forgave her, but her easy infidelities plainly had twisted something in him. They lived together in a tolerant partnership, and Josephine would appear a gracious empress. But she was seldom truthful and always in debt; Napoleon still loved her but hardly respected her. Worse, she could not give him a son and heir. Among history's might-have-beens is the wonder of what could have happened if some virtuous young lady of intelligence and charm had grabbed him first!

Napoleon's career after assuming command of the Army of Italy is well known: victorious in Italy, 1796–97; conqueror of Egypt, 1798–99; First Consul of France, 1799; victorious over Austria, 1800–1801. By 1802 he had achieved a successful peace throughout Europe. England, however, still held its basic national strategy—no one power should dominate continental Europe. There was war again in 1803. Secure behind the English Channel, Great Britian would raise coalition after coalition against Napoleon, hiring the kings of

Europe as casually as it had recruited German mercenaries during the American Revolution.

In 1804 Napoleon became, by a popular plebiscite, Emperor of the French. The next year he routed the Russians and Austrians at Austerlitz; in 1806, he broke Prussia; 1807 brought Russia to heel and into professed alliance with France. His first false move, the occupation of Spain, came in 1808. An English army arrived to assist the Spaniards; it was driven back to sea, but Napoleon had to leave his work in Spain only partly done—Austria had taken advantage of his Spanish troubles to launch a new, undeclared war. Napoleon crushed this effort by mid-1809, but another English army came into Spain, led by Sir Arthur Wellesley (soon to be the Duke of Wellington), to make a hard core for Spanish resistance. Spain would be an "ulcer" to Napoleon's military strength.

In late 1809, Napoleon regretfully divorced Josephine to marry an eighteen-year-old daughter of the Emperor of Austria, Marie Louise. A fair, blue-eyed, immature girl, weak-willed yet sincere, she fell in love with Napoleon—as he with her—and gave him a son whose destiny would be tragic and short.

Napoleon's "Continental System"—his economic war with England—had deranged European business. In Italy and Germany where Napoleon had swept away the repressive rule of kings, nobility, and clergy, a new spirit of nationalism stirred among the people. Russia had proved a false ally and a secret enemy, mustering armies for another war. In 1812 Napoleon decided to hit first, moving into Russia with over 500,000 men. The Russian plan to trap and destroy him just across their frontier collapsed. Napoleon reached Moscow, but Russia was too vast, its roads too poor, for Napoleon to destroy their main army. The Russians would not make peace; the dreaded Russian winter was at hand. Napoleon had to retreat, losing over 400,000 men, mostly to cold, hunger, exhaustion, and disease. In early 1813 Prussia joined Russia; Napoleon brought a hastily organized army into the field and repeatedly defeated their combined forces. But Austria and Sweden joined his enemies, and his German allies began deserting him. Finally defeated at Leipzig in October 1813, Napoleon fought his way back across the Rhine with the remnants of his army. In the first days of 1814 his enemies for once seized the initiative, pouring across the Rhine before he could reorganize his troops. He waged an amazing campaign of maneuver, winning a succession of small victories, but the odds against him were too heavy, his marshals weary of war. In April they forced his abdication.

Peace brought the Bourbons back to rule France. Napoleon was given the little island of Elba (ninety-four square miles), just east of Corsica. His father-in-law would not permit his wife and child to join him; the Bourbons would not pay the pension promised him. There were plans to move him to some more isolated island. Meanwhile, the Bourbons made a hash of ruling France, and the victors quarreled over the loot.

Napoleon suddenly left Elba with his little guard, slipped past patrolling English and French warships, landed at Antibes (just east of Cannes), and pushed north through the mountains to Grenoble and Paris. Not a shot was fired; troops sent to halt him joined him; the Bourbons fled. Napoleon offered peace, but all Europe marched against him. Again he struck first, outwitting and surprising Wellington and the Prussians, his strongest enemies, in Belgium. He came close to complete success—in Wellington's own words it was "a damn nice thing—the nearest run thing you ever saw in your life."[6]

Waterloo ended his military career. Faced with a variety of unpleasant fates, he surrendered to the English, who reacted with the grace of a keeper of a children's zoo presented with a live, fang-gnashing *Tyrannosaurus Rex*. There was a wild rush to shut him up somewhere far away—the unhealthy island of St. Helena, a thousand miles from anywhere in the South Atlantic Ocean. There, meanly treated, tightly guarded by thousands of troops and constantly patrolling warships, under the charge of a stupid, frightened governor, he won his last and greatest victory, writing and talking with poignant skill to apotheosize himself as the champion of freedom and the rights of man. And so he passed from life into a towering myth.

Napoleon had reigned as a true emperor, lawgiver, and builder. His Code Napoleon, which modernized and systematized French law in clear language, is still the basis of French law and has had worldwide influence. He built no new palaces but left a mighty heritage of harbors, highways, bridges, drained swamps, and canals. He planted trees along his roads; set up a government office to protect France's forests, lakes, and rivers; gave Paris better water and sewer systems, its first public fire department, an improved opera, and the modern system of street numbers. Wherever his rule ran, there was freedom of religion, basic human rights, better hospitals, orphanages, and public sanitation. During his ten-month rule on Elba he built roads, planted trees, organized a system of garbage collection, and pushed the building of private latrines. He encouraged vast improvements in French agriculture and built up an enlarged system of public and private education. Just as important was his emphasis on competence and honesty in his officials. All careers were open to men of talent who would serve loyally, regardless of family background or political orientation. Also, he balanced his budgets; even in 1814 France had practically no national debt. And he ruled as a civilian head of state, never as a military dictator.

Napoleon's achievements as a Great Captain had two sources—his training and his own genius. As a lieutenant of artillery, he had been painstakingly instructed by some of the world's best artillery officers, especially the brothers Baron Jean-Pierre du Teil and Jean du Teil. On joining his regiment, Napoleon was required to serve nine weeks as an enlisted man, to learn an artilleryman's basic duties. After that, he was given thoroughgoing instruction in the responsibilities of a regimental officer. As he later stated, "If there is no one to make gunpowder for cannon, I can fabricate it; gun carriages, I know how to construct. If it is necessary to cast cannon, I can cast them; if it is necessary to teach the details of drill, I can do that."[7] In addition, he studied Jean du Teil's *De l'Usage de l'Artillerie Nouvelle dans la Guerre de Campagne,* which advocated mobility, concentration of firepower at the decisive point, and artillery-infantry coordination.

An omnivorous reader, Napoleon went through everything on history or military science he could procure. As emperor, finding no satisfactory book on Marlborough's campaigns, he ordered one written. Among the more modern military authorities he studied—besides Saxe, Lloyd, Villars, and Folard—were Feuquieres, Bourcet, and Guibert.[8] Feuquieres had fought his way up from private to lieutenant general; his memoirs pulse with an aggressive, hit-first spirit. Bourcet was France's most expert staff officer and also an authority on mountain warfare. He seems to have been the first to use what is now termed an "estimate of the situation" as an aid in his planning, methodically comparing the advantages and disadvantages of the various possible courses of action. Guibert (1743–90) was rather a problem child. A young nobleman who had seen some service, in 1772 he wrote his very influential *Essai general de tactique.*[9] This called for an army of citizens rather than of mercenaries, and urged a war of maneuver and less reliance on fortresses. Armies should move light without long supply trains, living off the countryside. How much of this came out of his experience and how much from the joyously vague liberal ferment of that era is impossible to say. The book made him a lion of the boudoirs and salons, but did contain more wisdom than he himself probably realized. In 1779 he produced *Defense du systeme de guerre moderne,* which apologized for just about everything he had written in the *Essai.* "The vapors of modern philosophy heated my head and clouded my judgment."[10]

Excellently educated and self-educated, Napoleon had within himself the great captain's essential qualities of courage, decisiveness, steadfastness, and swift, lucid thought. His wars were mostly defensive. From 1800 to 1815 he was the aggressor only in his 1808 move into Spain and his 1812 invasion of Russia—and the latter was in the nature of a preemptive "spoiling attack" to cripple an open enemy. Yet, even when on the defensive, Napoleon usually managed to seize the initiative and hit first, surprising the enemy by the timing and direction of his offensive. He kept the greater part of his forces concentrated under his direct control. (In the terminology of the time, he "had only one line of operation.") He began his campaigns with an overall strategic plan that clearly defined his *objective*. The plan might be modified according to unforeseen circumstances—Napoleon was past master of the art of turning the unexpected accident to his own advantage—but he kept his objective always in mind. Victorious battles were only means of reaching it; battles that did not contribute to that desired end were a waste of men and time.

His objective was the main body of his enemy's army. He sought to catch it at a disadvantage and destroy it. Once that was done, everything else would be easy.

Napoleon's favorite strategic *maneuver* was to advance so as to bring the French into the enemy army's flank and rear, if possible cutting its line of communications. At the same time he was always careful to protect his own supply line. If confronted by allied armies, as in 1796 and 1815, he might strike suddenly at the junction of their forces, wedge in between them, and deal with them separately.

Though their combined armies might outnumber his, he relied on the speed and skill of his maneuvers to mass a stronger force at the decisive point.

> Don't give the enemy time to regroup; intercept him in his movements; and rapidly move against the different forces you are able to isolate; plan your maneuvers so as to be able in every fight to throw your entire army against portions of his. In that way, with an army half the size of the enemy's, you will always be stronger than he is on the battlefield.[11]

Such operations required a high degree of mobility and troops willing, if need be, to push forward and fight on short rations. "The best soldier is not so much the one who fights as the one who marches."[12] "The first quality of a soldier is consistency in enduring fatigue and hardship. Courage is only the second."[13]

He could get such service out of his men because he shared (portions of the 1812 campaign excepted) his men's dangers and hardships, riding just behind his advance guard, often taking what fortune might send in the way of food and shelter—a tumble-down farm building with some straw for his bed and rain and wind for company; a few potatoes, roasted in the embers of a campfire and shared with his staff, for supper. In action, he was fearless; after a battle he was concerned for the wounded. (Quite contrary to the usual concept of Napoleon, he was careful of his soldiers' health and had a surprising commonsense knowledge on that subject.) He rewarded good service generously, sought to be just and patient. And he won a legendary devotion, the *"Vive l'Empereur!"* that echoes yet across the centuries.

His mistakes—Spain, Russia, his Continental System—came in part from too much self-confidence, hardened by years of victories over stronger enemies, in part from his desperate need to keep the world at bay until France and his dynasty were solidly established. Also, his new system of war required subordinate commanders of unusual capacity, capable of handling independent commands. Such men are few in any generation, and only four or five of Napoleon's marshals—Massena, Davout, Soult, and Suchet, possibly Lannes—met the test. Lannes died of wounds in 1809; Massena wore out. Others of the marshals proved unfaithful in 1814, and Napoleon could never rely on his brothers.

In the end, like Charles XII, he found his task too great. He could build a new France, achieving a shotgun wedding between the best features of the French Revolution and the older Kingdom of France, and build it so strongly that fifteen years of restored Bourbon rule after Waterloo could effect no major changes. He failed in his effort to establish his own dynasty in France, but it was one of history's great failures.

> Seen like some rare treasure-galleon
> Hull down with masts against the
> Western hues.[14]

The major source for Napoleon's opinions on the art of war is his *Correspondence,*[15] which contains most of the orders and advice he showered on his subordinates. There also are smaller collections, such as *Ordres et Apostilles*,[16] containing material discovered since the publication of the *Correspondence*. Napoleon did dictate his memoirs[17] during his final exile on St. Helena, but the absence of essential reference material and his failing health combined to make them flat reading and unreliable history. Only the last volume, which deals with the art of war, shows flashes of the old imperial thought.

Even before his death the first of several little books, filled with maxims excerpted from available Napoleonic correspondence, appeared. One, *The Military Maxims of Napoleon,*[18] was useful—a sort of general officer's checklist (though Napoleon disowned it in his will). Portions of it are still applicable.

There was no secret to Napoleon's mastery of strategy and the whole art of war. To learn how he won battles and wars, simply study carefully and impartially what he did and the masses of information in his corrrespondence. That, however, is hard work, even when the necessary information is available. Many students found it simpler to consult the works of Antoine Henri Jomini, who announced himself Napoleon's one true disciple and interpreter.

Jomini (1779–1869) is a difficult man to know. His few biographies, unfortunately, are practically autobiographies, based on little more than his personal notes and the tales he told various admirers in his later years. We old soldiers are notorious for remembering "with advantages" our deeds in battles far away and long ago—hence the ancient phrase "Lie like a trooper."

Growing up in the town of Payerne in western Switzerland, young Jomini would have heard much of military life. Mercenaries were still one of Switzerland's major exports; there were Swiss regiments in the service of Sardinia, France, Spain, Holland, and Naples; the Pope and the King of France had Swiss Guards; Swiss soldiers of fortune had distinguished themselves in North America against the French and Indians. Apparently Jomini hoped for such a career. Frederick the Great was his first hero, and he read whatever books he could find on matters military. However, Jomini's father, magistrate of Payerne, was a practical man. His son's education was limited, but the boy had done well in mathematics; his father apprenticed him to a banker in Basel.

Jomini was soon complaining that his employer assigned him work unworthy of his intelligence. In 1796 he shifted to a Paris bank but quit that the next year to become an independent speculator. At first successful, he soon lost heavily. By 1798 he was ready to go home. (Switzerland meanwhile was being converted into a French satellite and given a new government.) Jomini clubbed together with a Swiss named Keller who, having won some credit in the French Army, expected to become Switzerland's new Minister of War, and promised Jomini a commission. The Swiss government did not want Keller, but Jomini got a job as secretary in the war ministry because of his excellent handwriting. He worked hard and neatly and was rewarded with a major's commission. But it soon was the Basel bank all over again. Though merely a desk soldier with no active service whatever, Jomini was bossy and tactless with more experienced associates, often insolent to his superiors. Moreover, he seems to have fiddled with Swiss politics. In 1802 it seemed best for him to resign his commission and go back to Paris.

It was not a propitious time for a would-be great captain. Europe was enjoying the unfamiliar pleasures of peace; the French Army was being reduced. Jomini was lucky to find employment with a dealer in military supplies. His solace was reading, both the works of Lloyd, Folard, Puysegur, Bülow, Saxe, and Guibert and various accounts of the campaigns of Frederick, Napoleon Bonaparte, and Archduke Charles.[19] He claimed that the latter were the more useful but nevertheless annexed most of Lloyd's ideas for his own books. Like Lloyd, Jomini possessed intellectual curiosity; wondering over the "whys" of military success and failure, he began writing *Traite de Grande Tactique* to explain them.

The renewal of war between France and England in 1803 brought Jomini increased frustration. The French Army did not want him; he could not find a publisher. He sought an interview with the Russian ambassador, only to be told that he was rather young to give lessons to Russian generals. Learning that Switzerland was providing four regiments for the French Army, he and another Swiss offered—in return for appropriate commissions—to use these troops to overthrow the existing Swiss government and force the annexation of Switzerland to France. This proposal was forwarded to Napoleon, who must have ignored it. (Jomini normally posed as a Swiss patriot, devoted to the welfare of his native land.) At the same time he was rewriting his *Traite,* seeking a more readable style and applying Lloyd's precept that the easiest way to prove a point was by the use of appropriate examples.

Then his luck improved. Learning that Gen. Michel Ney might be entrusted with the raising of those Swiss regiments, Jomini approached him and offered to dedicate the first volume of his new work to him. Ney—onetime hussar top kick, tall, redheaded, an excellent division commander, the "bravest of the brave," soon to be a Marshal of the Empire—had only a limited education and was a man of headlong enthusiasms. Mightily impressed, he not only loaned Jomini the money needed to publish the first volumes of his work but also took him into his staff as a volunteer aide-de-camp.

An aide-de-camp's duties were relatively simple, though often dangerous in combat. Jomini carried Ney's orders and reports, explained them when necessary, observed their execution, and generally served as an extension of Ney's eyes and ears. Ney was impressed by Jomini's energy and fluent explanation of military theory. Ney's staff, more often exposed to Jomini's everyday manners, was less happy. A foreigner and a civilian (he wore his former Swiss uniform and insisted on being treated as a major), Jomini was contemptuous of military routine and always ready to advise and correct.

During the 1805 Ulm campaign, Jomini proved himself useful: He spoke German, was full of initiative, and made a good secretary. Outside Ulm, Ney "blooded" him, probably deliberately, by involving him in a skirmish. His own later version of this campaign was that he functioned as Ney's chief of staff, preparing plans and orders. This, of course, was sheer invention. Marshal Berthier, Napoleon's chief of staff, aware that Ney had no experience with a large command, had detailed the competent Gen. Adrien Dutaillis to head Ney's staff. Jomini would also claim a glorious achievement, which has become part of Napoleonic folklore. Napoleon, so Jomini would tell it, left Ney's VI Corps (four divisions) north of the Danube River to block the roads leading northward out of Ulm. Marshal Joachim Murat, commanding the French forces around Ulm, ordered Ney to move south of the Danube, leaving those roads open; Jomini urged Ney to delay executing this order. Ney, thus sagely advised, left one division on the north bank. The Austrians did attempt to break out there, and this division checked them until Napoleon could arrive. It *is* quite a story—but a little serious study proves it fiction.

Once the Austrians around Ulm capitulated, Napoleon sent Ney south into the Tyrol to cover the right flank of his offensive toward Vienna. During the VI Corps' subsequent operations, Jomini was sent forward to imperial headquarters with Ney's reports. Jomini had long hoped to gain Napoleon's personal attention but on arrival—according to standard headquarters routine—was directed to Berthier. Accounts of this confrontation are vague, but it seems that Jomini astounded Berthier by demanding immediate access to the Emperor himself.

Louis-Alexandre Berthier (1753–1815), stocky, incredibly tough, of proven high valor, is one of history's great chiefs of staff. Of his many decorations he was proudest of his insigne of the Society of the Cincinnati, won in America under Washington and Count Rochambeau. He was noted for the courtesy with which he treated subordinates—but here was this strange little civilian of Ney's creating a scene! Apparently Berthier administered a brusque lesson on military courtesy; Jomini departed, a flea in each ear and a hatred flaming up inside him.

From then on Jomini would cherish Berthier as his special, malevolent personal enemy. His hatred grew pathological. Unable to conceive any fault in himself, he developed the quaint theory (which, unfortunately, a number of semihistorians would believe) that Berthier was jealous of his great talents and very much afraid that Napoleon might make Jomini his chief of staff in Berthier's place. His revenge was overwhelming; thereafter, he never missed a chance to blacken Berthier's reputation as a man and a soldier, not scrupling at an occasional lie.

On a second trip to imperial headquarters, Jomini added copies of the first two volumes of his *Traite* to the packet of Ney's reports. Having won the decisive battle of Austerlitz, Napoleon had time to receive and cross-examine him as to Ney's operations but tossed his packet aside unopened. Jomini made another unhappy departure. However, Napoleon had merely put the packet into his ''Hold'' file. Somtime later he had parts of Jomini's work read to him and found it impressive. Accordingly, he ordered Jomini given a French commission as ''adjutant-commandant,'' a staff grade equivalent to colonel. (Characteristically, Jomini assumed the title of colonel, with its implication of troop-command status, and resented any correction.)

With Austria defeated, Russia temporarily stunned, Jomini had time to write the third volume of his *Traite*. War with Prussia was threatening. On September 15, 1806 Jomini drew up a staff study for Ney, predicting such a war and the strategy Napoleon would employ. (It was accurate prophecy but relatively easy for anyone who had a map of northern Germany and knew the positions of the French army.) At the same time Jomini was bringing his *Traite* to the attention of the Prussian ambassador.

Late in September Jomini was summoned to Napoleon's headquarters in Mainz. There Napoleon himself informed him that because of the knowledge of Frederick's campaigns his *Traite* had exhibited, Jomini would be attached to the imperial headquarters for the coming campaign. As Jomini proudly recorded this interview, he promptly astounded Napoleon by rattling off an accurate forecast of the strategy the Emperor intended to use, and was at once warned to keep this knowledge secret—even from Berthier. This story may be largely true: Jomini had already forecast Napoleon's general plan; Napoleon was shrewd enough to know the best way to keep Jomini quiet was to make the whole business a ''secret'' between the two of them. What Jomini did not know, of course, was that Berthier and his staff had begun preparing plans for such an advance three weeks earlier!

Accordingly, Jomini accompanied the imperial staff into Prussia. At the battle of Jena, where Napoleon wrecked one Prussian army, he was sent with orders to Ney—as Jomini's chosen biographer phrased it, ''Jomini solicited permission to fly to [Ney's] assistance''[20]—and obviously spent a rough afternoon at the marshal's heels.

Once the French reached Berlin, Jomini hurried to Sans Souci, the former palace of his first hero, Frederick the Great. Entering the dead king's study, kept intact as a memorial, he bullied the keeper into letting him deck himself with Frederick's decorations, sash, and sword belt. His preening before a mirror was interrupted by the arrival of Napoleon and Berthier.

Jomini's fascination with things Prussian brought him further woe. He eagerly scraped acquaintance with a Prussian delegation sent to explore chances for peace and offered copies of his *Traite* to the King of Prussia. Possibly the Prussians used him for a cat's-paw. Told to prepare a study on the Prussian fortresses of Silesia as a guide for the French conquest of that province, Jomini produced a long paper on the general policy Napoleon should pursue toward Prussia. Prussia, he insisted, should not be punished but rather enlarged by the addition of any Polish territory Napoleon might seize from Russia. Poland was not worth liberating. On his next appearance before Napoleon, Jomini was greeted as "Mr. Politician."

He went on with Napoleon's advance into Poland. During the wild weather and wilder fighting at Eylau (February 1807) he got black looks from the imperial staff by expressing a wish to be in the Russian commander's boots for a couple of hours. Afterward, he would tell how, toward the battle's end, Napoleon had taken him aside and told him that—unless the Russians soon retired—he would have to retreat. Gen. Emmanuel Grouchy's dragoon division would form the rear guard, but supervision of the entire ticklish operation would be entrusted to Adjutant-Commandant Jomini, whose earlier advice the Emperor now regretted ignoring. (Jomini forgot, when he shaped this yarn, that Napoleon would not have used cavalry to cover a night withdrawal and that Grouchy had been badly injured during the battle.)

After Eylau, Jomini went on sick leave and did not rejoin in time for Napoleon's victory at Friedland, which ended the war. He did demand a definite assignment, either with Ney or the imperial headquarters. Dutaillis had lost an arm in the recent fighting, and Ney reportedly had asked for Jomini as his replacement. Dutaillis considered himself still capable of active service. Considering the known capabilities of the three men, Berthier named Jomini as Dutaillis's assistant. Instead of accepting this as an opportunity to learn staff procedures from a competent senior, Jomini threw a tantrum—he had rendered "great services," he "was Swiss," he absolutely refused to be "the subordinate of an invalid," he would resign if not given what he considered his just dues.

Napoleon seldom wasted potential talent. Now, he apparently decided to give Jomini a chance to prove himself. Dutaillis was found another assignment; Jomini named Ney's chief of staff. Ney went on leave; the VI Corps settled into comfortable billets in Silesia; Jomini courted a Prussian beauty and wrote more of his *Traite;* division and brigade commanders found their own pleasures; VI Corps discipline and training also relaxed.

Then came trouble in Spain. The VI Corps humped their packs and hit the road. Jomini led one division in a triumphal parade through Paris, riding at its head as if he had led it years in combat. In Spain, however, things were rough. Jomini was rebuked for adding officers to the corps staff and getting engaged to that Prussian girl without Ney's customary consent. Jomini complained that jealous officers were poisoning Ney's mind against him; other officers growled that Jomini considered himself Ney's brains (a hallucination common to ambitious young officers). Jomini was shaken by the brutalities of guerrilla warfare. (To his credit, he always tried to protect civilians.) He continually irked Ney, who suddenly sent him off with dispatches for Napoleon, then in Austria. Reaching Vienna just after Napoleon's victory at Wagram, he talked about Ney's supposed lack of aptitude for the Spanish war. (Later he would explain that he had hoped to secure Ney a more appropriate assignment.) Somebody wrote Ney, who picked a new chief of staff and left Jomini stranded. This also killed the promotion to brigadier general Jomini was to get as proper for his assignment.

Jomini clamored over this new injustice. Napoleon first thought of sending him to Marshal Davout in Germany (which would have been a kill-or-cure assignment) but finally left him on the imperial staff. Convinced this was all some of Berthier's devious plotting, Jomini determined to find a more congenial service and so began dickering with a Colonel Tchernitchev, head of a Russian liaison mission in Paris—and also of an efficient spy net. Then as now, Russia had a program for recruiting discontented specialists and potential traitors. Jomini's *Traite* had some popularity in Russia, and Jomini was supposed to understand Napoleon's system of warfare. He might be a handy catch for Russians secretly preparing war against their French ally. Tchernitchev offered him a brigadier general's commission. Jomini secured a long leave, sent in his resignation from the French service, and proceeded to Munich to pick up the promised Russian commission and passport. To his shock, he found the Russian diplomatic service there had never heard of him. Tchernitchev could not be contacted. He sought employment in Bavaria but was quickly ordered back to Paris. There he received a rather mild reprimand, asked Napoleon's grace, and was actually promoted to brigadier general and given financial support for his current publications. Moreover, Napoleon gave him the mission of writing a "thoroughly logical" history of his 1796 Italian campaign.

At once Jomini had new complaints: The war ministry's files were not conveniently arranged; he had to drudge through stacks of strength returns, orders, and reports. Soon he was convinced Berthier had ordered important documents kept from him. Also, a Swiss bank failure had hurt him just as he had married a French girl. Tchnernitchev reappeared with that Russian commission, jingling enticing sums of money. However, with war at hand, French counterintelligence cracked Tchernitchev's net, and Jomini decided to hold his Russian commission in reserve.

As Napoleon's army surged into Russia in 1812, Jomini went with the imperial staff as its official historian. By his own account, he was embarrassed over invading the territory of Czar Alexander I and so brought his apprehensions to Napoleon's attention at the first opportunity. It is a strange story, but Napoleon did detach him as military governor of Vilna, the first important city the French occupied.

Vilna became a major base for the French invasion, giving Jomini frequent opportunities to display any talent he might have for making bricks without straw. Wearied by his constant excuses, Napoleon rebuked him for sleeping and weeping, though Berthier showed understanding of his problems. Most of Jomini's energy seems to have gone into insubordination toward his immediate superior, General Hogendorp, governor-general of Lithuania. Their squabbles brought them imperial rebukes; Jomini was ordered forward to take over the governorship of Smolensk. There he did nothing noteworthy.

Shortly after his arrival the great "retreat from Moscow" began. Jomini's account of it is an epic of how he guided the French home, giving Napoleon wise advice and locating the vital ford on the Beresina River where three converging Russian armies tried vainly to trap the Emperor. In sad truth, he did none of these; up to the Beresina, he ran minor staff errands. After that, through a series of adventures, he survived. Shortly after his arrival at Stettin he was ordered to Paris. His memory of this was that Napoleon wished his advice on the creating of a new army and that only he himself and one other general were allowed the honor of returning to France. The truth is that all surplus officers and NCOs were recalled to cadre new units. Jomini, however, reported himself too sick for duty.

He rejoined the army in Germany in May 1813 and was reassigned as Ney's chief of staff, replacing an officer killed in action. Ney now commanded the left wing of Napoleon's army, five corps including his own—a mass of some 84,000 men. Almost three weeks later Napoleon caught the Russian/Prussian armies at Bautzen and called Ney in from the north to strike their right flank and rear. It would have been a battle of annihilation that would have reestablished Napoleon as master of Europe, but Ney and Jomini—unable to control their massive force effectively—turned the Emperor's planned hammer blow into a fumbling contest. The battered enemy escaped.

Shortly thereafter, an armistice brought a pause in the fighting. Ney recommended, as seems to have been his custom, many of his subordinates for decorations, cash bonuses, and promotions. Napoleon went through the list with a heavy hand. Jomini, recommended for a bonus and a decoration, got the decoration only.[21] Ney also recommended him for promotion to major general, possibly as much in hope of getting rid of him as to reward him. Elated over this anticipated advancement, Jomini suddenly found himself under arrest for allowing his administrative paperwork to lapse into arrears. Jomini and his admiring biographers have insisted this was a trivial incident, blown up out of all reason by Berthier's jealousy. Quite the contrary, Napoleon considered such status reports essential to his operational and logistic planning. He loved studying them "better than a girl loves novels."[22] Concurrently, Jomini managed to infuriate Ney by contradicting him in front of his division commanders.

The armistice would end on August 16, 1813. On the thirteenth, having put his affairs in order, Jomini abandoned Napoleon's "ungrateful flags" and deserted to the enemy.

Naively, Jomini expected to be accepted as Czar Alexander's military preceptor and a guiding genius to Napoleon's enemies. In his own mind he was not a French deserter but a free Swiss citizen seeking a kinder employer; the fact that he held a French commission was of no importance if it interfered with his own convenience. So began Jomini's new disappointments.

Alexander already had a mentor in Jean Victor Maria Moreau, a famous French general who had been exiled in 1804 after becoming involved in Bourbon plots to assassinate Napoleon. Jomini was assigned to the staff of Prince Karl Philipp Schwarzenberg, the fat, humorous Austrian field marshal who more or less commanded the allied Russian, Prussian, Austrian armies. Jomini offered advice, constantly. Any officer who had read his *Traite* must be, *ipso facto*, his ''student'' and so owed him proper respect. The English representative at Schwarzenberg's headquarters considered Jomini an unnecessary nuisance; one of his Russian ''students'' pungently defined him as ''not fit to serve during war.''[23] He was certain that only his genius had saved the allies from defeat and won their victories for them, yet no one else seemed aware of it. The postcampaign shower of promotions and decorations missed him altogether; only by personal complaint to Alexander could he secure a minor medal and appointment as one of Alexander's multitudinous aides-de-camp. Stunned by such ingratitude, Jomini was ready to leave the Russian service when his wife and child arrived, reminding him that he had given hostages to fortune. He continued with Alexander.

In 1814 Jomini attempted, with Alexander's permission, to make himself a place in the Swiss government but found all parties against him as a man not to be trusted. He had better fortune in France, wheedling Bourbon Louis XVIII into canceling the judgment for desertion still standing against him and restoring his pension as an officer of the Legion of Honor. His petition stated that Napoleon had ''imposed upon him an unjustifiable yoke. Therefore, there remained to General Jomini no other recourse but that of joining his efforts to those of a Europe armed for independence.''[24] In 1815 he rejoiced over Berthier's mysterious death; in Paris with the Russian occupation forces after Waterloo, he gave Madame Ney some extremely cautious assistance in seeking Alexander's intervention to prevent her husband's execution by frightened, vengeful Louis XVIII. (Alexander would not intercede.)

Thereafter Jomini followed Alexander to Russia, where he developed a new contingent of enemies. He organized the first Russian military academy (ambitious Russian comrades soon elbowed him out), was briefly military tutor to the future Czars Nicholas I and

Alexander II, created various diplomatic teapot tempests, and filled occasional staff assignments. After 1856 he lived largely in Paris, fighting inky wars in defense of his career and his theories, telling gaudy stories of his alleged deeds, and becoming increasingly touchy and imperious. When Clausewitz expressed doubt that the art of war could be boiled down into a set of general principles, Jomini repaid him in language once devoted to Berthier. But as passing years took the true Napoleonic veterans, Jomini came to be viewed as their unique spokesman. His glorification of Napoleon gradually led Frenchmen to forgive his conduct of 1813. In 1859 Napoleon III even consulted him concerning an impending campaign against Austria, but got nothing useful.

Jomini never served with troops. He did well enough as a free-lance aide-de-camp in 1805 but was a flat failure as a chief of staff in 1808–09 and 1813. As an official historian, he produced complaints but no books. He would not accept a subordinate position, a rebuke, or an uncongenial assignment for the good of the service, or even to gain experience and good will for his own future career. Thoroughly convinced of the magnitude of his own genius, he was certain that any opposition to his wishes was stupid and unjust. This vanity and arrogance, coupled with his chronic inability to keep his mouth shut, made him generally disliked. Fellow officers were especially irked by Jomini's view of himself as another Thucydides, with its implication that their historical reputations would depend on what he wrote. And, in an army where uncommon bravery was a very common thing, Jomini seems to have hazarded himself as little as possible. Only a belief that Jomini had *some* potential practical talent can explain Napoleon's patience and generosity toward him, but the Emperor failed in every effort to bring it out.

Through the Napoleonic Wars Jomini was a very minor figure, seldom mentioned in orders or dispatches, practically ignored in the memoirs of officers who had served with him. His fame came later as an author of several major works on military history and the military art. These books are methodical, rather well written, and truly inspired by his devotion to his subject. At the same time, they are flawed by Jomini's smoldering resentments. Never having achieved a position that matched his own opinion of his abilities, he sometimes rewrote history to show that—though always cheated of his due glory—he really had been an important figure and the trusted consultant to whom Napoleon would turn in time of trouble. Increasingly, he seemed to imply a mystic intellectual communion with the emperor.

Jomini's original *Traite de Grande Tactique,* later revised and rewritten as *Traite des Grandes Operations Militaires,* was basically a critical history of Frederick the Great's campaigns during the Seven Years' War (1756–63), comparing Frederick's operations with those of Napoleon and giving a summation of the general principles of the art of war.[25] Much of this work was unabashedly lifted from Lloyd, including his "fundamental principle" of war (see page 108). Jomini's contribution, besides his comparison of Frederick and Napoleon, was an attempt to define various features of the art of war. He elaborated Lloyd's concept of the *"line of operations"* and pointed out Frederick's successful use of *interior lines* as compared to the *exterior lines* his opponents were compelled to use.[26] (These were the items that impressed Napoleon in 1805.)

Jomini next produced a history of the wars of the French Revolution[27], a work of considerable scope but little depth, now practically forgotten. In 1827 came his famous *Vie Politique et Militaire de Napoleon.*[28] Since his Russian employers might feel that he was unduly glorifying Napoleon, he published it half anonymously as "General J. . . ." Its theme is unique: Following his death, Napoleon arrives at the Elysium Fields where the spirits of Alexander, Caesar, and Frederick question him as to his career; Napoleon decribes the whole of it through 1814, speaking in the first person, which makes it remarkably authoritative. It was a glib, informative work, easily read and extremely popular. But it is not accurate history. Jomini's desire to vaunt himself and defame Berthier, to glorify Napoleon and Alexander I, makes it frequently inaccurate. We are told that the character speaking is Napoleon, but the words suggest Adjutant-Commandant Jomini. Moreover, the work is flawed by errors of fact that a reasonable degree of research would have detected. (Research, one suspects, bored Jomini.) Jomini's critiques of Napoleon's various campaigns are usually sound—except that some of the incidents he mentions never happened.

In 1830, Jomini began an attempt to sum up his knowledge and opinions; eight years later this reached final form as his *Precis de l'Art de la Guerre.*[29] It outlined the relative scope of strategy, tactics, and logistics and set down principles, maxims, and rules of thumb covering almost every aspect of warfare. He insisted that there were a small number of fundamental principles of war that it was dangerous to ignore. A general should seek to seize the initiative and to *surprise* the enemy by rapid *maneuver* in order to bring the *mass* of his forces against a fraction of the enemy, striking the enemy in the "most decisive direction" (where it would hurt him most). This *Precis* was translated into practically every European language and was read by military men throughout the world as a handy compendium of what every general should know. There were no really new concepts in it, but it clearly explained, for the first time, the anatomy of warfare and how wars were waged.

Jomini's writings reveal deep inner uncertainties. Fascinated as he was by Napoleon's campaigns, he was troubled by the destructiveness of mass armies and great national wars. The development of new weapons distressed him. He wanted a return to the "loyal and chivalrous warfare" of the eighteenth century, waged by small armies for limited objectives with proper respect for civilian sensibilities. There really had been no such ideal warfare then, but Jomini probably got the idea from his fixation on Frederick the Great. Frederick's wars were small affairs, easily reduced to diagrams. Also, they were past history, their many sordid details hidden by the greater blood and thunder of the Revolutionary and Napoleonic wars—and so little known to Jomini. Somehow Jomini seems more comfortable with Frederick than Napoleon; Napoleon's campaigns he labeled "wars of conquest"; Frederick's 1744 seizure of Silesia (see page 93) was a "stroke of genius."

Jomini influenced the study of the art of war more than all of Napoleon's marshals lumped together. Yet he passed through the Napoleonic Wars with no apparent sense of personal commitment except to his own fortune. He gave himself to the study of war but had little taste for war itself. His interest was an abstract intellectual curiosity. War's emotional impact, its sense of challenge, its comradeship and dangers shared—which Clausewitz felt so deeply—slid off his armor of self-centered conceit. This detachment made him, at least when his own ego was not involved, a shrewd and impartial observer. But it also separated him from the personal conviction and leadership that were so characteristic of those wars. The men and armies who marched and fought and died, wind and weather, empty bellies and muddy roads, are only dim shadows across his methodical strategic studies. He could comprehend Napoleon the strategist but not Napoleon the battle captain who rode into the fire, lifting the greenest conscript into battle fury—or Berthier in 1814, over sixty and half sick, smashing into a group of Cossacks to rescue a captured gun.

To do him justice, Jomini was rather aware of this lack, as a blind man might understand colors. His books debate the importance of morale and acknowledge that war is an art that cannot be reduced to mathematical calculations. Yet no sense of personal participation puts life into those words.

We end with an ambiguity: Never really a soldier, a failure as a staff officer, frequently a careless or dishonest historian, Jomini still has been a useful educator to generations of soldiers. Basically, he was an observer and critic. Unlike Clausewitz, he was not interested in the nature of war. Instead, with care and clarity, he systematized and defined the art of waging it, bringing order into the study of that most disorderly species of human activity.

The best summation of Jomini comes in a letter Napoleon wrote on August 16, 1813 to his archchancellor in Paris, giving the latest military situation: Austria had declared war on France, the enemy had attacked before the armistice expired, Moreau was at the Russian headquarters. Then he wrote: "Jomini, Chief of Staff to [Ney] has deserted. . . . He is not worth much as a soldier; however, as a writer, he has gotten hold of some sound ideas on war."[30] And the Emperor wrote on about more important matters.

It was only natural that Carl von Clausewitz (1780–1831) would disagree with Jomini. His family had emigrated from Poland at the beginning of the eighteenth century and entered the Prussian service. It had not profited greatly thereby; Clausewitz's father, wounded and partially disabled while a lieutenant in Frederick the Great's infantry, was a minor official in the Prussian excise service. (Since Frederick hated wasting money on officers no longer fit for active duty, Clausewitz's father probably was lucky to secure that position, ill paid as it was.)

In Prussia there was only one honorable career open to young Carl. He was twelve when he found a place as a *Junker* (officer candidate) in an infantry regiment. There he served in the ranks through various odd grades such as free corporal and ensign until he proved himself worthy of a commission.[31] Since the wrangling government of Revolutionary France had attempted to solve its suicidal internal disputes by picking a fight with its neighbors, Clausewitz saw considerable combat along the Rhine frontier during 1793–95 and had plenty of opportunity to compare disorderly French improvisation with stodgy Prussian professionalism. (There was the occasion when bulky Prussian King Frederick William II complimented a wounded French grenadier for his courage but suggested that the French Republic was unworthy of such devotion. The soldier, politely enough, delighted most Prussians within earshot by replying, "Citizen William, we shall not agree on that point: let us talk of something else."[32]) This fighting was on the whole indecisive, but Clausewitz won promotion to lieutenant during the siege of Mainz in mid-1793. That meant that he would have a horse to ride and another to carry his baggage, plus a small tent to himself, an orderly, and further opportunity to master his trade by doing a large part of his captain's work for him.

After Prussia signed a separate peace with France in 1795, Clausewitz had five years of garrison duty. This normally could be a stultifying experience, but Clausewitz was determined to complete his scanty childhood education. He also had the luck to belong to an unusual regiment; its colonel in chief was Prince Ferdinand, a royal prince, and its officers as a group were sincerely interested in the education of their junior officers and even their enlisted men's children. It is obvious that Clausewitz both studied and taught. In 1801 he had the honor of being selected to attend the new War College, one of the schools that Scharnhorst (see page 178) had established for officers' training. He graduated in 1803 at the head of his class; Scharnhorst, having noticed his industry and intelligence, got him an assignment as adjutant (Prussian for aide-de-camp) to Prince August of Prussia, son of Prince Ferdinand. This duty left him in Berlin, in close association with Scharnhorst and other officers who hoped to use army reform as a lever to pry Prussia out of its increasingly ossified social system. He also found time to study the philosophy of Immanuel Kant, read Jomini's and Bülow's first books, and become engaged to a young countess, a girl of intelligence and charm, willing to wait until her shy young officer gained promotion and pay enough to support a wife.

In 1806 it seemed their chance had come. Prussia confidently declared war on Napoleon. But it promptly became evident that confidence was Prussia's one asset. In place of Frederick the Great's iron central control, there was a clutch of squabbling generals. Young King Frederick William III was brave but diffident and indecisive; his wife Louise, who wore the royal family's pants, hated Napoleon, wanted war, and was given to parading in uniform. Clausewitz watched Scharnhorst vainly urge that war with Napoleon required prompt action, while the Prussian generals held long councils of war in which the most impractical suggestions received full attention and nobody seemed to be in command.

While Frederick William's generals disputed, Napoleon concentrated and marched, sweeping northward with over 150,000 men in his massive yet highly flexible "battalion square" formation[33] behind a hard-riding, hard-probing cavalry screen. Clausewitz quickly learned the cost of Prussian unpreparedness and indecision. Near the little town of Auerstadt, just northwest of Weimar, he endured that terrible day when Marshal Davout's 26,000 French veterans smashed the 63,500-man main Prussian army despite its long-famed massed cavalry charges, infantry volley fire, and heavy cannon. (Simultaneously, Napoleon was wrecking another Prussian army just to the south at Jena.) After that he had two harried weeks with a disintegrating column of fugitives before Murat bagged them north of Berlin. Clausewitz saw this as a captain with a grenadier battalion, apparently showing himself a good combat officer, if all in vain. Captured, he set himself to working out the reasons for Prussia's defeat. The result would be a scathing work, *Notes on Prussia in Its Great Catastrophe of 1806,* which the Prussian government promptly suppressed for some time.[34]

Clausewitz came back to Prussia in early 1808, after internment in France and a stay in Switzerland. Scharnhorst, who now functioned as both chief of staff and minister of war, got him a major's commission in the Prussian General Staff. Clausewitz soon became one of his confidential assistants, accumulating a number of important responsibilities, which included teaching at the War College and acting as military instructor to the Crown Prince. Scharnhorst, seconded by Gneisenau (who became Clausewitz's firm friend), was the military leader of a group of reformers who understood that the French Revolution had changed the world and that Prussia must change, too, or dwindle into a French satellite. In France, any career was open to any individual with the necessary determination and intelligence, enabling Napoleon to draw on the brains, bodies, and enthusiasm of the whole French nation. The Prussian reformers had neither time nor inclination to create an actual revolution among the Prussian people. Their intention was to impose the equivalent of a revolution "from above" by changes achieved in their king's name. Their immediate aim was the creation of an effective army, trained for Napoleonic-style warfare, filled by national conscription, and officered by the most competent men available, regardless of their social background. Naturally, their efforts were opposed by most of the Prussian nobility, distrusted by the king, and periodically squelched by Napoleon, who kept a cold eye on Prussian military activities.

His improving fortunes enabled Clausewitz to marry his Countess Marie von Bruhl, who made him the most understanding and pleasing of wives. But Napoleon's decision to invade Russia in 1812 interrupted their life together. Not wanting to leave possible enemies in his rear, the Emperor called on Prussia for an auxiliary army to reinforce the north flank of his invasion. Clausewitz refused to serve Napoleon; with a number of other officers, he resigned his commission and entered the Russian service. (While this definitely was in keeping with Clausewitz's personal convictions, it also may have been planned by Scharnhorst's inner clique to give it contacts within the Russian Army.)

Clausewitz was a young major who could not speak Russian. His service was therefore restricted to various staff assignments, but he saw the major battles and experienced most of the hardships of that famed campaign. In the last days of December he took part in the Convention of Tanroggen, under which the Prussian contingent with Napoleon's retreating army coyly declared itself neutral. Clausewitz and others proclaimed this an act of pure patriotism, but as carried out by the Prussian commander Gen. Hans D.L. Yorck, it was little better than cheap treachery. From that, Clausewitz, turned to organizing troops in Russian-occupied East Prussia despite the fact that Prussia had still not declared war on France. After that event, in March 1813, he was with Scharnhorst's staff as a Russian lieutenant colonel until Scharnhorst got his death wound at Lützen. Therefore, he was chief of staff to a small polyglot army of Prussians, Russians, Swedes, Hanoverians, miscellaneous Germans, and English mercenary units made up of men of all nations that operated along the Baltic coast. It had an undistinguished existence, but Clausewitz seems to have been responsible for its one success. Following Napoleon's abdication in 1814, Clausewitz managed to obtain a colonel's commission in the Prussian army. (King Frederick William III had been irked by his Russian escapade.) During the 1815 Waterloo campaign he was a corps chief of staff; while Waterloo was fought, his corps waged a desperate rear guard fight to keep reinforcements from reaching Napoleon. It held on just long enough, at the cost of taking a bad beating. In 1816 Clausewitz became chief of staff to Gneisenau's command along the Rhine frontier. Two years later he was promoted to general and made superintendent of the Berlin War Academy. His duties there, however, were purely administrative; his influence on the curriculum and students could be only indirect. He thought of transferring to the diplomatic service, but his political views were unacceptable to the Prussian court, which was doing its best to return to the 1750s. He therefore concentrated on his writing. In 1830 he was transferred to Breslau as chief of a major artillery command, but this new assignment was interrupted when Poland (theoretically an independent kingdom under the personal rule of the Czar of Russia) rose up in determined rebellion and Prussia's Polish-inhabited eastern provinces seemed about to join in. Troops were hastily concentrated under Gneisenau, who again took Clausewitz as his chief of staff. The border districts were pacified, but the disturbances aided the spread of a cholera epidemic out of Russia. Gneisenau died first, Clausewitz on November 16, 1831, after he had returned to Breslau.

Clausewitz's usual reputation is that of an intellectual theorist who had limited practical experience and little influence on events during his lifetime. As a matter of fact, he was a combat infantryman at an age that finds the modern American boy still in grade school. He proved himself as a regimental officer, an aide-de-camp, a military instructor, and a chief of staff. He was also a trusted assistant to Scharnhorst and Gneisenau, raised and organized troops in 1813, and acted as military tutor to two Prussian princes. His service thus was far more practical and complete than Jomini's; if he had performed no great feats, his general reputation was that of a brave, devoted, and competent soldier who was also a thoughtful student of the art of war. Without family influence or wealth to aid him, he rose steadily and comparatively rapidly. But for his untimely death he undoubtedly would have received further promotion.

> "Free as he was of any petty vanity, of restless egotism and ambition," wrote his Marie after his death; "he nevertheless felt the need to be truly useful, and not let his God-given abilities go to waste. In his professional life he did not occupy a position that could satisfy this need, and he had little hope that he would ever reach such a position. Consequently, all his efforts were directed toward the realm of scientific understanding."[35]

These inner convictions went with an outward personality described as shy, sensitive, and very reserved. Such a man would experience frequent disappointment and discontent in the daily rough-and-tumble of military life, even while excelling at it. Like any professional soldier, Clausewitz hoped for promotion and decorations and, as his wife noted, especially for a command assignment that would allow him to put his own ideas into execution on his own authority and receive full credit for his accomplishments. His letters could be bitter, but this frustration never seemed to affect his performance of his duties or his home life.

Writing was Clausewitz's release from routine frustration and also his hope for future fame. A note found among his papers (undoubtedly referring to *On War*), told of his hope "to avoid everything common, everything that is self-evident, that has been said a hundred times, and is commonly accepted; for my ambition was to write a book that would not be forgotten in two or three years, and which anyone interested in the subject would be sure to read more than once."[36]

The military philosophy that guided Clausewitz's pen was a cross-breeding of two very different species of military experience. He had been brought up, trained, and indoctrinated in a Prussian Army very little changed from Frederick the Great's and in the acquisitive Prussian doctrine that war was a natural and ordinary thing, to be waged whenever there was an opportunity to seize some neighboring territory, with no particular concern as to the morality of the business. (Prussian national policy resembled that expressed by a certain Kansas farmer; "I ain't greedy. All I want is the land that adjoins mine.") Such wars were rather deliberate affairs, waged by professional, largely mercenary armies that existed more or less in isolation from the rest of their nation. Then, literally overnight, Clausewitz had been swept into Napoleonic warfare—swift, long-range campaigns, fought by whole nations-in-arms, that brought down empires in less time than Frederick had needed to conquer a minor province. As he later phrased it, "war itself" had lectured mankind. Comparing and analyzing these experiences brought Clausewitz to pondering the nature of war itself—its forms, its purposes, and its relationship to other human activities.

As an author, Clausewitz is commonly remembered only for his *On War (Vom Krieg),* but he was actually a prolific writer. During 1804–1808 he wrote and rewrote studies on contemporary strategy, commenting on the works of Bülow and Jomini, both of whom he found too eager to reduce war to sets of principles and theories instead of considering its psychological aspects—the combatants' morale, intelligence, and will to victory. He also wrote spare histories of Napoleon's 1796–97 Italian campaign and the campaigns of 1799, 1806, 1812, 1813 and 1814, and 1815 and on the wars of Gustavus Adolphus, Turenne, and Frederick the Great. Occasionally, as in his book on the 1813–14 campaigns, he slips into exultation over Prussian victories, but most of his work is careful and impartial. What happened, and how and why, was important to him because he believed only accurate history could be of any help in the development of military theory. He did not hesitate to criticize Prussia or Prussians. In his account of the 1815 campaign he roundly denounced the conduct of Field Marshal Gebhard von Blucher, the Prussian commander, during the occupation of Paris after Waterloo. (Blucher, a pugnacious old ex-hussar officer of elastic morality, proposed—among other capers—to blow up the Jena bridge, which Napoleon had built across the Seine to commemorate that victory. Fortunately, the Prussian engineers were all thumbs, and the Duke of Wellington was moved to put a British sentry on the bridge, so that the only real damage was to Prussian prestige.)

Comparatively few of Clausewitz's writings were published before his death. Most of his historical works were eventually translated and published in France during the last half of the nineteenth century. *On War,* his *Notes* on the 1806 campaign, and one or two other works have been brought out in English; except for *On War,* the average student finds them hard to come by.[37]

Clausewitz began the work that was to become *On War* in 1816 while stationed on the Rhine. Beginning with notes on various aspects of war that intrigued him, he expanded these short, usually unrelated essays. When he found that his next assignment, the War Academy, left him with plenty of spare time, he converted his wife's drawing room to a study and began developing and systematizing notes and essays into a connected whole. This proved extremely difficult—Clausewitz was constantly finding new aspects to his subject, and his basic honesty and thoroughness kept him from being contented with anything incomplete or superficial. His transfer to the Artillery in 1830 (very likely intended to fit him for further promotion, since he had no direct experience with that arm) forced him to put his writing aside. Out of eight projected books, he had written six and finished first drafts for the other two. None of them, however, came up to his growing appreciation of the complexity of his subject, and he had decided that his whole work must be completely rewritten. He had revised only the first chapter of his first book when his next orders came. "He arranged his papers, sealed and labeled the individual packages," wrote his Marie in her foreword to the first edition of *On War:*

> and sadly bade farewell to an activity that had come to mean so much to him. . . . When he returned [from the frontier] . . . he was cheered by the hope of resuming his work and possibly completing it in the course of the winter. God decided otherwise. . . . For twenty-one years I was profoundly happy at the side of *such* a man.[38]

Marie published *On War* in 1832. It was admired, but its sales went slowly. When a new edition, revised by her brother, Count Friedrich von Bruhl, was brought out twenty years later, some of the original 1,500 copies were still on the publisher's shelves. Von Bruhl apparently worked honestly, but his attempts to clarify Clausewitz's often rough original text resulted in changes in what now seems to have been Clausewitz's intended meaning. The first French translation appeared in 1849. Jomini considered it a "labyrinth"; in his *Precis* he denounced Clausewitz as "pretentious" and "arrogant" and accused him of plagiarism but expressed regret that Clausewitz had not lived to read the *Precis,* which might have convinced him of

Jomini's primacy as a military theorist. The first-known English translation was published in 1874.[39] Written in stiff and stilted mid-Victorian style, it was unnecessarily difficult to read; also, the translator was hardly a master of the German language. The first expertly translated and edited version based on the original 1832 text did not appear until 1976.[40]

Much of Jomini's distress over Clausewitz's works was caused by their neglect, as he saw it, of "good theories" and their denial that there was any hard-and-fast system for winning wars. Probably, like many later critics and commentators, Jomini had merely skimmed through Clausewitz's text. For Clausewitz *did* recognize the validity of certain basic principles of war. (He even recommended handy little service maxims such as "Cooking in the enemy camp at unusual times suggests that he is about to move" and "The intentional exposure of troops in combat indicates a feint."[41] He had been taught, with brutal clarity, the need for *unity of command* in 1806, with a postgraduate education from service in the feuding Russian high command in 1812. Selection of a decisive *objective* and the employment of *mass, surprise* and *maneuver* ("mobility" to Clausewitz) to achieve it were imperative. *Simplicity* and *economy of force* were also important. But to Clausewitz such maxims and principles were simply matters of common sense, which any competent commander always had in mind. Also, they might not always necessarily apply. War was infinitely variable—a "chameleon"—and there would be times when a wise commander might decide to throw the book away, dividing his forces in the presence of the enemy instead of massing them or otherwise conducting himself in unexpected ways. "What genius does is the best rule"[42] was Clausewitz's conclusion, which runs with Napoleon's observation that Jomini established principles for everything, whereas genius worked by inspiration.

However, geniuses being few, Clausewitz believed it *was* essential that all soldiers understand the basic nature of war. (When the average commander throws the book away, he normally seeks trouble and finds sorrow!) War was unleashed violence, but it was a form of political intercourse between nations in which they fought battles instead of exchanging diplomatic notes (see page 3). Battle was the decisive act of any war: "The decision by arms is for all operations in war, great and small, what cash settlement is in trade."

("The battle is the payoff," as some American said in World War II.) Therefore, every activity in time of war should contribute to victory in combat. "The end for which a soldier is recruited, clothed, armed, and trained, the whole object of his sleeping, eating, drinking, and marching *is simply that he should fight at the right time and the right place.*"[43] Battles are fought for one purpose alone: the destruction of the enemy army (which Clausewitz considered more a matter of killing its collective courage than of killing its soldiers). Statesmen and generals may seek to avoid decisive battles because of the losses and risks involved, but this is traveling "on devious paths," attempting to dodge the fact that war is inherently a serious, dangerous business. A sense of humanity may lead us to attempt half measures in the hope that they will suffice, but there is no guarantee that our enemy will reciprocate; in fact, if our enemy knows his business, he will exploit our indecisiveness.

> It is necessary to either wage war with the utmost energy, or not at all. . . . The maximum use of force is in no way incompatible with the simultaneous use of the intellect. . . . Woe to the government, which, relying on half-hearted politics and a shackled military policy, meets a foe who, like the untamed elements, knows no law other than his own power![44]

Clausewitz specified two types of war: general war, to completely defeat the enemy, forcing him to accept peace on our terms; or a limited war to gain minor pieces of territory or political objectives. (Grenada is an excellent example of the latter.) Before waging either one, the responsible government must determine what objectives it wishes to achieve, how it proposes to wage war, and—especially—whether its military forces are capable of winning that war. The civilian population must be ready to support military action. During the war it may be necessary for the government to change its policy. Unexpected casualties, logistical difficulties, or the intervention of a third power may force it to limit its aims, and thus its military operations; or a sudden collapse of the enemy's forces may inspire it to enlarge both. But whatever type of war is waged, it must be carried through as energetically as possible. Military operations therefore take on the characteristics of the national policy that launched and directs them. If this policy is set on complete conquest of an enemy, the war becomes absolute—for example, involving an indiscriminate exchange of nuclear weapons—or may, as presently being practiced by Russians in Afghanistan and Vietnamese in Laos and Cambodia, degenerate into a deliberate program of genocide.

In his discussion of standard military operations, Clausewitz is surprisingly conservative, some of his opinions being reminiscent of Lloyd's. He considered the defensive a stronger form of warfare than the offensive, since the offensive force grows constantly weaker as it advances from having to detach troops to cover its flanks and its line of communications—a process termed "strategic consumption." In this, Clausewitz was definitely influenced by Napoleon's 1812 Russian campaign: Napoleon had crossed the Niemen River into Russia on June 23 with over 400,000 men immediately under his hand; when he reached Moscow on September 15, he had only 95,000 men. A strictly passive defense, however, should be avoided in favor of the "defensive-offensive"—remaining on the defensive until the enemy is exhausted and disorganized by unsuccessful attacks, then seizing the initiative and launching a decisive counterattack. "The flashing sword of vengeance," Clausewitz termed it, which is more poetic than Napoleon's observation that a careful defense, followed by a vigorous offensive is one of the best possible maneuvers but that the passage from the defense to the offensive is a most delicate operation.

Clausewitz's most unusual contribution to the study of war was his insistence on the importance of moral forces in war. This includes his concept of "friction" to explain why simple strategic and tactical problems prove so difficult to solve in actual practice.

> Everything looks simple: the knowledge required does not look remarkable, the strategic options are so obvious that by comparison the simplest problem of higher mathematics has an impressive scientific dignity. . . . Everything in war is very simple, but the simplest thing is difficult. The difficulties accumulate and end by producing a kind of friction that is inconceivable unless one has experienced war. . . . Countless minor incidents—the kind you can never really

> foresee—combine to lower the general level of performance, so that one always falls far short of the intended goal. . . . A [military organization] is made up of individuals, the least important of whom may chance to delay things or somehow make them go wrong. . . .
>
> Moreover, every war is rich in unique episodes. Each is an uncharted sea, full of reefs.[45]

The sources of such friction are endless, ranging from the various cussednesses of the weather to the human factors, which include everything from a subordinate commander's inflamed ego to a staff officer's indigestion or some buck private mooning over the last letter from his girl friend to the exclusion of all else. From these we proceed to the mechanical—the broken-down tank blocking a narrow road, sudden engine failure in an airborne general's helicopter, and the innate perversity of electronic equipment. At times, the enemy can seem a minor nuisance.

A commander requires strength of character to resist this continual friction and to go ahead with full confidence in his own skill and judgment. He should expect friction and deal routinely and sensibly with its results, even when it seriously curtails his plans.

Friction, however, works upon both armies. Though at times it may be difficult to believe that the enemy's efforts are as fouled up as our own, they probably are. Friction thus throws military operations open to the workings of chance. (''War,'' wrote Napoleon, ''is composed of nothing but accidents . . . to profit from these accidents; that is the mark of genius.''[46])

This insistence on the importance of the moral factor in war is evidence enough of the depth of Clausewitz's experience as a field soldier. He went well beyond Saxe (see page 83) in searching out the effect of war on the individual soldier. ''War is the realm of physical exertion and suffering. These will destroy us unless we make ourselves indifferent to them . . . War is the realm of uncertainty . . . War is the realm of chance. . . .'' He noted how danger, exhaustion, and uncertainty combined to increase friction and wear down the physical and mental strength of soldiers and commanders alike and also how difficult it is for a commander to sort out conflicting reports.

> As a rule most men would rather believe bad news than good, and rather tend to exaggerate the bad news. The dangers that are reported may soon, like waves, subside; but like waves they keep recurring without apparent reason. The commander must trust his judgment and stand like a rock on which the waves break in vain. It is not an easy thing to do.[47]

Much of *On War* covers various types of operations such as retreats, defense and attack of a river line, encampment, marches, and the like. Most of this, naturally, is out-of-date; nevertheless, it does make an excellent reference for the routines of military life and operations during the early nineteenth century, containing material which this author has seen nowhere else. Clausewitz's discussion of guerrilla warfare (which he termed "The People in Arms"), however, remains worthy of consideration. He had seen a good deal of irregular operations in Russia and Germany during 1812–13 and had quite modern ideas as to how it would be most effectively employed.

There is nothing in *On War* on naval operations or the nineteenth-century version of economic warfare. Prussia then being a nation of rather hard-up landlubbers, such things were completely outside Clausewitz's experience. As concerns land operations, it is notable (and generally overlooked) that Clausewitz's concept of war—except for his preference for defensive warfare—is far more in agreement with Napoleon's than is Jomini's.

Clausewitz had hoped that *On War* would not be forgotten. It has not, but the influence he hoped it would exert has frequently been of quite another sort than he intended. Somehow Clausewitz—the practical professional soldier who believed in defensive warfare, the man who urged greater democracy for Prussia's citizens, the serious military philosopher who taught that war was hazardous—has been transfigured into a champion of all-out offensive warfare.

Part of this, of course, can be attributed to his somewhat pedagogic style of writing, part to the fact that *On War* had to be published in an unfinished state with sections of it still only in draft form. Also, the major revision Clausewitz had begun left a number of contradictory ideas unresolved. Finally, Clausewitz loved a striking phrase, and *On War* (especially its earlier translations) is filled with sentences like "War is an act of violence, pushed to its utmost bounds,'' ''Blood is the price of victory,'' and ''The bloody solution of the crisis, the effort for the destruction of the enemy's forces, is the first-born son of war.''[48] Since a proper understanding of his work requires thought and effort, many of his alleged students seem merely to have skimmed along from one forceful statement to the next and remembered little else.

Until Von Moltke became Chief of the Prussian General Staff in 1857, Clausewitz remained ''Well-known but little read.'' Moltke (see page 179) considered Clausewitz his professional guide, through he tended to ignore Clausewitz's insistence that war was only one instrument of national policy. Schlieffen probably understood him better than Moltke, but German generals of Schlieffen's time, faced with a probable two-front war waged by mass armies, could see no possible use for limited warfare or an overall defensive strategy—once the balloon went up, they must hit out with all the force they could muster, to take out France and then turn on Russia.

Meanwhile, the French, after their 1870 defeat, became fascinated by Clausewitz's views on the moral aspect of war. Blended with Du Picq's teachings, these inspired the wild-eyed theories of Foch and Grandmaison (see pages 189–190), which doubtless had Clausewitz turning over in his grave. Meanwhile, the Japanese Army, trained by German officers and studying their own translation of *On War*, won the Russo-Japanese War of 1904 and politely acknowledged Clausewitz's contribution. More important, *On War* early interested leaders of the developing Communist movement. Friedrich Engels, the German-English manufacturer who was Karl Marx's angel, coauthor, and sometimes ghostwriter, was much taken—naturally—by Clausewitz's comparison of a decisive battle to a ''cash settlement in trade.'' Murderous Vladimir Ilyich Lenin particularly appreciated Clausewitz's statement that ''a conqueror is always a lover of peace; he would like to make his entry into our country unopposed.''[49] Communist strategy has applied that maxim repeatedly in its various ''liberations.'' Clausewitz's work also passed into China, either from Russia or through the German officers who trained the pre-World War II Chinese Army. On the whole, the Communists probably have read more deeply into Clausewitz than the average Western soldier.

Clausewitz was largely ignored in England until just before World War I, when it was suggested that it would be wise to learn something concerning the military philosophy of a putative enemy. One Englishman to profit thereby was the British historian Sir Julian Corbett, who brilliantly applied Clausewitz's work, especially his concept of limited war, to naval operations. (Unfortunately, in the United States Corbett's *Principles of Maritime Strategy* remains completely overshadowed by Mahan's works.[50]) Following World War I, Clausewitz was furiously denounced by such British writers as Liddell Hart and J.F.C. Fuller for the ''mausoleums of mud'' and vast casualties his teachings—they claimed—had caused. They would admit that Clausewitz might have been misunderstood by some of his disciples, but their criticism plainly revealed that they themselves had never really studied *On War*. Hart especially decried Clausewitz's failure to consider limited war!

The United States was the last of the world powers to discover Clausewitz; his first popularity here developed around the time of the Korean War. (There probably was some knowledge of *On War* among more studious army officers even before World War I but hardly enough to affect American policy.) His present students are both civilian and military; their interest is the interrelationship of national and military policy and of the armed forces and the civil government, as well as his version of the brutal facts of international relations.

Clausewitz taught no ''system'' of war. Like Napoleon, he urged his readers to educate themselves by a thorough study of military history, which would teach them the various forms war might take and various methods of waging it successfully. Self-education and experience increased the officer's self-sufficiency, the moral reserves he could summon against times of unexpected crisis. He was no warmonger, but he knew war as an entirely natural thing, not to be entered into casually, not to be cravenly avoided.

Clausewitz wrote *On War* with painstaking care, to ''iron out many creases'' in the minds of strategists and statesmen as to the true nature and uses of war. In it, he and his Marie produced one of history's really great books—a work that has given wise counsel to men of many nations for a century and a half and promises to continue through unguessed years to come. It is a book that must be studied. Most of the ideas it contains are common sense plainly put, but they are concentrated and sometimes packed in irregular fashion. Like hardtack and jerky, *On War* yields its flavor and nourishment only after some mastication; then it is rich in both.

Probably Clausewitz's most potent advice is simple:

> The first, the supreme, the most far-reaching act of judgment that the statesman and the commander have to make is to establish . . . the kind of war on which they are embarking; neither mistaking it for, nor trying to turn it into, something that is alien to its nature. This is the first of all strategic questions and the most comprehensive.[51]

VIII

PROPHETS AND PONTIFICATORS

Dear Lord God Almighty!
I value your criticisms and have accepted about 50% of them, but, in spite of your omniscience, I cannot accept them all.

J.F.C. FULLER TO LIDDELL HART, 1929[1]

Youth comes back again no more.
It was spent in the army corps

OLD GERMAN SONG

Bright morning over a valley deep in the Black Forest. A joyous bride, slim and blonde, steps out upon her bedroom balcony, gazes ecstatically at the view, and calls back over her shoulder to her new husband, a promising young officer of the German "Great General Staff."

"Oh, Hermann—come see how beautiful the sunrise is on the stream and the hills over there."

"That stream," replies her doting bridegroom, "is too shallow to slow down an enemy attack; that ridge is too low and narrow to be a satisfactory defensive position."

The German "Great General Staff" (usually termed simply "General Staff") has been one of the most feared military organizations in world history. Its very title evokes apparitions of bemonocled, immaculately uniformed professional militarists—ramrod straight, icily efficient, utterly ruthless. And yet it was simply another army staff. It differed (and probably still differs) only in that its members were more carefully selected and more highly motivated than the average staff officers of other armies. Restricted in numbers, constantly tested, repeatedly culled, routinely overworked, they formed a strategic machine that again and again shattered Europe's armies. Their inspiration and purpose was never the blind "Master, order us where we may die" of Frederick the Great's damned-and-drilled dogfaces. Rather, they followed the example of Frederick's famous cuirassier general Friedrich Wilhelm von Seydlitz, who—when a panicked Frederick ordered him to make a suicidal charge at Zorndorf (1758)—quietly informed the aide-de-camp who brought him the order: "Tell His Majesty that my head will be at his disposal after the battle, but as long as the battle lasts, I intend to use it in his service."

This Great General Staff had its beginnings—naturally—in raw, semifeudal Prussia. Prussia's Junker nobility and gentry, blended out of generations of German adventurers, Czech mercenaries, Polish and Lithuanian renegades, Huguenot refugees, and Scots exiles, were first-class fighting men, stout of heart and arm but with little time or interest for intellectual exertion.[2] They had small use for foreign ideas, but there always was a certain basic frontiersman's pragmatism in their makeup. Finally convinced that some new idea was a military necessity, they would, however reluctantly, master it or hire some foreign expert to show them how.

Professional staff officers were the last of these experts. Preeminent among them was Gerhard Johann David Scharnhorst (1755–1813), son of a retired sergeant turned farmer, who had won distinction in the Hanoverian service. A rather unmilitary scholar in uniform, he labored through the Napoleonic Wars to modernize the Prussian Army. Dying after a lost battle from a French bullet and incompetent Prussian surgeons, he nevertheless left a thoroughly established general staff and a military educational system to provide it with qualified officers. Among the younger officers he trained and inspired were Carl von Clausewitz and August Neithardt von Gneisenau.

Gneisenau (1760–1831), son of an impoverished Austrian nobleman turned artillery officer, was born amid the night retreat of the defeated Austrian army after the battle of Torgau. He grew up a soldier of fortune. During the last years of the American Revolution he was a lieutenant in a German mercenary light infantry battalion the British hired for garrison duty in Canada. Shifting to the Prussian service in 1806, he rose rapidly, becoming chief of staff of the Prussian field army on Scharnhorst's death. (It must be confessed that his performance during the 1815 Waterloo campaign rather resembled the 1814 American militia fiasco at Bladensburg, just outside Washington, D.C.) A willful, intelligent officer, he believed that Prussia as a whole—and not just its army—needed modernization and supported Baron Heinrich vom und zum Stein and his fellow reformers in their effort to extricate Prussian society from the late Middle Ages. While the Napoleonic Wars lasted, they had some success; those wars won and Napoleon no longer a danger, the King of Prussia and his *Junkers* concluded that the old ways were more comfortable. Reformers were banished from the refulgence of royal favor. Gneisenau was promoted but shelved. In 1830 a new French revolution fused disorders all across Europe; recalled to active duty, Gneisenau died of cholera on the Polish frontier.

In 1821 the General Staff was placed directly under the king, making it independent of any other authority, civil or military. But not until Karl Bernhard von Moltke's (1800–1891) accession did it have a chief capable of using its full powers and capabilities.[3]

Moltke had grown up in Denmark and attended the Royal Danish Military Academy but transferred from the Danish to the Prussian army in 1822, apparently in hope of faster promotion. A very quiet military intellectual, much like Scharnhorst in appearance and manner, he sought wisdom in his Bible, Homer's *Iliad* and *Odyssey,* and Clausewitz's *On War*. At least one Junker officer expressed emphatic doubts that Moltke ever would make a proper Prussian soldier, but his keen mind and industry brought him through all tests, including a frustrating three-year detail as military adviser to Sultan Mahmud II of Turkey. And, like Scharnhorst, he was accepted, promoted, and ennobled in a military service that never has been given credit for sensitivity. Despite his retiring personality (even in 1866 some Prussian generals did not recognize his name), he was accepted into the General Staff in 1839, became its chief in 1857, and held that position until 1889. He made the Prussian Army (after 1871, the

German Army) Europe's finest and shaped the General Staff into that efficient organization that remains a popular bogeyman. His planning and organization were largely responsible for Prussia's victories over Denmark (1864), Austria (1866), and France (1870–71). That last victory inspired the combination of the twenty-six German states into the German Empire. In all of these he had the cooperation of Otto Eduard Leopold von Bismarck, the "blood and iron" Prussian chancellor who arranged the wars that Moltke won, but he bluntly rebuffed all of Bismarck's attempts to exert any control over the army.

Unlike Scharnhorst and Gneisenau, Moltke had few interests outside the military. He did, however, very much appreciate the nineteenth century's technological progress, especially the possible military uses of railroads, improved highway systems, and the telegraph. It is possible—even if he did not exactly say so—that he regarded the American Civil War as "two armed mobs chasing each other around the country, from which nothing could be learned",[4] yet in 1864 he added a "Railway Section" to the General Staff, and some two years later the Prussian Army was given a railroad construction unit like that employed by the Federal forces.

To Moltke, strategy was

> a system of expedients . . . a science applied to everyday life . . . the art of acting under the pressure of the most arduous circumstances. . . . In war it often is less important what one does than how one does it. Strong determination and perseverance in carrying through a simple idea are the surest routes to one's objective.[5]

Moltke's application of his theories was unusual, a complete break with Napoleonic command practice. Napoleon had moved into battle with his forces concentrated in his tightly controlled "battalion square." Moltke preferred to operate after his own principle "First reckon, then risk."[6] After long study he would estimate his enemy's probable strength, location, and intentions (which he did quite skillfully), then—using all existing railroads and highways—send his forces forward in several columns, sometimes widely separated. Once they neared the enemy's expected position, he gave their senior commanders general directives as to what results he wanted and turned them loose. Thereafter he functioned as a detached brain, observing but seldom intervening as the campaign developed. When one of his columns struck resistance, the others closed in toward the "sound of the guns," achieving their concentration of force and effort on the battlefield itself. In theory, this would envelop and crush the hostile army, as it did the French at Sedan in 1870. In fact, it offered the enemy commander an opportunity to defeat those converging columns separately. A Robert E. Lee, Stonewall Jackson, Napoleon, or Chu Teh would have made monkey meat out of such an offensive. Moltke tried to avoid this risk by careful preplanning but always found himself troubled by fouled communications, misunderstood orders, and bullheaded subordinates. In the 1866 Seven Weeks' War against Austria, this risk was intensified by Moltke's misuse of his cavalry. Forgetting the effectiveness of the Napoleonic cavalry screen, Moltke advanced without one. The Prussians located the Austrians by the primitive process of blundering into them. (Moltke could learn; in the Franco-Prussian War he had his cavalry well out to his front and flanks.) Also, he might lose control of the offensive since his army and corps commanders were an aggressive, ambitious lot, sometimes inclined to ignore Moltke's carefully drafted directives in favor of the barroom-brawl principle "If you see a head, hit it!" However, if one of them sought trouble and bit off more of it than he could chew, he could count on help from the nearest Prussian units—a comradeship that the backbiting French generals of 1870–71 frequently did not display. Moltke accepted such insubordination, if it was successful, even if it scrambled his strategic plans.

In retrospect, with all proper appreciation of Moltke's genuine talents, it is probable that his fame comes largely from the fact that Bismarck picked wars only with nations obviously weaker than Prussia, after having made certain by forehanded diplomacy that no other European power would interfere. But Prussia's victories impressed the world, and there was widespread imitation of all things pertaining to the Prussian army. In a few cases this was beneficial: The United States began to improve its antediluvian staff system. Generally, it was trivial: the British and American armies adopted German-style spiked helmets. Occasionally, it had dangerous long-term results: The Japanese fired their French military advisers and requested Germans, and French officers began skimming through Clausewitz's *On War*.

The General Staff continued to grow in power and influence, and for very good reasons. German senior commanders were a more mixed bag than ever before, the usual indifferently educated Junkers mixed in with princes of the royal family and princelings and nobility from the smaller states of the new German Empire. The first, having always commanded Prussia's armies, believed they had an established priority despite all the newfangled additions to their profession. The second might or might not be good generals, but their whims had to be considered. The third, *noblesse oblige,* kept up their family traditions of military service, though some loathed it as penal servitude. Such commanders required an infinite variety of restraint, professional advice, or encouragement. This the General Staff provided, tactfully, calmly, firmly, following Moltke's directive: ''Accomplish much, remain in the background, be more than you appear to be.'' Steely-minded Gen. Hans von Seeckt, who rebuilt the German Army after World War I, declared, ''General Staff officers have no names.'' One of them left a postscript: ''We have no time to be weary.''[7]

The General Staff was an army staff only, with next-to-no interest in naval affairs. Its members provided only a small portion of the various staffs of the whole army, especially at the lower echelons. In 1914 there often was only one General Staff officer in the typical German division—always the division chief of staff. The relationship between these highly trained officers and the commanding officers of the organizations to which they were assigned was unique. The General Staff officer was responsible both to his commanding officer and to the chief of the General Staff. Moreover, he was the commander's collaborator, sharing both his responsibility and authority. This exact relationship was not spelled out in printed regulations but was established through custom; it varied from division to division and corps to corps according to the characters of the two officers involved. If the commander were able and authoritative, the General Staff officer might function much as the chief of staff in other armies. If he were not, his chief of staff would assume the necessary responsibilities, often being the commander in all but title. These chiefs of staff often—as in the case of adjoining divisions, corps, or armies—dealt directly with one another, informing their respective commanders after the fact. Even more than their commanders, they were rewarded for victory or punished for defeat; in the latter case, the chief of staff might be replaced, but the commander given another chance. In any other army such a command system would be, as the old American folk saying goes, "one hell of a way to run a railroad." With the Germans, it worked.

It was said of Moltke that he smiled only twice in his entire adult life—once when he saw how obsolete the Swedish coastal defenses were and again when he was told his mother-in-law had died. However, compared to Alfred von Schlieffen (1833–1913) who became chief of the General Staff in 1891, he must have seemed genial indeed. Schlieffen, though of the Prussian nobility, was another studious officer, nearsighted and retiring. From his first service it was evident that he was a natural staff officer. He was also devoted to his young wife, who died early. Thereafter he had no family, interest, or happiness except his military duties. He saw combat in both 1866 and 1870–71, emerging a practical solider without illusions. Never sparing himself, he had no pity for human weakness. German Army legend had it—apparently truthfully—that his idea of a Christmas gift was a staff problem, to be solved immediately. Every scientific advance was to be probed for possible military value, all world events examined for their effect on German military strength. Sardonic, aloof, imperturbable, ascetic, yet gifted with a strength of character and will that made able soldiers his devoted disciples, he came to be a high priest of war.

Schlieffen taught and planned under a haunting foreknowledge of an impending *Gotterdammerung*. The European power balance was swinging against Germany as France, Great Britain, and Russia drew together. Germany's allies were only the Austro-Hungarian Empire and Italy: the first was twisted by internal dissensions and militarily inefficient; the Italians would probably hang back until they could rush to the rescue of the victor. (There was an unkind Polish proverb that God's only reason for creating the Italians was to give the poor Austrians somebody they could whip!)

Faced with the prospect of a two-front war, Moltke had proposed to stand on the defensive against France and throw most of Germany's strength against Russia. Schlieffen reversed that priority. Longer-headed than Moltke, he realized that the size of the Russian army and the extent of Russian territory would make a quick knockout there impossible; also, he realized that a long war of attrition would ruin Germany. His solution was to leave a small force in East Prussia to fight a delaying action against any Russian offensive and mass the rest of his army in the west for the quick destruction of the French. Since the French frontier was strongly fortified, he would leave only approximately one-eighth of his available troops there facing the main French forces. If the French attacked them, so much the better. Meanwhile, the remaining seven-eighths of Schlieffen's army would wheel suddenly through neutral Holland, Belgium, and Luxembourg, enveloping the French right flank and crashing into their rear. (Schlieffen had been much impressed by Hans Delbruck's description of Hannibal's battle of annihilation at Cannae and hoped to emulate that tactical masterpiece on a strategic scale.[8]) This Schlieffen Plan (actually a series of plans, as the original 1894 concept was modified to meet changing conditions) was necessarily complex, covering everything from initial mobilization to the general scheme of combat operations. Given the size and complexity of his planned offensive, Schlieffen realized that once the actual fighting began, he would have to depend, like Moltke, on the aggressiveness and competence of his corps and army commanders. However, so far as can be judged from his writings, he proposed to function as a commander in the Napoleonic sense, maintaining as tight a control as possible over the action.

To prepare for *''Der Tag''*—the expected, if not exactly anticipated, day when Europe would again burst into war—Schlieffen sought to equip the Germany army with the latest technical developments: medium howitzers for its field artillery, mobile siege guns, field radios, improved railroads and railroad units, and aircraft. While he emphasized offensive warfare, he also trained his troops in defensive operations and the construction of field fortifications. His cavalry was organized, trained, and equipped to fight either mounted or dismounted; his infantry and artillery to work together. It seems that he gave little thought to political/diplomatic matters. His planned violation of Dutch and Belgian neutrality was sure to invite English intervention, but this was probably a calculated risk. England had relatively few troops available; a quick demolition of the French army would leave them impotent. And there is the interesting fact that he recommended, if his grand offensive should fail, Germany ''should at once seek a negotiated peace'' rather than risk the attrition of a prolonged war.[9]

Schlieffen, however, found his planning dogged by opposition at all levels. German officers, especially the ultraconservative Junker types, mistrusted the new technologies and the officers (mostly expert middle-class specialists) who handled them. When *Der Tag* did arrive, they frequently misused them, as young Heinz "Hothead" Guderian, then commanding a mobile "Heavy Wireless Station" attached to the 5th Cavalry Division, would emphatically testify.

Of more immediate importance was the German Reichstag's[10] refusal to provide sufficient troops and money. The result was that Germany could train only 52 percent of its available conscripts each year (France trained 82 percent) and the German army on the western front in 1914 was twenty divisions short of the strength Schlieffen had considered essential.

The last and most frustrating problem was embodied in the person of Emperor Wilhelm II of Germany—the horrendous "Kaiser Bill" of the author's childhood. He was in fact no evil ogre but something possibly worse—a monarch who had been poorly educated for his job, who meant well enough but remained a restless, erratic adolescent all his life. The pomp and pageantry of imperial rule fascinated him, but he was incapable of conceiving, or even comprehending, a sensible national strategy. He wasted immense sums building a high-seas navy that would be of next-to-no use in a European war but alarmed England. One of his first acts was to dismiss Bismarck, who, after 1871, had devoted himself to keeping Europe peaceful. Thereafter, Wilhelm meddled in every possible crisis, displaying an alarming talent for stupid oratory. Schlieffen, a loyal old-school Prussian monarchist, could not bring himself to oppose Wilhelm's follies, which included capsizing carefully planned army maneuvers so that he might "win" them by personally leading thousands of assorted cavalrymen in wild charges against infantry, machine guns, and artillery. (There *is* an old tale of the Kaiser riding proudly up to the chief umpire—Gen. Paul von Hindenburg or some comparable Junker—after such a display, to meet the grimly respectful verdict "All dead but one, your Majesty.")

Wilhelm's final folly was his worst. When Schlieffen retired in late 1905, the Kaiser made Gen. Helmuth von Moltke, nephew of the late Count von Moltke, Chief of the General Staff, apparently on the theory that his name would impress Germany's putative enemies. The younger Moltke did not want the assignment and accepted it only out of personal loyalty; his previous service had been largely of the "carpet knight" variety as aide-de-camp to his uncle and the Kaiser;

he had not been shaped in the General Staff's hard mill, and his health was indifferent. He had intelligence but no backbone, could be wheedled or browbeaten into actions he knew were unwise, and was oversensitive. Steadily, he weakened Schlieffen's plan. His decision not to move through Holland was probably wise, since it might reduce the risk of English intervention. His continual reassignment of troops from his offensive right wing to his defensive left made limited sense. The French army was increasing in size and power, and the German Rhineland and Saar industrial areas must be protected, as the aged Schlieffen himself agreed. But Moltke overdid it. Regardless of Schlieffen's last deathbed words, "see that you make the right flank strong,"[11] he continued his whittling until the original seven-to-one preponderance of the right wing over the left had been reduced to less than four to one.

Schlieffen's plan had been an outstanding example of the application of the principles of *objective, offensive, mass, economy of force, maneuver,* and *surprise*. Even with Moltke's tinkering it came amazingly close to success. A modified version of it would bring Hitler a smashing victory in 1940. Its 1914 failure had several causes, but the most potent of these was Moltke's attempt to imitate his uncle.

The elder Moltke's concept of the great strategist as a detached, omniscient brain had enjoyed amazing popularity among soldiers who ought to have known better. A queer school of military fiction evolved in which commanding generals gave their orders and then went fishing, returning in the evening with full creels to be hailed as conquering heroes. Napoleon was dragged in to support this myth, with quotations from such works as the faked memoirs of his false friend Louis-Antoine de Bourrienne. (It is always easier to read alleged memoirs than research original material in the archives.) In these Napoleon was portrayed as laboring for a night over his pin-studded situation map and thereafter uttering detailed, all-comprehending orders for the entire forthcoming campaign, including the exact spot where he would win its decisive battle!

So, once the 1914 German offensive was launched, young Moltke settled down in comfortable quarters far to the rear to function just like uncle. He promptly lost contact with his right flank and control of the whole war. What decisions were wrung out of him were almost invariably wrong. His principal opponent, Gen. Joseph J. C. Joffre, was never accused of intelligence, a distinguished ancestry, or sensitivity. Energy, a brutal courage, and plenty of low animal cunning were his strengths. He began the war with suicidally stupid plans, strategy, and tactics. He was thoroughly whipped in all the opening battles. But he literally rode the French army, roughing out new plans, breaking defeated generals, reorganizing shattered commands, driving the French armies back into combat. The plans often failed, the new subordinates might be defeated, but—Moltke having practically abdicated responsibility—he somehow stopped the German offensive. And so Schlieffen's planned quick war of annihilation became the long war of attrition he had hoped to prevent.

Not only Germany found the elder Moltke's strategy a curse. Some English soldiers thought it congenial, in particular Sir Ian Hamilton, who was sent in 1915 to command the ground forces in a hastily prepared attempt to take the Dardanelles area and thus open a short sea route to Russia. Trained in England's colonial wars, an expert catcher of Boer commandos, high command left him, as he said, "so wrapped in cotton wool . . . [that] it was not for me to force" foot-dragging, blundering subordinates to *do* something.[12] And so a vital campaign that might have shortened World War I by at least a year was lost because Hamilton would not drive—or, better still, relieve—several incompetent old blisters wearing generals' insignia.

The French, meanwhile, operated by their own concept of strategy. One of its originators was a little-known young officer with the most appropriate name of Charles Ardant du Picq (1831–70). A graduate of St. Cyr (France's West Point and somewhat exclusive), he was captured at Sebastopol during the Crimean War (1853–56) and had considerable experience in North Africa and Syria. Early in his service he became fascinated by the question of what motivated individual soldiers and whole military organizations in action. Saxe's *Reveries,* with its emphasis on the "human heart," probably was his first inspiration, but he also drew on Guibert and especially on Marshal Thomas Robert Bugeaud—redheaded "*Pere* Bugeaud" of the famous *casquette,* who had conquered Algeria with the Roman strategy of forts, roads, and flying columns.

Du Picq attempted to poll his fellow officers as to their experiences during the Crimean War and the so-called Austro-Sardinian War (1859) when their troops first made contact with the enemy. Apparently this was not too successful, for he turned to ancient history for further examples. He certainly was no student; his *Combat in Antiquity*, published privately just before his death, is a stew of misinformation and brash assumption, liberally spiked with pure-quill chauvinism. He seems to have done a great deal of writing, but probably only part of it was collected and published as *Etudes sur le Combat* in 1902.[13] However, literary endeavors were not in good repute in the armies of Napoleon III. One of its senior marshals, the brave-enough blockhead Marie-Edme-Patrice de MacMahon, Duke of Magenta, threatened, ''I shall remove from the promotion list any officer whose name I have read on the cover of a book.''[14] MacMahon is officially remembered for having allowed Moltke to bag him at Sedan, but there was also the tale of his profound observation while visiting a hospital: ''Ah, typhoid fever, I have had it; it either kills you or leaves you an idiot.''[15]

Du Picq's service was largely colonial warfare in which the organization, discipline, and fire power of French regular troops routinely shattered undisciplined native hordes. From this, he derived the theory that France should develop a relatively small army of long-service professional soldiers officered by military aristocrats who were to have a life of ''money, little work, and leisure.'' This army must be carefully trained, drilled, and educated until it became a ''collective man,'' ready to march and fight, proudly and efficiently, under all conditions. Its uniforms and equipment must be simple and practical, designed for combat. He violently opposed diluting this elite regular force with large numbers of reservists, maintaining that citizen soldiers were ineffective and loath to fight. Being happy to pontificate on any subject, he cited the American Civil War as an example of the failure of volunteer armies. Its battles, he declared, were affairs ''between hidden skirmishers, at long distance . . . a melee of fugitives'' who never risked close combat.[16]

To du Picq, war appeared a contest of wills and morale. His research, he announced, proved that troops almost never met in hand-to-hand combat, that one side or the other ran away before the moment of contact. If an attack were well planned and the attacking troops came on boldly, the defending force facing them—no matter how strong its position—would bug out. Therefore the key to victory was to inspire your troops with the will to conquer and to lead them forward: "he will win who has the resolution to advance. . . . No enemy awaits you if you are determined. . . . No one stands his ground before a bayonet charge."[17] It is a pity this cock-a-hoop young theorist could not have completed his education at Gettysburg, Wilson's Creek, or the Bloody Angle.

The element of truth in this harangue was just sufficient to make it dangerous; troops, whether attacking or defending, frequently *do* flinch at the critical moment, but usually as much because of enemy fire power as any deficiency in their own will power. Possibly du Picq's colonial experience led him astray. A pocket of Arab tribesmen behind a sand dune would have been much more amenable to his theory than a company of Pomeranian infantrymen defending a hedgerow. Du Picq was never to learn if there were a difference. He died in action at the head of his regiment a few days after the Franco-Prussian War began. His slim book killed Frenchmen by the thousands all through World War I.

In the metaphysical hothouse of the French War College after 1871, du Picq's teachings were brewed with hasty readings from Clausewitz and selected quotations from Napoleon to make strong medicine that would enable the French to defeat the stronger, better-armed German army. A gaggle of enthusiastic young colonels worked over the solution that finally took shape in the 1912–13 *Regulations for the Conduct of Large Units:* "The teachings of the past have borne their fruit. The French Army, reviving its old traditions, no longer admits for the conduct of operations any other law than that of the offensive."[18] The French army was trained and armed accordingly. Anything that might slow up its advance—heavy artillery, medium artillery—was not worth having; since French troops would always attack, there was no need to teach them how to entrench or to modernize the frontier fortresses; it was not necessary to know the enemy's plans or intentions, because the attacking French would impose *their* will on him. In short, *attack—always attack—attack all out!*

The high priests of this strange military creed were supposedly France's best and brightest army officers; some of them passed for military scholars. One or two were walking encyclopedias of Napoleon's campaigns but somehow missed the Emperor's views on the value of fortresses, entrenchments, and defensive operations. High among them was one Ferdinand Foch (1851–1929), who had been a student, a professor, and finally the commander of the War College. There is no particular evidence that he was a real student of the art of war, but he certainly was an impassioned advocate of the offensive and the will to victory. The officers he taught were even more fanatic.

And so the French went bravely off to war with a strategy and tactics very much like those of the wild dervishes and "fuzzy-wuzzy" spearmen who followed their Mahdi and emirs across the Sudan through 1882–98 until English fire power made hash of their Moslem zeal. So deep was the evocation of France's traditions that French soldiers marched in the famous red trousers they had worn since 1830. (An attempt to put them into a sensible green field uniform in 1912 had been squelched as thoroughly subversive to French morale.) Du Picq had disliked those red trousers, but the War College priesthood overlooked his practical ideas.

"En avant!" They went forward, ranks tight behind slanting lines of long, needlelike bayonets, bands playing, new *sous-lieutenants* flaunting traditional white gloves. Over the rumble and mutter, the snap of the first shots, long infantry *clairons* sang the old call:

> There is a glass to drain up there!
> There is a glass to drain!

"A la baionnette!"

The German machine gunners, their sights on the lines of red trousers, opened fire.

French generals, however, could take comfort from Foch's maxim: "a battle won is a battle in which one will not confess oneself beaten."[19] They could feed in more men and try again. That seldom worked, but Foch and others remained convinced they were unbeaten—given sufficient manpower, they would win. They did concede they would need more artillery. Meanwhile, it was very rough on the boys in red trousers.

Ferdinand Foch was a frizzy-moustached, bandy-legged bantam rooster of a general, utterly convinced that he was a transcendent military genius. He also was a devout Catholic, with an apostolic zeal and effectiveness in presenting his opinions and an outrageous, scalding temper when his ambition was challenged. He had never heard the proverbial "shot fired in anger" until 1914, when he went into World War I as a corps commander, and it is not certain that he was ever really under fire except by accident. He was no field soldier; he insisted on headquarters with steam heat, and his headquarters commandant caught blazes whenever its temperature dropped below eighty degrees Fahrenheit. When the fighting went against him, his kepi was set squarely on his head; when it went successfully, the kepi was cocked "well over toward his right ear," and he carried his cane over his shoulder.[20] He spoke with many gestures, in a unique mix of images and clipped, often cryptic expressions. As a general, he was hardly competent. He would push offensives long after they had reached their maximum possible penetration and could achieve nothing more except an increased butcher's bill; he did his best to sabotage Gen. Henri Pétain's excellent "elastic" defense system in 1918 because he thought it shameful to give up an inch of ground without fighting desperately for it—an attitude that also lengthened French casualty lists and greatly facilitated that year's German offensives.

Frequently, he dramatized himself, launching resounding phrases (often unacknowledged borrowings from Napoleon) that looked impressive in the newspapers. He somehow managed to appear as the real hero of the 1914 Battle of the Marne—where, in fact, his ears had been beaten down into his socks. Later he would confess that the French reliance on morale had been foolish; his *Memoirs* maintain that he was aware of the weaknesses of the 1914 French army and took sage measures to correct them. However, he continued to preach the validity of French offensive doctrine and scold British and American commanders whenever they seemed insufficiently willing to expend their troops according to his desires.

All that said, the ironic fact remains that Foch was the only general who could have been placed over the Allied forces in early 1918, when German offensives had almost crippled the French and British armies and the Americans were only beginning to arrive in strength. (Since the French Army was the largest of the Allied forces, any supreme commander *had* to be a Frenchman.) Most French generals were discouraged; Foch was endlessly and noisily pugnacious, at least by proxy; the Allies needed an aggressive, self-confident leader. Foch's massive counteroffensive exhausted the French and British armies, but the surging inflow of big American divisions gave him the cannon fodder to keep it moving. He could remain rather condescending about it: "The young American army was excellent and full of enthusiasm but, naturally, inexperienced and immature. . . . I could not treat it as though it had been fighting for us for years.''[21] However, he wanted as many American divisions as possible under French generals for use as shock troops.

The great victories are not always won by great generals. (An excellent example is the decisive battle of Saratoga during the American Revolution, fought between two incompetents, Horatio Gates and John Burgoyne.) Having little basis in fact, Foch's reputation as a great captain has evaporated. There remains his record of loyal service, after his own fashion, to France in a time of deep peril.

World War I produced no great strategist. Its great strategic schemes, such as the Schlieffen Plan and the Allies' Dardanelles campaign, foundered because of insufficient preparation, dumb execution, the innate cussedness of terrain and weather, and the general amateurishness of the armies involved. Europe had not seen a really major war since 1815; the last war between major powers had been in 1870–71. (The Russians and Japanese *had* fought in 1904–1905, away off in Manchuria, but the Russians learned nothing much useful from the beating they took.) From buck private to field marshal, except for a comparative few salted in colonial squabbles, western Europe's soldiers were professional innocents. Their commanders and staffs had prepared for war the best they knew how, but the early locking of the western front into continuous lines of fortifications from the English Channel to Switzerland denied them any real chance of strategic achievement. The Allies—except for the few with Gen. Edmund H. H. Allenby in Palestine who heard the sound of many horses running furiously to battle across the plains of Armageddon—ended with little strategic inspiration beyond the meat-grinder experiences of trench warfare. They did have the sweet justification of a final victory and an obsessive belief in the power of the defense, which begot the Maginot Line and the belief that the tank and airplane were only handmaidens to the infantry and artillery—a school of strategy sometimes impolitely termed the Sitzkrieg.

In the East, however, there had been a war of vast distances where the largest armies had maneuver room. Cavalry scouted and fought much in the old way, and enterprising commanders could find opportunity for initiative and ingenuity. There, always outnumbered, fighting a left-handed war, the Germans squashed Romania and broke Russia, meanwhile propping up their Turkish, Bulgarian, and Austro-Hungarian allies. (As sideshows, they also took out Serbia and crippled Italy for the Austrians.)

There were successful and unsuccessful German commanders in those eastern battles, the most famous being the tight team of Field Marshal Paul von Hindenburg (1847–1934) and his chief of staff, Erich von Ludendorff (1865–1937). A carnivorously ambitious son of a middle-class family, Ludendorff was the team's brains and savage driving power, a tireless, imaginative trainer and organizer; during 1916–17 he masterminded the development of entirely new and highly effective German offensive and defensive doctrine.[22] He had a remarkable personal bravery, but prolonged strain and ill fortune might drive him to erratic decisions. Hindenburg was the old-fashioned Junker officer and gentleman, responsible, competent, and unshakable. His simple loyalty and austere dignity made him a natural leader in times of trouble. This team had an icily efficient "chief of operations," the sardonic Col. Max Hoffmann (1869–1927), who was left—probably to Germany's misfortune—in the East when the Hindenburg/Ludendorff team was promoted to the command of the entire German Army in August 1916. (Gifted with quick and deep intelligence, a photographic memory, *and* a taste for night life, Hoffmann had not always slaved over his studies like a proper General Staff officer. Before the war, German Army gossip accused him of dressing his soldier-orderly in his uniform, leaving him, apparently deep in study, at his well-lighted desk, and slipping out the back door in civilian clothes for an evening on the town.)

In the end, the Germans had defeat and a broken nation to ponder. In Article 160 of the 1919 Treaty of Versailles the Allies demanded the abolition of the General Staff and its school system, which naturally proved an exercise in trying to wring pots of gold out of a rainbow. Surviving members of the staff went incognito in various assignments and set about rebuilding their army and nation. They knew how a war of maneuver might be won and sought the weapons, communications systems, and vehicles needed to wage it. Hitler would cut back the General Staff's authority, but the skill with which it once more guided Germany's battle against a world in arms during World War II brought new demands for its eradication as "a sacred and rigid duty, dedicated to those who have fallen while combating the historic and evil efficiency of the Great General Staff."[23] But the

General Staff survives in all its old efficiency in the West German Army today—and, unfortunately, probably in the East German Army as well, however clogged by Communist commissars.

The naval warfare of World War I, on both sides, was much inspired by an American, Albert Thayer Mahan (1840–1914), son of Dennis Hart Mahan and the first American strategist to achieve truly international fame. Wilhelm II, Emperor of Germany and King of Prussia, described himself as "devouring Captain Mahan's book, and . . . trying to learn it by heart."[24] The universities of Oxford and Cambridge decked him with honorary degrees, officers of the new Japanese navy exclaimed "Ah-so!" over his books, and French sailors read his strictures on their naval history with respect, if not with joy. He was even honored in his own country. For Albert T. Mahan was that most fortunate mortal—a prophet born into his own time.

He had refused to be a soldier, but after two years at Columbia College and much to his father's displeasure, he entered the U.S. Naval Academy, graduating in 1859. Most of his Civil War service was on blockade duty. In 1867 he was sent on a two-year cruise to the Far East. Like Halleck, he spent his spare time in study, a practice he continued through a series of assignments at sea and on shore. A methodical man, he sought out the best foreign books and professional periodicals. Two books he found particularly stimulating—William F. P. Napier's vivid and contentious *War in the Peninsula and in the South of France* and Theodor Mommsen's *History of Rome*—the first for its insistence on the sequence of cause and effect in military operations, the second for its accounts of Roman naval warfare. Then or later, he also read and valued Jomini.

In this, Mahan was part of an intellectual groundswell building up within the U.S. Navy, much to the unsettlement of many senior officers whose hearts still lingered in their early years of stately sailing vessels. The U.S. Naval Academy, established in 1845 in an obsolete coastal fort borrowed from the Army, had fallen into ill days and ways during the Civil War. The war done, efficient Secretary of the Navy Gideon Welles set Rear Admiral David Porter over it as a new-broom superintendent. Porter was all ambition, brag, bluster, and impossibly tall tales—but also intelligent and energetic. He had served wholeheartedly with Grant and Sherman in the West; like Sherman, he believed in education. At Annapolis he dealt with unsatisfactory instructors and midshipmen as he had with Confederate gunboats and batteries and rebuilt a rowdy prep school into a functioning college. In 1873 the Naval Academy staff founded the present U.S. Naval Institute to promote study and discussion of professional and scientific problems. Mahan served for a year as an instructor and made himself noticed. When, in 1884, Capt. Stephen B. Luce, Porter's right-hand man in the Naval Academy's reform, finally wangled authority to establish a Naval War College for the postgraduate schooling of naval officers, he requested Mahan as an instructor. (Luce's idea was violently opposed by Congress and senior naval officers; the best site he could find for it was a vacant poorhouse near Newport, Rhode Island.) Mahan was on sea duty off South America and was not made available until 1886. He spent the two years in intensive preparation for his coming assignment, concentrating on the history of naval warfare. Mommsen's comments on Hannibal's reasons for marching out of Spain, across southern France, and over the Alps into Italy—instead of going directly by sea—at the beginning of the Second Punic War (218–201 B.C.) particularly caught his imagination, since the basic reason was the Roman naval superiority in the western Mediterranean. Gnawing this mental bone, he decided that there had never been a methodical study of the importance of sea power. Of necessity, he drew his historical studies from wars waged by wooden navies, propelled by oars or sails. But he had active service enough on distant seas to appreciate the problems and possibilities introduced with steam power, armor, long-range guns, radio, and—in his last years—submarines. As his principal area of study, he took the military history of the eighteenth and nineteenth centuries—an era of world-girdling campaigns that had decided the fate of North America and offered splendid examples of the interplay of land and naval warfare and of the fate of nations.

Coming to the Naval War College in 1886, he found Luce ordered off to sea and himself the college's new president. The college still had many enemies; in 1890 they succeeded in abolishing it. But that same year Mahan published his lectures at the college under the title of *The Influence of Sea Power upon History, 1660–1783*. The book was an immediate success, worldwide. Two years later, he followed up with *The Influence of Sea Power upon the French Revolution and Empire*. The Naval War College was justified and saved; Mahan's reputation established.

Mahan had come into the U.S. Navy when a combination of technical progress and the long American westering was changing its mission and its needs. The title of his autobiography, *From Sail to Steam,* only hints at the magnitude of these changes. Steam propulsion had freed ships to move against wind and storm, to maintain steerageway and tight maneuverability in the cross chops of harbor entrances and narrow straits. It also hobbled their voyaging. Sailing ships could range free across the world. Their maintenance was relatively simple; supplies they needed could be found in almost any seaport. The new steamers needed good coal and friendly ports in which to refill their bunkers; on occasions enough either the coal or friendship might be lacking. Steamers needed shortcuts to avoid long, coal-eating voyages around Cape Horn and the Cape of Good Hope. From this need would come the Suez and Panama canals, with all their attendant political complications. With steam came iron ships, stronger built and often swifter but requiring more elaborate shipyards for repairs. Armor and powerful new guns compounded that problem.

Moreover, the United States was now firmly established from east to west and bound together by railroads. But the sea route between its coasts went around the stormy Horn for 15,000 miles, more or less. (In the Spanish-American War the battleship U.S.S. *Oregon* needed sixty-six days to make that run.) Burgeoning American commerce across the Pacific went in danger of Malay pirates and foreigner-hating Japanese clans; various Latin American nations were apt to confiscate American shipping on any pretext they could imagine. And by 1875 a hurriedly modernizing Japan had a navy and was beginning an empire. But the U.S. Navy had dwindled from its proud Civil War days into a maritime curiosity shop, uncomfortably inferior to the Chilean navy, increasingly manned by chance-enlisted foreign seamen who knew barely enough English to fumble through their duties. It had no bases beyond our shores, so its warships must still trust to sail on distant voyages, hoarding their small supplies of coal against emergencies. Beyond a catch-as-catch-can willingness to try anything, it had no concept of naval strategy. Against the mighty Royal Navy during the Revolution and the War of 1812 it had attempted, without much success, to defend our coastline while raiding British merchant shipping. Against the North African Barbary pirate states, Mexico, and the Confederate States of America, none of which had effective navies, it had blockaded their coasts and raided their harbors.

Porter, Luce, and Mahan managed a marriage of modern technology with a revival of the old, ocean-troubling American ship mastery that had outsailed, outadventured, and outfought all comers. Mahan gave it a doctrine and sense of mission. He saw naval power as the true measure of national power, the one effective means by which the United States could extend its influence outward from its coastline to protect its own citizens and commerce, or halt an enemy far from our shores. In times of international crisis American diplomats might be clever and persuasive—but the sight of the long gray hulls of American warships lying off the other nation's coast could add the necessary emphasis to their words. Earnest, peaceable Americans—never quite able to accept the fact that much of their world was, and remains, an irrational place—would talk of "gunboat diplomacy." (Few would seem disturbed a half century later when Mahan's theories got a landlubber's echo in Mao Tse-tung's "Power grows out of the muzzle of a gun."[25]) In war, Mahan taught, a navy's overriding objective was the destruction of the hostile fleet. Its strength should be concentrated for that purpose and not scattered in attempts to defend seacoast towns or chase enemy merchant shipping. Strategically located bases were essential to give the fleet the necessary reach and flexibility. Originally, Mahan had not believed in territorial expansion, but long service and study made him something of an imperialist, eager to annex Hawaii, collect useful bases in the Caribbean, and build and control a canal through the Isthmus of Central America.

America's imperialism, however, was relatively modest for the late nineteenth century. Hawaii, the Philippines, Guam, Wake Island, some of the Samoan Islands, Puerto Rico (along with the Virgin Islands, purchased from Denmark), sated it. There was no attempt to build a navy comparable to England's; one that could handle either Germany or Japan, both of which were moving aggressively into corners of the Pacific, would suffice. But the Panama Canal would be essential; without it the United States would have to maintain major fleets in both the Atlantic and the Pacific. A canal under the control of some foreign nation (a French company began an unsuccessful attempt to dig one in 1881) would be an abiding threat to the United States. Exorbitant tolls might be levied on our shipping, and the passage of American vessels might be delayed or even blocked during international crises.

Mahan retired in 1896, but two years later the Spanish-American War plucked him back to active duty with the Naval War Board, which directed the Navy's operations. His authoritative hand and voice were much needed, the new Navy being an adolescent organism with a wondrous talent for stepping on its own trailing shoelaces. The Army had been allowed to lapse into an under-manned museum of military antiquities, and the almost-instinctive Army/Navy cooperation of the Civil War had vanished. Their joint effort against the Spanish-held port of Santiago de Cuba produced a row over whether the Navy would risk a ship to force the harbor entrance, or the Army suffer heavy casualties in an attempt to storm the city's land-side fortifications. When the Secretary of War appealed to Pres. William McKinley, the Secretary of the Navy loosed Mahan upon him. Mahan bluntly informed him that "he didn't know anything about the use or purpose of a navy."[26] (The Army thereupon considered fitting out improvised warships—armed transports with bales of hay for armor—and crashing the harbor entrance on its own. Unfortunately, the Spanish surrendered before any such unseemly doings could be perpetrated.)

In 1899 Mahan was one of the American delegation to the Hague Disarmament Conference where his realistic appraisal of international relations often ran crosswise to the pacific hopes of American civilian delegates. Promoted to rear admiral on the reserve list in 1906, he performed various odd jobs for the navy, wrote his memoirs, and continued with his studies. Magnificently bald, with heavy-lidded, appraising eyes and an aggressively curved nose above a graying moustache and short beard, he was a formidable presence. Despite the visible success of his teachings, he doubted the continued will of the United States to maintain the strength of its armed forces, democratic governments being in his opinion lacking in foresight. His last book, *The Harvest Within, Thoughts on the Life of a Christian,* testified to the religious conviction that had buttressed his life.

The amazing popularity of Mahan's works and theories doubtless derived in part from the fact that they fitted exactly into the spirit of the times. To the English, they explained and upheld what the English had been doing instinctively for centuries; to the Germans and Japanese, they justified vast naval construction programs and imperial expansion. The Germans, unfortunately, misread Mahan, skipping over his warnings that no nation could be both a great land power and a great sea power. (Modern Russia has managed that feat, but only by using what Mahan would have considered an impossible portion of its wealth for that purpose.) Germany also lacked two other qualifications Mahan had listed—a favorable geographic position with ready access to the high seas and large numbers of citizens in maritime occupations. "Kaiser Bill's" navy was an artificial creation—swift, deadly ships and some daring sea captains—too small to defeat the Royal Navy but large enough to draw off money, steel, and men that could have made Schlieffen's army invincible.

Mahan's greatest weakness was that he was completely a sailor; his works therefore always contain an element of special pleading, since his descriptions of naval operations lack any deep comprehension of concurrent army and diplomatic activities. He wrote eloquently of how the Royal Navy saved England from French invasion, but gave little attention to the limited effect it had on a self-sufficient land power like France. His assertion that sea power was the standard of national power would be challenged by a school of "geopoliticians" who studied the potential effects of political and economic geography on world power. One of their founders, Halford J. MacKinder, an English educator (1861–1947), stated their general premise: The "pivot area" or "Heartland" of the world (roughly, eastern Europe and European Russia) was inaccessible to sea power. If sufficiently provided with railroads, roads, and highways, it could shift troops and goods in any direction from its central position. He followed that with his famous assertion "Who rules Eastern Europe commands the Heartland. Who rules the Heartland commands the World Island (the connected land mass of Europe, Asia, and Africa). Who rules the World Island commands the world."[27] The Western Hemisphere and Australia/New Zealand he considered merely outlying islands, lacking the population, area, and resources to challenge the World Island. His thesis was militarized and popularized by Karl Haushofer (1860–1946), a former Bavarian General Staff officer and artillery instructor to the Japanese army, turned professor of geography and military science, who influenced Adolf Hitler. (Haushofer went into the Nazi doghouse when he disapproved of Hitler's impulsiveness in attacking Russia before he had settled with England.) MacKinder's basic thesis may seem overstated, yet it does present an irrefutable limit to the capabilities of sea power.

Mahan's world has gone down the winds of history. Britannia no longer rules the waves; control of the Suez and Panama canals has been relinquished to unstable minor nations; Russia is a major sea power. Mahan's teachings, however, retain considerable validity. The surface of this planet Earth is mostly water. Its oceans are vast, lonely reaches across which fleets and convoys can pass, screened by storm and fog. Submarines can prowl deep within them, sheltering under subsurface currents that muffle probing detection devices. Ships still remain the most efficient method of moving the oil, minerals, and other heavy, bulky goods that our economy needs in peace or war. Also, the United States still prefers to do any necessary fighting overseas in the enemy's front yard, and for the foreseeable future it shall still be dependent on shipping for the logistic support of any sizable military operation short of all-out nuclear war. Mahan's vision of battleship navies competing for command of the sea has now become a struggle to control the surface of the oceans, the sky and space above them, and the depths below—three-dimensional, split-second warfare. But the objective still is the freedom of the seas, the ability of our shipping to pass safely in peace or war.

Though World War I produced no Great Captain, it did beget military prophets and pontificators aplenty, with an accompanying fringe of cranks. Out of its latter years certain officers had deduced that future wars could be won quickly and decisively by tanks or airplanes, or a combination of both. Considering the short range, dubious mechanical efficiency, and limited killing power of the World War I versions of those weapons, such assumptions had to be something of an act of blind faith, if touched with prophecy.

Of these harbingers of new doom the airplane enthusiasts were easily the loudest. Just possibly they might have been more modest had they paused to compare their own achievements during the war with the prewar prognostications of various early aviation zealots. Prominent among these had been the British soothsayer-novelist and pseudohistorian H. G. Wells, who direly pictured a dirigible fleet—"cheap things of gas and basketwork"—demolishing cities and navies with dispatch, accuracy, and impunity.[28] Naturally, nothing of the sort had happened during World War I.

It is only fitting and proper that the first major apostle of air warfare should have been an Italian, Giulio Douhet (1869–1930). After all, Italy had been the first country to suffer aerial bombardment (in 1849 the Austrians released a flight of unmanned, bomb-carrying free balloons over the rebellious city of Venice with no apparent results) and also the first to employ aerial bombardment against another nation. (During 1911–12 Italian planes and dirigibles dropped small bombs on Turkish troops in Libya, achieving dramatic successes—but only in the Italian newspapers.)

Giulio Douhet was an artillery officer with a speculative mind. After championing the possible military uses of motor vehicles, he became interested in aircraft. By 1909 he was beginning to crank out air-war scenarios that rivaled those of H. G. Wells. Transferred to the Italian Army's Aviation Section in 1912, he became its head for a short period during 1913–14 but was soon in hot water, apparently as much from his disagreeable disposition as his advanced ideas. In 1916 he got himself court-martialed for passing material critical of his superiors to a politician in the Italian cabinet and was put out of the service for a year. He was recalled in 1918 but retired as a lieutenant colonel at the war's end. In 1921 he was promoted to general on the retired list; when Benito Mussolini came into power in 1922, he appointed Douhet a Subsecretary of Aeronautics. Douhet, however, soon quit his position, choosing to devote himself to expounding his theories in books and articles. These final years seem to have been lonely and somewhat embittered (we really know very little about Douhet as a private person), but he wrote in a calm, studious style and avoided polemics and personalities in his controversies. His principal work, *Command of the Air,* seems to have attracted little attention when first published in 1921, but the revised and expanded second edition of 1927 certainly raised a row within Italian military circles. Possibly because of Italy's inept military performance during World War I, general European interest in Douhet's theories did not develop until after his death. A somewhat abbreviated French version appeared in 1932; this was translated into English the next year and mimeographed for officers of the U.S. Army Air Corps. German, Russian, and British editions followed, and there finally was a complete English translation in late 1942.[29]

Douhet's central theses were simple: air power was irresistible, and any nation could be quickly defeated by merciless aerial attacks on its major cities. Aviation therefore was the most important part of Italy's armed forces: Italy must maintain an air force strong enough to seize command of the air as soon as hostilities began. This air force must be independent of the Italian Army and Navy, which would have only the minor mission of holding Italy's frontiers during the short period its air force would need to win the war. Command of the air would be achieved by destroying the enemy air force (if possible, by surprising it still on the ground) and its ground installations. That accomplished, Italian aviators would begin the methodical destruction of the enemy's population centers, using high explosive, gas, and incendiary bombs. Within two or three days the enemy would be pleading to be allowed to surrender; the enemy's civilian population, crazed by fear, might even revolt against their own government to force it to capitulate. So sudden would this overthrow be that the hostile army and navy would not even have completed their mobilization. This formula for victory, of course, could be used by any nation willing and able to build up sufficient air power.

Douhet had read reports of panics in London and Italy following enemy air raids and felt these proved his theory that civilian morale was the weak link in any nation's military strength. He had no hesitation over the morality of killing large numbers of women, children and other noncombatants. "The only principle to be considered is the necessity of killing to avoid being killed."[30]

In contrast to his cold and deadly theorizing, Douhet himself was a rear-area operator, the sort of air force officer British aviators dubbed a "Kiwi" because he couldn't, or wouldn't, fly.[31] There is no record of his ever qualifying as a pilot; his doctrine shows no evidence that he ever went along on a real combat mission. In fact, one may doubt that he had any experience whatever with the shooting end of World War I. He shows an amazing lack of technical knowledge, claiming that combat aircraft were cheap and easy to build and that it would be possible to create a superior air force in secret from limited resources. His imaginary air force would operate unhindered by enemy fighter aircraft and antiaircraft artillery, both of which he scorned, though both, and especially the first, were visibly increasing in effectiveness all through World War I. As one excellent airpower historian has observed: "Douhet's bombers always flew; his pilots always found their targets."[32] It could be added that the weather was always favorable for their missions.

Nevertheless, Douhet undoubtedly was an effective prophet for air power. He laid down in clear words a doctrine for its organization and use that appealed mightily to ambitious air force officers of many nations. J.F.C. Fuller, who came to abominate Douhet's works, dubbed him "a wonderful salesman, and like many such people—a prophet of the ridiculous."[33]

William ("Billy") Mitchell (1879–1936), Douhet's American counterpart, was a very different person who came to much the same conclusions on the uses of air power. His father was a U.S. senator; he left college at the outbreak of the Spanish-American War to enlist as a private in the infantry, serving in Cuba and the Philippines. Given a direct commission in the Signal Corps, he was involved in the construction of the Alaskan telegraph line system. Intelligence, drive, and zeal brought him promotion. When the Signal Corps began purchasing its first fragile aircraft in 1908 (they were considered to have possibilities for courier service), Mitchell became interested in their employment but did not learn to fly until 1915–16. Meanwhile, he had graduated from the Army Staff College and in 1912 was the youngest officer ever appointed to the U.S. General Staff. During World War I he rose to the command of American air operations, winning special distinction during the St. Mihiel offensive, September 12–16, 1918, where he directed some six hundred American, French, British, Italian, and Portuguese planes in support of the American ground forces. Mitchell flew his own plane in action and studied the air services of both allied and enemy nations. Returning home with many medals, he was promoted to brigadier general and appointed Assistant Chief of the U.S. Air Service.

Mitchell had been impressed by the British Independent Air Forces's operations under Maj. Gen. Sir Hugh Trenchard and was familiar with Douhet's theories. He had been eager "to blow up Germany" but was thwarted by the U.S. Air Service's complete lack of bombers.[34] Though the post-World War I years were starving times for the U.S. Army, Mitchell launched himself into a campaign to establish a large independent air force, coequal with the two older services and able to do its own strategic planning. He was particularly hostile to the U.S. Navy, which *was* getting the major share of available military appropriations, besides insisting on maintaining its own air force.

Proclaiming that air power had made surface warships obsolete, Mitchell "proved" his point by sinking four old battleships in demonstrations carefully staged to achieve maximum publicity. (The ships, of course, were anchored in coastal waters; service talk had it that Mitchell's flyers could not locate moving ships some miles out at sea.) As another demonstration he sent a flight of three aircraft around the world. He was a tireless speaker and writer, constantly throwing out new variations on his gospel of air power. He warned of Japan's growing strength, recommended the development of transpolar air routes, claimed that it would soon be possible to fly around the world on a single fill-up of gasoline, and predicted the fast-approaching day when "future wars again will be conducted by a special class, the air force, as it was by the armored knights in the Middle Ages."[35]

In all this he bore himself as the anointed prophet, divinely sent to bring salvation to the stupid and ignorant. Neither higher military rank, longer service, nor superior technical knowledge entitled anyone to disagree with his apocalyptic assertions, and any disagreement was high treason and personal insult. Eventually, his noisy insubordination wearied the Army; in 1925 Mitchell found himself a colonel at a minor post in Texas. Later that year the crash of the big navy dirigible *Shenandoah* inspired him to charge the Army and Navy with "incompetency, criminal negligence, and almost treasonable administration of the national defense, and with . . . almost always [giving] incomplete, misleading, or false information about aeronautics" to Congress.[36] A court-martial found him guilty of making statements to the prejudice of good order and military discipline, and sentenced him to five years (later reduced to two and a half) suspension from the service with forfeiture of pay and allowances. He was guilty beyond any doubt but never realized it, regarding himself a martyr and persuading a large number of people that he actually was one. Leaving the Army in early 1926, he became a Virginia farmer and continued his agitation for an independent air force.

World War II found the major powers' air forces following a variety of doctrines. Russia's and Germany's were relatively "short-legged" organizations, designed to operate in close cooperation with their ground forces. Though Douhet's theories had impressed many Frenchmen, including Foch and Pétain, France had failed to maintain an up-to-date air force and had no clear idea of how to use what it did have. Italy had an independent air force, but its planes were poorly designed and its high command resembled a debating society. England's very independent Royal Air Force (RAF), following its own doctrine worked out late in World War I by such officers as Trenchard and John Slessor, intended to bring down Germany by strategic bombing—long-distance raids by heavy bombers that would destroy the enemy's "centers of production, transportation, and communications."[37] There was to be no *intentional* destruction of civilian lives and property, but any that was inflicted in the process of destroying industrial targets could be welcomed as a means of wrecking civilian morale. In the United States, doctrine was a mix of Douhet theories and Trenchard practice, spiked with Mitchell's impatient, aggressive opinions. The old Air Service became the Air Corps in 1926 and gained its own headquarters, directly under the general staff, in 1935. It concentrated on the development of a heavy bomber, the future B-17 Flying Fortress, presented to an isolationist Congress and public as designed primarily for coast defense against imagined invaders. Gradually, it evolved a strategy of precision daylight raids against key enemy industrial systems, which were to be surgically excised, in place of Douhet's blind bludgeoning. The B-17 was to fly high and fast and mount enough machine guns to drive off enemy fighters.

All this was theory. The English and American emphasis on bombers led them to neglect the development of fighter aircraft. The Spitfire fighters that—with the equally new radar—beat back the German air assault on England in 1940 had started coming into the service only the year before. In the Far East, Japanese Zeros made sad havoc of the slow, unwieldy crates in which devoted American fighter pilots had to go up against them. Both air forces also neglected (possibly "avoided" would be more accurate) the development of aircraft and techniques for close cooperation with their respective ground forces. As a final blow, their much-touted bombers proved unsatisfactory, requiring extensive modification or replacement.

As the war began, both sides tiptoed around each other, limiting their bombing to purely military objectives, seemingly out of a mutual reluctance to risk bombings and counterbombings of civilian populations. Eventually, in May 1940, the British began strategic raids against the war industries of the German Ruhr, with very little effect. The following August, when the German Luftwaffe (air force) at last was beating the RAF Fighter Command into a corner, British Prime Minister Winston Churchill made a bold strategic decision: British bombers would strike Berlin. Churchill's intent was to bait Hitler into switching his attacks from the gasping Fighter Command to vengeance raids on London and other major cities; after the sixth raid on Berlin, Hitler hit back, throwing away his chance of victory over the RAF. His light, relatively short-range bombers could not carry enough bombs to do vital damage and were vulnerable to British fighters. Thereafter, life was harsh and uncertain for European civilians.

Churchill's next strategic decision, in February 1942, was impure Douhet: British strategic bombing would be concentrated against the morale of enemy civilians, especially industrial workers. In practice, this simply meant the destruction of German cities; if any armament factories were hit in the process, that would be a bonus. Churchill's agent in this destruction was the RAF Bomber Command's new chief, Air Chief Marshal Arthur Harris, "Bomber Harris" to the press, "Butch" (for "Butcher") to his men. A gruff, aloof martinet (he would not allow the wives of his bomber crews to live within fifty miles of their husbands' bases but kept his own wife with him), he was also a smarmy personality-boy salesman with politicians and newsmen. As a strategist, he had perfect tunnel vision. He intended to destroy German cities, and he went about that mission with gusto. "Victory," he pontificated, "speedy and complete, awaits the side that employs air power as it should be employed"—meaning as Harris did it.[38]

Harris's "area bombing" heaped destruction on the center of German cities but usually missed the industrial districts on their outskirts. Repeatedly, the Luftwaffe's planes and antiaircraft guns inflicted heavier casualties than even Harris would accept.

The Americans disappointed Harris by sticking to their doctrine of daylight precision-bombing raids against specific targets such as German aircraft factories and fuel-oil installations. They had a number of notable early successes until October 1943, when the Luftwaffe snubbed their offensive up short. It did not resume until satisfactory long-range fighter escorts could be provided. Even then, European weather often made precision bombing impossible. Consequently, much American bombing was done "blind" through the clouds by primitive radar sights, little different from Harris's primitive methods.

It was not until September 1944 that the Anglo-American command at last set definite target priorities. Most important was the German oil industry, an objective previously stressed by Lt. Gen. Carl Spaatz, commander of the American air forces in Europe; the next was Germany's communication system of railroads, canals and rivers, and highways. Harris did not approve and cooperated only partially. But Allied air strength had now increased amazingly. Swamped by thousands of Allied planes of all types, the Luftwaffe went down fighting. Massive raids raked Germany from end to end as the Anglo-American and Russian armies closed in from west and east, overrunning oil-production facilities the air offensive had not knocked out. By early 1945 there was a shortage of worthwhile targets, and English and American air forces turned more and more to people-killing strikes, ostensibly to break German morale. Thus it was that an undefended city such as Dresden, crowded with refugees and of little military importance, could be considered a fitting target for repeated air raids. (Dresden, be it noted, suddenly reminded Churchill of his religion. He shocked his RAF command by suddenly opposing further "bombing German cities simply for the sake of increasing the terror, though under other pretexts.")[39]

The war in Europe finished at last, the Anglo-American "bomber barons" were ready to accept full credit for winning it, though their initial claim that they had made a "decisive" contribution was relatively modest. Later they would assert that they could have done it all themselves, and that there would have been no need for the 1944 Anglo-American landings in Normandy had they only been given all the bombers they required *and* complete freedom to employ them exactly as they saw fit. They would also make similar claims concerning Japan.

The knowable facts, however, demonstrate that they had not made good the air-power prophets' predictions. Air power had not brought a quick victory. Civilian morale in England and Germany had only grown tougher as the bombs rained down. Under the lashing of Anglo-American air raids, German production of tanks, vehicles, planes, and weapons actually increased steadily into July 1944 and remained high into November. Postwar surveys showed that it had required three tons of bombs to kill one German of whatever sex or age. (A B-17 carried between two and nine tons of bombs, depending on the distance of its target.) The maintenance and supply of enormous bomber forces was a major problem, as was their manpower requirements. By late 1944 the U.S. Army Air Forces in Europe had approximately 3,200 heavy bombers, with two crews for each bomber to prevent undue fatigue; the U.S. Army had its last divisions in line and a crucial shortage of combat infantrymen.

Air power might have been more effective had it been more intelligently employed. Spaatz's campaign against the German fuel-oil industry was highly effective. Hitler's efficient Minister of Armaments Albert Speer remembered the first major raid on May 8, 1944 as the "day the technological war was decided."[40] However, this campaign proceeded by fits and starts (during April–September 1944 many of the Anglo-American strategic bombers were diverted into preparations for and support of the Normandy invasion) and really did not get rolling steadily until late in the year. Then the weather turned sour. But a shortage of fuel was one of the major reasons for the failure of Hitler's surprise December offensive of the Ardennes. Had such an operation been pushed earlier, and had Harris been forced to commit his entire Bomber Command to doing something more useful than redistributing the rubble in already-ruined cities, the war might have ended sooner. This speculation, however, is only another historical "What if?"—impossible to answer with any certainty. For one thing, Hitler's response to such an air offensive would have been unpredictable. He had new weapons—the Me 262 jet fighter, utterly superior to any Anglo-American aircraft and an almost-perfected ground-to-air guided missile—and he might have been moved to employ them properly. (As it was, his maniacal whim was to decree that the Me 262 was to be used as a bomber and to sideline the antiaircraft missile in favor of the strategically useless V-2 supersonic rocket.)

The strategic employment of air power was complicated by a human factor: as custodians of a new—and, they were certain, all-decisive arm—many English and American air force officers developed a viewpoint much the same as that of any jingling, clanking French cavalry officer of Turenne's day. They were far above the grubby surface battles; army and navy officers obviously lacked their vision and arcane knowledge of modern war. This was especially true of the strategic bomber barons. They resented any need to support the land forces that pulled them from their own war. One sour example was the First Canadian Army's attack on Walcheren Island in November 1944 with the vital mission of clearing the sea approaches to Antwerp harbor. Requesting heavy bomber support to take out some strong German coastal defenses, the Canadians met a firm refusal: The RAF felt that such action "was not essential . . . when its only purpose is to save casualties [and] must eventually lead to the demoralization of the Army" if the "poor, bloody infantry" were so humored.[41] This problem would continue into the postwar years, affecting both strategic planning and actual operations, and undoubtedly still persists.

In the matter of prophets and pontificators after World War I, England was doubly blessed. Besides Trenchard, Slessor, and other air-power dervishes, it possessed a band of tank enthusiasts who were equally convinced of their own virtue, invincibility, and indispensability and quite vociferous about it. This did not preclude their belaboring one another over points of doctrine, but they did agree generally on the urgent need to rebuild the British Army completely, with the tank as its major weapon. They were, almost without exception, men of notable intelligence and imagination, tenacious and pugnacious, keen soldiers, if often unorthodox.

Their doyen was Lt. Col. Ernest D. Swinton of the Royal Engineers, military historian and student of the art of war. (These two qualities are by no means synonymous.) Service in the Boer War (1899–1902) and work on the British official history of the Russo-Japanese War (1904–1905) had brought him to consider the need for a bulletproof, self-propelled fighting vehicle capable of cross-country maneuver. In late 1914, seeing how the western front had locked into two opposing trench systems that stretched from the English Channel to Switzerland, he proposed the construction of "Armored Machine Gun Destroyers" that would assist infantry attacks by crushing enemy barbed-wire entanglements and knocking out machine guns. Swinton had the technical knowledge to design such a vehicle and the professional skill to plan its use in battle; his gentle humor made him a persuasive advocate. Winston Churchill, the First Lord of the Admiralty and always a patron of military cranks and novelties, gave him vigorous support. Since Swinton saw his "destroyer" simply as a supporting weapon for the infantry and not as a new, independent branch of the army, even old-school generals were willing to give it a try. And so, from late 1916 on, tanks became the spearhead of British offensives. ("Tank" was originally a code name, adopted to conceal the purpose of these odd, cumbersome vehicles; soldiers liked it and kept on using it.) Swinton, however, never commanded tanks in action. Promotion took him to staff assignments back in England. By World War II he was too old for active service.

After Swinton came more demanding characters. The staff of the newly organized ''Heavy Branch of the Machine Gun Corps'' (soon rechristened ''Tank Corps'') included two formidable officers, Giffard le Quense Martel (1889–1958) and John F.C. Fuller. Martel was from an old army family and a combat engineer of headlong bravery. Fellow officers complained that ''his idea of pleasure was to get into a shelled area and dodge about to avoid the bursts,'' depending on his instinct for where the next shell would land. His ''deep, hoarse laugh . . . had a most peculiar note of good-humored ferocity.''[42] Small and loose knit and privately a gentle person, he was a famous boxer, repeatedly welterweight boxing champion of the British Armed Forces. Friends called him ''Slosher.''[43] For relaxation he worked on a lathe. In the summer of 1916 he had been sent back from France to lay out a replica of a section of the battlefront that would be used to train the first tank crews—a project carried on with almost as much secrecy as the construction of the first atom bomb. Martel was quickly indoctrinated; that same year he wrote a paper titled ''The Tank Army,'' which described future armies consisting entirely of tanks and other combat vehicles operating out of fortified base areas and waging an overland version of naval warfare. This ''All Tank Army'' was a radical strategic concept; even Fuller at first had trouble accepting it.

John F.C. Fuller (1878–1966) was a clergyman's son. His beloved mother was of French blood but German raised. He went into the Royal Military Academy at Sandhurst apparently because his mother's father wished it. Small (at nineteen he stood five feet four), he developed into what American soldiers would call a ''runt with a runt's complex''—impatient, quick to take offense, always with a ready pin for any available balloon. Usually short of money (a junior officer could barely exist on his pay, and Fuller's father could give him only a small allowance), he was debarred from much army social life. Deep reading on assorted subjects and solitary tramps and hunting were his pleasures. Also, from boyhood his prying mind had worked into matters occult, macabre, mystic, and demonic. Strange cults fascinated him. And somehow, somewhere, he became an expert poker player, occasionally thereby supplementing his pay.

Fuller made, as he dubbed himself, an unconventional soldier. Hating regimental routine, he sought odd jobs outside it, chasing Boer guerrillas with a detachment of Kaffir scouts or serving with reserve units. In 1906 he married an attractive German-Polish girl, Margarethe Karnatz, whom he called Sonia, possessed of a terrible skill at insult. Their childless marriage was a tight partnership, but Sonia was never the traditional army wife. In 1913 Fuller won admission to the Staff College; there his less imaginative instructors found him a bother. For one thing, he had read through Napoleon's *Correspondence* in search of the emperor's guiding principles of war (see appendix), which he then applied to his study of military history. That, with his small size and imperial attitude, got him his Army nickname of "Boney."[44] He had already begun writing—mostly excellent pamphlets on training, tactics, and railroad movements.

World War I brought Fuller a series of staff jobs in England. (Canadians' ability to get drunk and disorderly between their transports and the nearest railroad station inspired him to language that would have shocked the legendary Colonel Blimp.) He finally got to France by the simple, if risky, process of telling off his commanding general. Eventually he came to Heavy Branch headquarters, a major, prematurely balding, his lean face thrusting forward behind winged eyebrows and a jutting nose. He had swift, sardonic wit, keen intelligence, and incredible energy, plus unusual side interests that surfaced unexpectedly. Assisted by Martel, he became the Tank Corps' planning and training officer, the brain behind its victories at Cambrai (1917) and Amiens (1918). He thereafter insisted that Amiens was World War I's decisive victory, but though Ludendorff called it a "Black Day," the German lines closed up again, largely because accident or German fire took out most of the tanks. Thereafter, Fuller planned for a great 1919 offensive employing thousands of tanks, some of which would strike directly through the German lines and—with air support—break up the German command and communications systems.

At the war's end in November 1918, England had the world's finest tank force and the keenest tank officers. At once, however, it was uncertain whether the Tank Corps would survive in the small peacetime army a war-impoverished Great Britain could support. Fuller, who customarily relied on challenge and exaggeration to make his points, intensified the dispute by proposing that tanks replace the cavalry and infantry and most of the engineers.

Unfortunately, there was no easy solution to England's strategic problem. First and foremost, there was too little money. As a verse of unremembered origin explained:

> Tanks is tanks, and tanks is dear.
> There shall be no tanks this year.

Next came the unbudging fact that the British Empire covered much of the world. Some of its subject peoples and their wilder neighbors tended to be restless. Consequently, there was always a frontier squabble somewhere, usually in backcountry areas where tanks were of little use. Cavalrymen who had recently ridden over enemy machine guns in Palestine refused to believe themselves obsolete. Finally, the tank advocates disagreed over whether they should have small, cheap tanks to work with the infantry or big, expensive machines for armored units.

This cross-feuding went on for years, enlivened by books such as Martel's *In the Wake of the Tank*[45] and Fuller's neck-or-nothing *Tanks in the Great War* and *The Foundations of the Science of War*.[46] Fuller's picture of a double-Douhet future war, given in the passage that follows, upset H. G. Wells.[47]

> Fast-moving tanks equipped with tons of liquid gas . . . will cross the frontier and obliterate every living thing in fields and farms . . . fleets of aeroplanes will attack the great industrial and governing centers. . . . All these attacks will be made against the civil population to compell it to accept the will of the attacker.

Slowly, with the help of a popular columnist named Liddell Hart (to whom they "leaked" confidential information in return for favorable publicity), Fuller, Martel, and a clique of energetic younger officers saved the Tank Corps and pushed the development of a "mechanicalized" army. This slow progress was not enough for Fuller. His multitudinous books and articles grew shriller and sometimes impractical; though he had not commanded troops since 1913 and had seen no real combat since his Boer War skirmishing, he too often treated veteran fellow officers with a grating intellectual arrogance. Offered command of an experimental mechanized force in 1927, he stirred up a teapot tempest over some organizational trifles, threatened to resign if not humored, and so cut his own professional throat.

He remained in the army until 1933, was promoted to major general, but never had another important assignment.

Once retired, Fuller turned to writing, largely as an outlet for the frustration and prophecy seething within him but also to supplement his retirement pay. He remained an international authority on armored warfare, though his books on that subject—for example, his *Lectures on Field Service Regulations, II* and *III*[48]—favored an all-tank army and were in no sense a forecast of the Blitzkrieg. Increasingly, he wrote history and biography. His major historical work was his repeatedly revised *Decisive Battles,* culminating in *A Military History of the Western World.*[49] *The Generalship of Ulysses S. Grant* and *Grant and Lee*[50] must have contributed to his reputation for contrariness: In the teeth of accepted British opinion, he pronounced Grant by far the greater general! Two of his last works were *The Generalship of Alexander the Great* (whom he lauded) and *Julius Caesar: Man, Soldier, and Tyrant* (of whom he disapproved).[51]

Fuller's histories and biographies are excellent reading, spiked with his wit and personal views on practically everything. They are not models of research and accuracy: Fuller worked too rapidly and had too many irons in his intellectual fire. Intent on justifying his own ideas, he often scamped the laborious sifting of associated facts—for example, his version of the 1777 Saratoga campaign in his *Military History* is a mishmash of old fables and utter nonsense. As he aged, his apocalyptic theories on future war imperceptibly reverted to something characteristic of a British officer and gentleman of the beginning of the century. The bombing of Dresden struck him as "Juggernaut gone mad."[52]

Yet, at the same time, Fuller moved farther into his personal ideological wilderness. Always scornful of democracy, he became fascinated by fascism and Mussolini, whom he saw as a Napoleonic figure rather than a tragicomic little Italian politician in an overornate uniform. He praised General Franco's pacification of Spain, was an honored guest at Hitler's fiftieth birthday parade, expressed strong anti-Semitic opinions, and became an associate of Sir Oswald Mosely, leader of a British Fascist party. Naturally, he was an object of official suspicion during World War II, but he died forgiven and honored as the prophet of armored warfare—an intellectual tail twister to generations of soldiers.

Fuller's place as a professional irritant within the British Army was largely filled by Percy P. S. Hobart. Born in India, commissioned into the Royal Engineers, a pig-sticking, tiger-shooting, polo-playing pukka sahib of the imperial "Mad dogs and Englishmen" breed, Hobart was a frontier soldier of unlimited energy and independence, both a roistering bachelor and a man of culture. His creed was "The secret of success in the Army is to be sufficiently insubordinate, and the key word is sufficiently."[53] During World War I he once not only refused to make a last-minute change to orders already issued but seized possession of the only available field telephone so that no one else could. In the Tank Corps he was "Old Hobo," an unsparing, skilled trainer of armored units, noted for speaking his mind to anyone. As an instructor at the Indian Army Staff College, he persuaded the wife of one of his students to divorce her husband and marry him—conduct much frowned on in most civilized armies even before the incident of David, Bathsheba, and Uriah. It definitely put him on the blacklist of British army wives. His readiness to make enemies ended in his retirement in 1940. Briefly a home-guard corporal, he was recalled by Churchill.

Hobart wanted a more or less independent armored army that would require several years to organize. Since time and considerable good will were lacking, he ended as commander of the oversized new 79th Armored Division, which was to consist of specialized armored vehicles—amphibious tanks, flame-throwing tanks, minesweeping tanks, tanks designed to lay trackways over patches of slippery clay, tanks carrying bridges, demolition tanks—for use against German fortifications.[54] "Hobo's" mission was to push the development of these "funnies" and train his troops in their employment. They formed the jagged point of the Allied strategic assault on Hitler's Fortress Europe, from the Normandy beaches to the Elbe River, saving thousands of infantrymen's and tanker's lives. (Oddly and unfortunately, neither Dwight D. Eisenhower nor Omar Bradley, the senior American commanders, were much interested in Hobart's odd machines. American soldiers would pay for that lack of professional curiosity.) Hobart was luckier than Martel, who, after a game attempt to stop the German panzer offensive in 1940 by sending his undergunned infantry tanks in against German "88s," got into Churchill's black book and was exiled to Russia as a military attaché, to return an incensed anti-Communist.

Probably the most publicized of these prophets/pontificators was

Basil H. Hart (1895–1970). He is, of course, not popularly known by that name: In 1921 he changed it to the more striking B. H. Liddell Hart (''Liddell'' was his mother's family name)—and woe betide any insensitive lout who later indexed his name as ''Hart, B.H.L.''!

Another minister's son, Hart was interested in military matters from boyhood. In 1914, despite uncertain health, he secured a temporary commission in The King's Own Yorkshire Light Infantry Regiment. His service in France was brief but convincing; in 1914, after a short front-line tour, a nearby shell burst gave him a concussion that sent him back to England; in 1916 he was gassed during the opening phases of the Somme offensive. He thereafter served as an instructor and specialist in infantry tactics until 1927, when he was retired as a captain on account of ill health caused by wounds.

With his own living to make, he turned to writing on military affairs. A very tall, imposing man of pleasing address and utter self-assurance, very skillful in bringing himself and his writings to the attention of influential people, he managed very well as a journalist and author. Around 1922 he came under the influence of Fuller with whom he maintained a mutually beneficial association, broken by occasional spats, as when Hart took it on himself to correct Fuller's version of history. His influence grew; in 1935 he became a military correspondent to the *London Times* and in 1937–38 unofficial handyman to a new-broom Secretary of State for War, Leslie Hore-Belisha. Then suddenly things soured. He clashed with the *Times* editors. Hore-Belisha was vigorous and modern minded but could not get along even with generals who supported his reforms; Hart found himself increasingly mistrusted by formerly friendly officers. Worse, with war at hand, he began preaching that England should adopt a strictly defensive strategy, relying on air and sea power and economic pressure; RAF units might be sent to support France, but no ground troops should be committed lest they be squandered in Somme-style mass attacks. Any large-scale offensive by either side was doomed to fail because modern weapons gave the defensive forces a decisive advantage. And then the Blitzkrieg swept across Poland.

Hart's prestige shriveled. His wife had just divorced him; overwork brought on a heart attack. Like Fuller, he spent 1940–45 in limbo, opposing unrestricted bombing, urging a compromise peace without victory, and warning of future Russian expansion.

Also, like Fuller, after 1945, Hart came back into prominence, advocating a doctrine of limited war and opposing the American theory of massive nuclear retaliation. Gradually, he restored his public image as a military sage and was knighted in 1966 and decked with academic honors.

It seems indisputable that Hart was always ridden by his memories of the Somme and determined to do what he might to prevent another war of that kind. In his personal life he was a sensitive, kindly person who fought for better treatment of German prisoners of war and aided and encouraged young students. He corresponded endlessly with friends, was somewhat foppish in his dress, and played a wicked game of croquet. Almost entirely dependent on his writing for a living, he neglected no gimmick that would enhance his repute. He was advertised as "the captain who teaches generals"[55] and made a profitable business of camouflaging ancient military truisms under striking names and presenting them as if they were new discoveries. One was his "Expanding Torrent," which merely meant that an attacker found a soft spot in the enemy's defenses, broke through, and fanned out to exploit his success—something Hart borrowed from the German infantry tactics of 1917–18. His pet strategic thesis was the "Indirect Approach," a catchall phrase elastic enough to contain any historical example Hart wanted to exploit. Its basic sense was that an aggressor should avoid frontal attacks, seek surprise, utilize enveloping maneuvers, and strike at the enemy's command system rather than the mass of his troops—all of it already ancient when Egypt's pharaohs loosed their chariots against the Hittites.

Though a champion of a mechanized army, Hart did not favor Fuller's all-tank version, believing—with Hobart and others—that some infantry would still be necessary. Like Fuller, he went through a phase of believing that unrestricted aerial bombardment with poison gas would shorten wars, but he later wrote approvingly of the limited warfare of the eighteenth century, when opposing armies supposedly fought genteel wars, into which no passion entered, for minor territorial gains (see chapters IV and V). He opposed conscription and scolded Napoleon—whom he labeled "the Corsican vampire"—for having introduced it.[56] (Napoleon didn't.) Clausewitz was another favorite bugaboo, routinely belabored as responsible for the mass slaughters of World War I in terms that merely make it obvious that Hart never studied *On War*.

Hart was an effective, highly readable author, an apt coiner of phrases and catchwords. His colorful, dogmatic style implied vast erudition and unquestionable authority; his theories ostensibly were based on deep study of old wars and their commanders and were lavishly supported by historical examples. In fact, however, Hart scorned "the pedantic burrower in documents," preferring to rely on "creative imagination" to determine the true lessons of military history.[57] He normally worked from secondary sources and—for modern examples—a large-scale correspondence with military men, not all of whom appreciated his uses of it. Considering war an affair of competing generals, Hart gave little thought to social, technical, and political factors that might influence their strategy. (The facts that he never completed college and normally wrote in a tearing hurry may help explain his frequent lack of background information.) He wrote on armored warfare without knowledge of tanks except what Martel and other officers leaked to him, and his history of the Royal Tank Regiment[58] is flawed by bias and deliberate omissions. Possibly his most useful books are *The Rommel Papers,* the English version of which he edited, and *The Other Side of the Hill,* a collection of interviews with captured German generals.[59]

Taken all together, his writings are mostly sermons, urging all men (including, one occasionally feels, long-dead generals) to harken to the true doctrine of Liddell Hart and thereby be saved.

Like Jomini, Hart went armored in almost-impenetrable intellectual vanity. He considered himself a genius, fully capable of leading armies or swaying affairs of state. And he blandly credited himself with having been the true creator of the Blitzkrieg even though in 1939 he had considered it impossible. Through his acquaintance during and after the war with captured German officers, he managed to collect testimonials as to the vast effect his writings allegedly had had on their strategy and tactics. Typically, one of his major vaunts—"General Heinz Guderian, the creator and leader of the panzer forces, has stated that he derived this new technique from my writings"[60]—trips over the uncomfortable fact that this proclaimed tribute "appears only in the English editions of *Panzer Leader,* for which Liddell Hart wrote the foreword, and not in the original German [edition]"[61] In fact, Guderian's first published work *Achtung! Panzer* (1937) lists books by Fuller, Martel, and De Gaulle in its bibliography but nothing by Hart, who has only a passing reference in its text. The truly surprising thing about this rampant egoism is that many people were eager to buy Hart at his own price.

Liddell Hart's long-term importance in the development of strategy probably will be slight. He *was* an effective propagandist and educator in the value of a mechanized army, though he presented himself as St. George battling entrenched stupidity rather than as a gadfly newspaper columnist attacking harassed officers who were trying to stretch a tiny budget to meet major strategic and organizational problems. It is doubtful that he originated any really new ideas on tactics or strategy; for all his talk of ''true facts'' and ''purely scientific spirit,'' he was neither an objective nor accurate historian. His pontificating did move other men to examine their own reasonings and beliefs, and this probably was his greatest service.

Meanwhile, in Germany, the Great General Staff still was at work, even if incognito. Von Seeckt (see page 182) wanted a highly mobile army. Among officers put to studying this problem was Capt. Heinz Guderian (1888–1954), known to fellow officers both as ''Hot-Head'' and ''Heinz the Hustler,'' a Prussian of Prussians, trained as a jaeger (light infantry) officer under the strict hand and eye of his colonel father. He became a specialist in the new radio-signal equipment, learned English and French, and made an early reputation as a dedicated officer. He also married young, and very wisely; Margarete Goerne—Gretel to her Heinz—was a dark-haired beauty, cool, sensible, a perfect soldier's wife who would both encourage and restrain her soldier-husband through his more furious moods. Their sons followed him into the army and to the wars.

After his misadventures in 1914 (see page 185), Guderian served as an intelligence officer. On occasion he flew as an observer in reconnaissance aircraft. In February 1918, thoroughly tested in a series of staff and line assignments, he became a General Staff officer. Serving with a corps on the Western Front, he experienced the effects of massed Allied tank attacks. From that he was transferred to Italy, just in time to see the Austrian armies crumble, then came home to a Germany where everything was falling apart as the war ended. He was ordered up into the Baltic States as staff officer to the grim ''Iron Division'' of ''Free Corps'' volunteers who held the eastern marches against Bolsheviks and Poles. It was a wild time: Guderian's sympathy was more with his Free Corps comrades than with Germany's new republican government which, under Allied pressure, eventually disowned them. But his seniors realized his potential; he was recalled and sent back to troop duty to cool down. Von Seeckt's little 100,000-man force, top-heavy with officers and noncommissioned officers, was a military laboratory as well as a cadre for a future German Army. Guderian passed through a series of educational experiences—service with motorized troops, instructor in tactics and military history, logistical staff duties.

There was considerable interest in tank warfare among German officers, especially those who had been on the receiving end of tank attacks. Guderian became a sparkplug among them. In visits to Swedish tank units, by training and experimentation in secret bases in Russia,[62] through study of all available foreign material on tanks, the German panzer organization and doctrine took shape. By 1929 it was solidly established, though much infighting remained before this theory was converted into actual combat troops in 1934. Hitler—like Churchill a collector of military novelties—favored the panzers but gave them no special support until they proved themselves in the 1939 Polish campaign.

The panzer division created by Guderian (many other able men were involved but have received little public recognition) was a balanced force of all arms—tanks, infantry, artillery, engineers, reconnaissance troops, antitank and service units—equipped and trained to operate together as a smooth-running team. The idea was not exactly original: Hobart had been trying to form something of the sort in England; the French and Russians had experimented with similar organizations; in the United States Gen. Adna Chaffee (1884–1944) was working himself slowly to death with ramshackle equipment and little encouragement to create a similar force. But the Germans built largely after their own ideas, whatever they might have learned from Fuller and others. Moreover, remembering the World War I breakdown of signal communications, Guderian equipped his panzers with highly effective radios.

Guderian commanded a panzer corps against Poland in 1939 and a bigger corps in the 1940 drive through the Ardennes and across northern France to the English Channel. In the 1941 invasion of Russia his group was increased to three panzer corps plus attached infantry divisions. He shared in the victories at Minsk and Smolensk that trapped hundreds of thousands of Russians; turned southward by Hitler (see page 324), he had a major part in the great victory at Kiev;

sent northward again with his weary soldiers and worn-out vehicles, he joined in the too-long-delayed drive on Moscow with initial success. When the winter, for which Hitler had made no preparations, came down upon the overextended Germans, Guderian defied Hitler's stand-fast order to pull his troops gradually back into better positions. Hitler put him on the retired list but gave him a farm in East Prussia. Guderian cheerfully tried his luck there until recalled to active duty in 1943 as Inspector General of Armored troops to clear up the appalling mess Hitler had made of tank development and production. Working closely with Albert Speer (see page 267), he accomplished marvels but failed in his attempts to bring a little common sense into Hitler's command system. Apparently he knew of the July 20, 1944 assassination plot against Hitler, refused to join in it, but did not warn Hitler or betray the plotters. Subsequently, Hitler appointed him Army chief of staff, in addition to other duties. This made him responsible for the defense of the Eastern Front, subject to Hitler's whims. These he resisted or evaded at considerable personal hazard in the forespent hope of saving Germany from Russian invasion and occupation. In late March, with the Anglo-Americans across the Rhine and the Russians on the Oder River forty miles from Berlin, Hitler ordered him to take six weeks' sick leave. On May 10 the Americans captured him. He was not released until 1948; the Poles had declared him a war criminal, and UN zealots ached to try all former members of the General Staff and the German Armed Forces High Command. The latter part of Guderian's imprisonment was spent discussing armored warfare and staff organization with American interrogators. He also learned to play bridge and worked in the camp vegetable garden. Released, he gardened and wrote: his *Panzer Leader,* translated into ten languages, was a best-seller.[63]

While never a strategist in the basic sense of that term—he did not plan any of the German Army's major offensives—Guderian was one in practice during the 1941 invasion of Russia. Commanding what was actually a panzer army of some 300,000 men, he moved off into the Russian spaces against vastly superior numbers of men and tanks, shifting between Kiev and Moscow, Russians all about him—and always won, even when he might have to personally lead a field bakery company or other scratch force to plug a gap.

One of Guderian's friends described him as a "coiled spring,"[64] and driving energy was one of his strengths. He was always well up at the head of his command, available at any crisis, day or night. ''*Nicht kleckern, sondern klotzen*'' was his motto, translatable as ''Not a drizzle, but a downpour'' or ''Don't tickle them—slug them!''[65] He maneuvered swiftly and skillfully, always was superior in fighting power at every decisive point. Most of all, he was a gifted trainer and leader of men, able to get the last quarter ounce of energy out of his soldiers—and then a little more. He was quick to praise and encourage, ready with a smile and a soldier's jest, careful to explain the ''why'' of his orders, tireless in looking after his men. Hitler rebuked him, in terms Sun Tzu would have used (see page 225), for sympathizing too much with them and being influenced by their hardships.

Guderian was also a difficult man to command, especially for superiors who did not understand armored warfare. On occasion he treasured a grudge. Since he led from the front, he needed a steady, like-minded chief of staff at his rear headquarters. He was not politically gifted and trusted Hitler too much and too long. More conservative generals found the wheeling and dealing he employed in creating and building up the panzer force unworthy of a General Staff officer.

But Guderian was the greatest of World War II's armored generals, the only one to conceive, create, organize, and train an armored force—and then lead it successfully in battle. Add that he was a German patriot and a man of honor who would refuse to publish Hitler's 1941 orders offering pardon in advance to soldiers who might mistreat Russian civilians. ''Be the day dark, be the sun bright, I am a Prussian and a Prussian I will be.''[66]

IX

SUN TZU SAID

Thanks to the power of the Everlasting Heaven, all lands have been given to us from sunrise to sunset. How could anyone act otherwise than according to the Command of Heaven? . . . If you fail to act in accordance with it, how can we foresee what will happen to you? Heaven alone knows.

KHAN GUYUK OF THE MONGOLS TO
POPE INNOCENT IV, 1247.[1]

If you know the enemy and know yourself, you need not fear the result of a hundred battles.

SUN TZU WU[2]

It was early in the Time of Warring Kingdoms (ca. 500–221 B.C.) and Ho Lu, King of Wu, was sorely beset by bloody-minded neighboring states. He summoned to him, from out of the militant northeastern Kingdom of Ch'i, a general named Sun Tzu Wu, author of a famous book on the art of war, and requested a practical example of his system of drill and training. For a demonstration unit, he gave Sun two companies of his palace women, each commanded by a favorite royal concubine. Probably swallowing a few adjectives much used by Ch'i first sergeants, Sun Tzu briefed the women on what they were to do. They assured him that they understood, but when he gave his first command, they broke into a fit of giggling instead of obeying it. Sun Tzu went through his explanation again, to make certain his instructions were clear and comprehensible to them and again got no response to his word of command except more girlish giggles. He thereupon ordered the captain/concubines executed. The king, watching from a high pavilion, at once sent a courtier scurrying with an order to end the demonstration and spare his favorite bed warmers.

Sun Tzu sent back a grim answer: "Having once received His Majesty's commission as general of His forces, there are certain commands of His Majesty which, acting in that capacity, I am unable to accept."[3] The execution proceeded. Two pretty heads rolled, Sun appointed two new captains, and the demonstration went off without hitch or bobble. Finally, comprehending what sort of soldier Sun Tzu was, Ho Lu made him his commander in chief. Sun put the fear of Wu into its neighbors and Wu, Ho Lu, and Sun Tzu flourished together.

Much of this undoubtedly is old legend. Even Sun Tzu Wu's actual existence has been questioned. But the slim book, *The Art of War,* attributed to him, certainly exists. Despite its early date (approximately 500 B.C.) and the fact that its original text may have been somewhat altered in the centuries since it was written, it remains the most respected of all known Chinese books on war. Sun Tzu's wars were waged by armies of "a thousand swift chariots, as many heavy chariots, and a hundred thousand mail-clad soldiers (mostly infantrymen)."[4] (At this same time in the West, Persia was invading Greece with an army of comparable size, including effective cavalry which seems not to have appeared in China until around 307 B.C.) Parts of it are still perplexing to Western students, Sun having put his thoughts into lean prose and aphorisms. His coverage of mountain warfare, for example, is "Pass quickly over mountains [which are barren of fodder] and keep in the neighborhood of valleys. Camp in high places. Do not climb heights in order to fight." So much for mountain warfare.

However, his book does cover the whole art of war. To Sun Tzu, war was "a matter of life and death, a road either to safety or to ruin." Before risking it, wise generals and kings must compare their strengths to those of the hostile kingdom. Which ruler had the "moral law" (the complete support of his subjects) on his side?

Which nation had the better generals? How did their armies compare as to size, training, discipline, and morale? How would the prevailing weather and the terrain aid either army? War was expensive and placed a great burden on the civilian population. "Thus, though we have heard of stupid haste in war, cleverness has never been associated with long delays. There is no instance of a country having benefited from prolonged warfare. . . . In war, then, let your great object be [speedy] victory, not lengthy campaigns." Sun's major thesis, to which he returns repeatedly throughout his book, is:

> All warfare is based on deception. Hence, when able to attack we must seem unable; when using our forces, we must seem inactive; when we are near, we must make the enemy believe that we are far away; when far away, we must make him believe we are near. Hold out baits to entice the enemy. Feign disorder, and crush him. If he is secure at all points, be prepared for him. If he is superior in strength, evade him. . . . Pretend to be weak, that he may grow arrogant. If he is inactive, give him no rest. If his forces are united, separate them. Attack him where he is unprepared, where you are not expected.[5]

A wise general invaded enemy territory and seized his food and supplies there.

The great triumph in war, Sun Tzu said, was to "take the enemy's country whole and intact; to shatter and destroy it is not so profitable. So, too, it is better to capture an army entire than to destroy it . . . supreme excellence consists in breaking the enemy's resistance without fighting."[6] Like Turenne, Sun Tzu did not approve of long sieges. He also did not approve of rulers meddling in military affairs through ignorance, politics as usual, or favoritism. His strategy and tactics alike were based on securing the initiative at the beginning of hostilities and keeping the enemy constantly off balance and bewildered through swift maneuver, stratagems, and secrecy. "Appear at points which the enemy must hasten to defend; march swiftly to places where you are not expected." As a result:

> We can form a single united body, while the enemy must split up into fractions . . . which means that we

> shall be many in collected mass to the enemy's separate few, amongst his separated parts. And if we are thus able to attack an inferior force with a superior one, our opponents will be in dire straits.[7]

His campaigns were to be like running water: "avoid what is strong to strike what is weak. Water shapes its course according to the ground over which it flows; the soldier works out his victory in relation to the foe whom he is facing." But no campaign could be successful unless you were prepared to wage it: "The art of war teaches us to rely not on the likelihood of the enemy not coming, but on our own readiness to receive him; not on the chance of his not attacking, but rather on the fact that we have made our position unassailable."[8]

Sun Tzu obviously was a stickler for discipline; he held that soldiers must be treated "with humanity, but kept under control by iron discipline." He would reward courage and good conduct, and look on his men as his "own beloved sons," but he had no intention of allowing them to become "spoiled children." In fact, he listed "over-solicitude for his men" (along with recklessness, cowardice, a hasty temper, and a "delicacy of honor that is sensitive to shame") as one of the most dangerous faults to be found in a general because it exposed him to "worry and trouble." He would not overtax his men until a time of crisis, but except for one mention of selecting healthy campsites, he left nothing on the problem of sick and wounded soldiers. The morale problem of his day seems to have differed from that in Greek and Persian armies. "On the day they are ordered out to battle," he wrote, "your soldiers may weep, those sitting up bedewing their garments, and those lying down letting the tears run down their cheeks." (Probably many of his infantry were impressed peasants who wanted only to be back on the farm.) His solution was to "plunge your army in deadly peril," carrying them into a situation where it was fight or die anyway. And he would not allow any taking of omens or fortune-telling in his camps, since this might endanger the soldiers' morale.[9]

Knowledge of the terrain and the use of spies were two important factors in Sun Tzu's strategy. The latter "called 'divine manipulation of the threads' . . . is the sovereign's most precious facility," since it was the only way of getting sure information of the enemy's dispositions. Sun especially valued "converted spies"—enemy spies that had been caught, "turned around" by large bribes and kind treatment, and converted to his service. He also used "doomed spies"—agents given false information about his future movements, then sent into the enemy lines on a mission that was sure to result in their capture. Caught, they would tell the enemy what they knew, which would get them executed in some painful fashion once the enemy learned they had been flimflammed. (It would be interesting to know just how these unsung heroes were selected.) There were also "surviving spies" who were supposed to get home again, "inward spies" who were suborned enemy officials, and "local spies" who were picked up as the army went along. All in all, the organization seems very modern, especially as Sun Tzu used his spies for assassinations as well as intelligence gathering.[10]

The Art of War is a multitude of intellectual riches in a few printed pages. A large part of it is ageless, as applicable today as in the Time of Warring Kingdoms. It was the basis of all subsequent Chinese military doctrine; most of Mao Tse-tung's rules of strategy are nothing more than a simplification of Sun Tzu's teachings. Just when Westerners became aware of Sun Tzu is uncertain. One account has it that Jesuit missionaries brought back a translation to France in the eighteenth century and that Napoleon, that most omnivorous of readers, had seen a copy. This is so far splendidly unsubstantiated—yet Napoleon is credited with the statement "Let China sleep: when she wakes up the world will be sorry," and his comments on whether a general should obey every order he receives from his government have a bite very like Sun Tzu's refusal to spare the unfortunate concubine captains.[11] Possibly the first good English translation appeared in 1910. However, the Chinese armies that Americans and Europeans saw during the nineteenth and early twentieth centuries were seldom impressive, sometimes even comic. This was a period of weakening emperors, intriguing palace women and eunuchs, and a corrupt bureaucracy, the members of which were selected largely for their literary skills and knowledge of Confucian philosophy. Such a system was loftily scornful of military skills, even when expert, loyal soldiers were badly needed. Then came revolution in 1911, Dr. Sun Yat-sen's attempt to form a republic, counterrevolution, and a collapse into anarchy, with the country split up among constantly battling warlords. It seemed obvious that nothing useful could be learned from a study of the military institutions of a nation that disdained its own soldiers or of warlords who were seldom more than large-scale bandits. Some Westerners meanwhile had read Sun Tzu's *Art of War,* but it did not become readily available or widely studied until after the Japanese began, in 1931, their attempted conquest of China. Such interest was intensified by the success of Mao Tse-tung's Communist armies after World War II and still more by the defeats they inflicted on U.S. forces in Korea between late 1950 and early 1951—some of the worst in American history.

China's story has been a series of cycles: small states battle each other; eventually, some strong ruler unifies them into an empire; the empire grows mighty and engages in foreign conquest; the ruling family degenerates; the empire breaks down into quarreling small states or is overrun by foreign enemies who finally are absorbed or driven out. This process has now been under way for approximately 3,750 years, give or take a few centuries; today we are probably watching the consolidation of a new empire after a long time of troubles.

The one foreign enemy the Chinese really dreaded was a succession of northern nomad tribes. Over the centuries these were many different peoples—Huns, Tatars, Turks, Mongols, Manchus, and others less famous but quite as deadly in their time. Such nomads always have haunted the outward edges of established civilizations. Begging from the strong and wary, pilfering small booty from the strong but careless, squabbling among themselves, they bide their destinies. And, in rare moons, a great captain fights his way up to power among them, while the civilizations weaken from their wars or from inward decay.

Such raiders harried China well into the eighteenth century. They were skilled horsemen and archers, hardened by a life of herding, hunting, and raiding, mounted on tough ponies that could go for days on little feed and water. Chinese armies of infantry and chariots seldom caught up with them. When they did, the nomads' normal strategy was to retreat slowly, luring the Chinese out into the wastelands where fatigue, hunger, and thirst would make them easy prey. The Chinese tried many counterstrategies: fomenting wars between different nomad nations, hiring nomad mercenaries, paying tribute in silver and silk. A hostile chieftain might be persuaded to become an ally by loading him with royal honors and giving him an emperor's daughter for a wife. (What with the average emperor's collection of wives, concubines, and palace women there usually was some nubile girl available for that sacrifice.) More aggressive emperors, especially the T'ang Dynasty (A.D. 618–907), would rely on mobile armies with their own elite mounted archers and launch preemptive campaigns through the north to keep the tribes cowed. Like the U.S. cavalry hunting Sioux and Comanche, the Chinese learned that nomads could be surprised in their winter camps. Also, they were at a disadvantage in the early spring when their mares were foaling, and their raids could be hindered by burning off the grass north of the Great Wall for a hundred miles or so in the autumn. But few emperors or generals had such hardihood, and those grew old and foolish. There always were new tribes coming down out of the wilderness, eager for the loot of Chinese towns. And Japanese pirates and Tibetan hillmen would join in the plucking of a weakening empire.

The most common strategic solution was to fortify China's northern frontier with an increasingly elaborate system of walls and fortresses. During the Time of Warring Kingdoms, some of the separate states began to fortify their respective borders. When wolf-hearted King Cheng of the western state of Ch'in brought all of these contending states under his rule (221–210 B.C.) as Ch'in Shih Huang-ti (First Emperor of China), he had most of the interior walls demolished and those along the northern frontiers linked up and improved to form the first section of what would become the famous Great Wall of China. Ch'in Shih was a thorough, larger-than-life tyrant but a man of remarkable intelligence and force. Aided by able advisers, he established a centralized state linked by good roads and canals, launched irrigation works, freed serfs, burned a great number of philosophical and historical books (copies were preserved in the imperial library), and executed several hundred scholars who talked too much. Their successors blackened his name until we cannot be certain just how deplorable a character he really was. Mao Tse-tung and modern Chinese leaders have hailed him as one of them, a "progressive" ruler and unifier. He did build himself a splendid tomb where recent archaeological work has uncovered an entire army—approximately 7,500 life-sized clay figures, spearmen, archers, charioteers, cavalrymen, scouts, and horses—all interred in battle array, ready to serve their emperor in the world to come.

Legend says that the building of the Ch'in Shih portion of the Great Wall cost a million lives. Its extension and improvement went on by fits and starts. The Han Dynasty (202 B.C.–A.D. 220) added a system of beacon fires to warn of enemy raids. (Chinese poets thereafter would lament that those fires never seemed to cease burning as wave after wave of barbarians crashed against northern China.) As noted, the T'angs neglected the wall, preferring offensive action. Other rulers extended the wall westward as the area under Chinese control spread. The Ming Dynasty (1368-1644) brought the wall to its greatest length and complexity, running along the high ground from the Yellow Sea some 2,000 miles westward into the southern edges of the Gobi Desert.[12] The Mings had as many as a million men garrisoning the fortresses that studded its length. For various reasons, however, the Chinese aptitude for military service was dwindling; in 1644 another people—the Manchu—came out of the north through the wall and swiftly conquered China.

The wall turned back hundreds of raids during its existence as a functioning military installation. But it was extremely difficult to keep its garrisons properly supplied, especially in the more isolated western stretches, or even to find troops enough to man it. Emperors attempted to meet this problem by creating colonies of soldier-farmers along its length in the hope they could feed themselves. This, however, frequently proved impossible for lack of arable land in many places. The wall seldom checked the major invasions, but this should not be too surprising, since these normally came only when the empire behind it was in disorder and decay. Often, it was not military force or skill that took invaders through the wall but bribery or the treachery of the commanding officer at a key gateway. Even Genghis Khan had considerable trouble with some of the wall's inner fortifications in 1211.

Genghis Khan (ca. 1167–1227) possessed the first characteristic of a great captain: he could win battles and wage successful campaigns. His victories, the loot they won, the terror they inspired, brought him followers in swelling numbers. He was also a man capable of being taught by experience. Son of a powerful Mongol chieftain, who named him Temujin, he grew up in the bitter nomad world of raiding, clan feuds, treachery, biting cold, scorching heat, and lurking death. When he was twelve, some Tatars poisoned his father during a supposedly friendly banquet. Unwilling to follow a boy, his clan abandoned their dead chief's family. For years, Temujin was pursued by his father's enemies, reduced to fishing and hunting small game to feed his mother and her younger children. Gradually, by sheer courage, cunning, and toughness, he hacked his way up to power, taking the new name Genghis Khan. In 1202 he had his vengeance on the Tatars: the battle over, every captive male Tatar who stood higher than the hub of a cartwheel was slaughtered. The Tatar women and children were issued out through the Mongol army. Three years later the last surviving Tatars were brought to bay. This time, Genghis Khan ordered total extermination; except for a few small children Mongol women saved, his sentence was carried out. "The merit of an action," said Genghis Khan, "is in finishing it to the end."[13]

In 1206, at some forty years of age, Genghis Khan was hailed as the Khagan (Great Khan) of the Mongol and Turkic nomad tribes, "all those who live in felt tents."[14] We have certain contemporary descriptions of him; unlike the typical squat, flat-faced, black-haired Mongol, he was tall with "cat's eyes" (probably green or blue) and reddish hair. (Whether that difference was due to Turkish ancestry or was a heritage from certain legendary tribal leaders to whom fable ascribed almost superhuman powers will probably never be settled.) One certain thing is that Genghis Khan was a man of iron determination and foresight, the natural leader for a nomad empire.

The tribes he had come to rule were tough soldier material—herders, hunters, horse traders and horse thieves, petty bandits, occasional mercenaries in the great Chinese empires to the south. They had ridden, endured, and killed since childhood, and there was no mercy in them. Their weaknesses were those natural to nomads; their loyalties were the small ones of family and clan, and yet uncertain even there—young Temujin killed a half brother who stole a fish from him. Their bravery was real but of the grab-and-run variety; they had little sense of order and less of sustained effort.

Out of them Genghis Khan created a professional army, organized as tightly as any Roman legion. Military service was compulsory, with an active campaign to be expected every year and a great ring hunt—carried out on a military basis—every winter to keep officers and men in training. The army was organized by tens; its basic unit being a troop of ten men. Ten troops formed a squadron, ten squadrons a regiment, and ten regiments a division, or *tumen*. Two or more *tumens* constituted an army. All were mounted archers, approximately two-thirds lightly armored; the rest were heavy cavalry for shock action, with lances and shields as well as bows. Like American Indians, each soldier would have several horses, changing from one to another as necessary. The larger armies were accompanied by wagon trains of disassembled siege engines. Having learned the hard way during his first campaign into China that siege warfare required specialists, Genghis Khan had drafted Chinese officers for that purpose. Later, more skillful Moslem engineers replaced the Chinese. Light engines, either in carts or knocked down for transport on pack animals, marched with the cavalry. The pick of the army, of course, went to form Genghis Khan's bodyguard, which in 1206 was increased to some 10,000 men. Besides ensuring the Khan's safety, it served as a special reserve in days of battle and as a school for future senior officers.

This army had a flexibility and mobility seldom equaled in military history. Its discipline was strict and savagely enforced; in an army made up of a dozen unruly nomad peoples ranging from semicivilized barbarians to semisavages, it could not be otherwise. Also, the Mongol empire was rawly in the making; no first king of any dynasty sits easily upon his throne. Genghis Khan governed through a code, the *Yasaq* (or *Yasa*), which he imposed on his subjects to serve them as Constitution, Ten Commandments, Uniform Code of Military Justice, and common and international law. It decreed a quick and bloody end to theft, disobedience, perversions, and neglect of duty. The rear-rank trooper who failed to retrieve and return anything dropped by a man in front of him, the soldier who broke ranks to plunder before the battle was over, the nobly born commander who failed to have his *tumen* ready to mount up when summoned some stormy night, all died promptly. So did men who gave a prisoner of war food without the captor's consent or entered a senior officer's residence without permission. Death would come wholesale to soldiers of a unit that failed to rally after being defeated.

There was nothing much new in Genghis Khan's strategy except the scale on which he applied it and the determination with which he carried it out. Asia had been full of skilled commanders of mounted archers, kingdom breakers and kingdom builders, for centuries. But Genghis Khan thought in terms of empires.

The *Yasaq*'s basic assumption was that Heaven—The Everlasting

Blue Sky—had made Genghis Khan its representative on earth and had given him dominion over all nations. Resistance to him therefore was sacrilege and must be so punished. So came the formula (given at the opening of this chapter) pronounced to a ruler who refused submission: "As for us, what do we know? The Everlasting God knows what will happen to you."[15] Once such a warning had been given, an enemy had no more hope of peace or pardon.

Mongol campaigns were planned, at least to outward appearance, in true barbarian fashion. A general council would be called to which all the senior officers of the tribes not on far-distant active duty would be summoned. While the actual decisions undoubtedly were made by Genghis Khan and a small group of trusted advisers, the other officers might at least furnish information and receive their orders, at the same time being reindoctrinated with the Great Khan's power and benevolence. Failure to attend was deliberate suicide: like "a stone that is dropped into deep water, or an arrow [fallen] among the reeds,"[16] the absent ones would vanish. The campaign was considered in detail, available forces and supplies assessed, and orders given for the necessary road building and stocking of advanced depots. It is difficult to determine the approximate strength of Genghis Khan's army, but it probably numbered between 100,000 and 150,000, with several times that number of horses and thousands of servants and camp followers. Preparations had to be made accordingly. Also, the imperial courier service, which kept the campaigning Khan linked with Mongolia, must be extended behind the advancing army. These administrative matters would be handled by men from the more civilized tribes of the Mongol Empire, Khitans and Uigurs who could read, write, and cipher. Genghis Khan appreciated their usefulness; a Khitan prince, Ye-lü Ch'u-ts'ai, became his administrative deputy and principal counselor.

Possibly the most important part of this planning was the collection and evaluation of information concerning the enemy and his country. Genghis Khan and his lieutenants had outstanding skill in military intelligence operations. Long before the first Mongol outrider crossed the hostile frontier, clouds of spies had been at work. Contact had been established with dissatisfied groups within the enemy kingdom. Apparently the Mongols' best agents were merchants of all nationalities and faiths, men whose lives and fortunes depended daily on their knowledge of roads, passes and fords, and the availability of water and forage along their routes. Merchants would know the great cities, the nature of their inhabitants, and also which of their officials might be more interested in bribes than in duty well done. Prospective traitors might betray some minor gateway in a strong fortress. Genghis Khan and most of his officers might be illiterate, but they had the nomad's instinctive ability to remember every turning of trails once ridden. Those who had served Chinese rulers as mercenaries or allies (Genghis himself had won the honorary title ''Commander Against Rebels'' for such service, ca. 1198) knew the roads and bypaths of the north China Kin (Chin) Empire better than its own generals did. And spies' detailed descriptions of foreign lands were methodically stored in nomad minds accustomed to memorizing long epics for winter evenings' entertainment.

The campaign begun, the Mongols advanced rapidly on a broad front with several armies more or less abreast, each of them covered by a screen of scouts and light cavalry. This strategy not only magnified their apparent strength (pages 76-77) but left the enemy uncertain as to which army was the strongest and the major threat. Meanwhile the scouts slipped across the countryside, stalking and locating the enemy forces. Hard-riding couriers linked the whole Mongol horde, making it swiftly responsive to the Khan's will and keeping him aware of the progress of every part of it. Once the enemy's main army was located, part of the Mongols moved up into contact with it, while others converged on its flanks and rear, cutting its communications. (This tendency to attempt to envelop the enemy, even though they might be more numerous, seems characteristic of Asiatic strategy and tactics alike. American forces encountered it in both Korea and Vietnam.)

If the enemy proved in good heart and alert, instead of risking a battle, the Mongols might employ some variant of what Sun Tzu Wu had termed ''indirect methods.'' A favorite ruse was a feigned retreat to draw the enemy into broken or barren country where his forces would soon become disorganized and exhausted. Another variant was for the retreating Mongol army to lure its pursuers into a large-scale ambush set up by another Mongol army. A third was for the Mongols to retreat so rapidly that the enemy would lose all contact with them; then, several days later, to change to their best horses and make a forced march back to catch their enemy off guard. If a surrounded enemy army stood and fought, an apparently safe line of retreat might be left open to lure it into a Mongol trap or simply to take advantage of the confusion of its retreat. A defeated enemy was pursued and slaughtered systematically; special efforts were made to hunt down the enemy commanders.

Mongol tactics were a small-scale repetition of their strategy. Their light cavalry attempted to wear the enemy down or to lure him out of position by a series of partial attacks and feigned retreats, constantly showering him with arrows. Once the enemy was sufficiently shot up and entangled, the Mongol heavy cavalry charged with lance and saber. It was observed that Mongols had no particular appetite for hand-to-hand fighting in which their enemies—often better armed and armored—could meet them on equal terms. They seldom engaged in individual heroics. In short, Genghis Khan fought a strictly professional war to destroy his enemy as expeditiously as possible. The very appearance of his armies increased the terror that rode with them. Armored in black-lacquered leather, filthy and stinking (the *Yasaq* forbade them to wash in running water),[17] they appeared suddenly out of nowhere, maneuvering in silence in response to signals passed from their commander's standard-bearer to standard-bearers of smaller units and closing in confidently and steadily.

Usually the Mongols avoided sieges until the enemy's main army had been destroyed, though they might drop off an occasional *tumen* to contain the garrisons of larger cities until that had been accomplished. Thereafter, they would herd together captives from a city already taken or from the rural population and drive them into assaults against those cities that still held out, killing them if they flinched. These improvised attacks wore out the defenders and exhausted their supplies of missiles. Then, once their unwilling allies were expended, the Mongols moved in on the "softened" defenses. During Genghis Khan's conquest of the Khwarizmian Empire (centered in what today is Iran) in 1219–22, great city after great city was beaten or tricked into submission and their populations (except for handfuls of useful craftsmen and choice women) butchered. Details of Mongols would remain behind to hunt down any survivors like rats; the surrounding farmland would be left too ruined to feed any that might finally escape. During the Mongol raid into Hungary in 1241, fleeing inhabitants were promised mercy and protection if they would return to their homes and get in their crops. The crops once gathered, the Mongols killed the men, used the women, then killed them. At the end of each campaign the captives picked up for various purposes were sorted over, and all but the most useful were casually cut down. Mercy was forbidden without express authorization from Genghis Khan himself.

Such massacres usually were methodical, dispassionate affairs, inflicted on unwarlike civilians who had sought only to submit with as little fuss as possible. Word of them sent terror sifting across the world like a psychological fallout. It was horrifyingly effective. Throughout Asia multitudes of people would submit to comparative handfuls of Mongols; serve them in numb, cringing terror; and finally, on command, tie one anothers' hands and wait to have their throats slit. It did take longer to kill the population of a major city by hand than it does to drop one nuclear bomb, but the resulting mess and demoralization were much the same, and the percentage of actual "kills" was infinitely higher than in Hiroshima or Nagasaki. Besides, the week or so thus joyfully spent allowed plentiful opportunities for leisurely rape and looting. Genghis Khan spoke for all Mongols in his expression of the greatest joy possible to mankind: "to cut my enemies to pieces, drive them before me, seize their possessions, witness the tears of those dear to them, and embrace their wives and daughters."[18]

There were emissaries who went bravely from the Pope and European kings to the Mongol court. Unbelievingly, they observed that, among themselves, the Mongols were obedient, generous, and honest, in keeping with the teachings of the *Yasaq*. All religions were respected; traders' caravans passed safely; there was rough justice in conquered areas. The endless tribal wars and petty rebellions that had raked all Asia had ceased. As was said of other mighty rulers in other lands and times, so perfect was the internal peace and order of the Mongol Empire that a virgin decked in gold could ride safely across its whole length. But for those nations beyond the Mongol frontiers, those that had not harkened to the decree of Heaven and submitted themselves to the Great Khan's rule, there could be only war. They were to be treated without ceremony or mercy; any treachery or deceit was permissible in dealing with them.

Genghis Khan was over forty when he conquered the title of Great Khan, master of all the wild tribes of Mongolia, lord of vassal kingdoms stretching into the far reaches of Korea, Siberia, and Tibet.

Thereafter, the speed of his conquests increased. Some seven years sufficed to cripple the great Kin Empire; the vast Khwarizmian Empire was wrecked in less than three. (Both had larger armies but weakling rulers; the Sultan of Khwarizm was hated by his subjects and betrayed by many of his mercenaries.) Before Genghis Khan died, his sons and generals had forced their way through the Caucasus Mountains to raid southern Russia and the Crimea. He died at the age of sixty in 1227, on campaign against a disloyal vassal king, his last commands being for the murder of that king (who had surrendered under promise of amnesty) and the eradication of everyone in the king's capital. He was buried in a secret place, with many fine horses and forty beautiful girls—so said the old tale—from noble Mongol families to serve him in the hereafter.

To the outer world Genghis Khan had been the ''Accursed,'' a destroyer and a mighty man of blood. Among his own peoples he was mourned as a wise and just ruler. Born into a ruling Mongol family, he had been a natural king and commander, responsible, openhanded, and grateful for good service. If he saw his world through a nomad's eyes, he had plenty of common sense and the rare willingness to recognize—and accept—good advice. He honored a brave enemy; several of his chief lieutenants had fought him long and daringly before they passed into his service. His was the genius to make a professional army out of half-wild barbarians, thus achieving an unparalleled blend of efficiency and enthusiasm, and to adapt their traditional hunting and raiding skills to a grand strategy that swept over half of Asia. But, in the end, he was still a nomad chieftain with little more than a nomad's values. He sought the rule of as much of mankind as he could achieve in a lifetime of war so that he and his descendants might ''wear garments of gold . . . eat sweet, greasy food, ride splendid coursers, and hold in their arms the loveliest of women.'' (And sometimes he wondered whether those descendants would remember how such things were won.[19]) To him, cities and agriculture were of little or no importance. They took up land that could be better used for grazing Mongol flocks and herds, and especially Mongol warriors' horses. (On at least one occasion, Ye-lü Ch'u-ts'ai had to intervene to prevent the forcible conversion of immense areas of Chinese farmland into pasturage, with its inhabitants utilized for manure. His winning argument was the amount of tribute and foodstuffs those Chinese could furnish if spared.)

Genghis Khan had gone across Asia like a prairie fire in dry grass, consuming old dynasties and boundaries. However, as after such a fire, new growth struggled up behind him. His single-minded, deadly work unintentionally opened a clear path across the earth as had not been done since the days of Alexander. Behind him there was a mixing of races, religions, and knowledges such as followed Alexander's march from Greece to India. This change was all the more effective because Genghis left a firmly established dynasty, divided among his four sons, who methodically set about enlarging their patrimonies in China, western Persia, Russia, and Korea and raided across Poland and Hungary to the Adriatic Sea. Their sons, one of them the famous Kublai Khan (1216–94) of Marco Polo's memoirs, still went forward. Kublai conquered the Sung Empire of southern China and northern Burma, though his amphibious expeditions against Java and Japan failed and he had limited luck in modern Vietnam. Another, Hulagu Khan, took Bagdad and conquered most of Asia Minor. He also wiped out the dreaded Assassins (or Isma'ilis), an aberrant Muslim sect with a headquarters fortress near modern Teheran. Under the command of a mysterious chief, "The Old Man of the Mountain," its hashish-drugged killers had terrorized the Near East for over a century.

The Mongol strategy, like the Vasa's, was designed for rapid offensive operations by carefully trained, well-armed, tautly disciplined soldiers in which their superior mobility and aggressive leadership would make up for their enemy's superior numbers. The Mongols, however, had greater, swifter armies and campaigned across the vaster distances. They conquered kingdoms; the Vasas struggled to seize provinces. There would be nothing really like the Mongol conquests until Adolf Hitler's panzer armies broke out across Europe in 1939. Hitler also would come close to equaling their use of deception, their careful preliminary gnawing away at an enemy's alliances, their search for useful traitors. But even Hitler would not match their use of terror. Stalin did somewhat better at that task.

Western study of Mongol strategy and military organization came slowly. The Poles, Germans, and Hungarians who went down before the expertly maneuvered Mongol *tumens* in 1241–42 were at too low a stage of military development to appreciate just what had hit them. The average European thought of the Mongols as "Tartars," an immense ravaging horde that carried everything before it by savage fury and sheer weight of numbers. This concept lasted well into the last of the nineteenth century for lack of reliable histories of the Mongol era. Students eventually produced these from Persian, Arab, Chinese, and Armenian records, but the warfare they described was foreign to European experience. For two centuries European armies had been mostly infantry; their clashes in eastern Russia, India, and China with the remnants of the Mongol horse-archer tradition had not impressed them. Their colonial wars might employ devastation and terror, but selectively and usually only after what they considered sufficient provocation; like Alexander, they wanted profitable conquests and contented subjects. For the rest, the Mongol's use of *surprise, maneuver,* and *offensive* action was nothing particularly new to competent soldiers anywhere.

It was not until after World War I that bits and pieces of military studies began to suggest, now that Western armies had reliable armored vehicles and aircraft, it might be possible to revive the Mongol art of waging war at high speed over great distances. In 1927 Liddell Hart made mention of Genghis Khan in his *Great Captains Unveiled*[20] as an example of the successful application of mobility, offensive power, and his pet ''indirect approach.'' Harold Lamb, an American popular historian, brought out a better-researched, if high-colored, biography of Genghis Khan that same year. One practicing soldier who certainly studied Mongol warfare was Douglas MacArthur. His annual report as Army Chief of Staff for 1935 included this assessment:

> More than most professions the military is forced to depend upon intelligent interpretation of the past for signposts charting the future. . . . Were the accounts of all battles, save only those of Genghis Khan, effaced from the pages of history . . . the soldier would still possess a mine of untold wealth. . . . The successes of that amazing leader, beside which the triumphs of most other commanders in history pale into insignificance, are proof sufficient of his unerring instinct for the fundamental qualifications of an army. . . . It is these conceptions that the modern soldier seeks to separate from the details of the Khan's technique, tactics, and organization, as well as from the ghastly practices of his

> butcheries. . . . So winnowed . . . they stand revealed as kernels of eternal truth, as applicable today . . . as they were when, seven centuries ago, the great Mongol applied them to the discomfiture and amazement of a terrified world.[21]

MacArthur's strategy in the southwest Pacific through 1940–45, with its bold, deep offensives, was in many ways an updating of Genghis Khan's campaigns.

Genghis Khan's grandsons and great-grandsons turned against their own kin. The breed dwindled, conquered peoples rebelled, and new nomad hordes came down across the northern horizon. Out of this welter of rebellions, wars, invasions, and treachery came Timur (1336–1405), a young Turkish nobleman from the area south of the Aral Sea, which now is a part of southern Russia. He was known particularly as Timur ''lenk'' (''the lame'') from an early wound, which has been corrupted into the better-known ''Tamerlane.'' Well-educated, cultured, a fanatical Muslim when that served his purposes, very brave and enduring, he excelled both in Mongol-style warfare and personal combat. He protected and encouraged poets, historians, and artists and lovingly beautified Samarkand, the capital of his empire, which covered modern Iran and Afghanistan and portions of Turkey, Iraq, and southern Russia.

He also was without honor or faith and an enthusiastic sadist. His trademark was towers of chopped-off heads, built at the corners of cities he had captured. He could carry on long, casuistic discussions with captured dignitaries concerning poetry or the finer points of Islamic sacred law while his soldiers collected the essential building materials and put those towers together. If angered, he might build his towers by stacking living prisoners in a mud mortar. Christians he buried alive or tossed down handy wells. Genghis Khan, the barbarian nomad, had been deadly; Tamerlane, the civilized city dweller, was vicious.

A very competent strategist when it came to planning his campaigns, Tamerlane seems to have been lacking in any sense of ''national'' strategy. His realm remained a jerry-built affair, battered by revolts and ground down by his vengeances—he would even wreck the irrigation system of a rebellious province. For no good reason he destroyed the east-west caravan trade Genghis Khan had protected.

Many of his campaigns ended as nothing more than great raids that swept up slaves and incalculable loot and made his name a terror and glory, but left only chaos behind where Genghis Khan and his advisers might have established a profitable tributary kingdom. His one strategic/tactical innovation seems to have been the utilization of war elephants, large numbers of which he had captured during his raid into India. Their unexpected appearance shook the morale of both Mamelukes and Turks. He proclaimed himself the champion of Islam, yet his heaviest blows were directed at his Mohammedan neighbors. Through 1387–96 he defeated the Mongol Golden Horde in southern Russia; in 1398 he plundered the Muslim Sultanate of Delhi. (To justify his claim to be waging a holy war in order to acquire merit in the eyes of Allah, he did flay a good many of his Hindu captives alive.) His defeat (1400–1401) of the Mamelukes, that strange aristocracy of white slave-soldiers who ruled Egypt and Syria, ended with the destruction of the great Muslim trade centers of Damascus and Aleppo. In particular, his crushing victory over the formidable Turkish conqueror, Sultan Bajazet, "the Lightning," near Ankara in 1402 saved the dwindling Byzantine Empire from impending extinction and drew off the mounting Turkish pressure on Hungary and central Europe. He died at seventy-one, preparing to conquer China and forcibly convert its population to Mohammedism.

His sons and grandsons promptly went for each others' throats. The youngest son, Shah Rukh, saved something from the wreck and ruled justly and well for forty years. Shah Rukh's son was a scholar, poet, and astronomer but no soldier; his own son deposed and murdered him. Another wave of nomads came ravaging, and Tamerlane's empire passed like dust along the wind.

Mao Tse-tung was a peasant's son, born in 1893 in the inland South China province of Hunan. In 1949 he became the first "Chairman of the Republic of China," worthy heir to all of China's emperors. How many millions of Chinese died during his ascension to supreme power, how many perished once he had achieved it, will never be told with any accuracy. His career, however, was nothing unusual in Chinese tradition; both the famous Han (202 B.C.–A.D. 220) and Ming (1368–1644) dynasties were founded by commoners in times of national disaster. And Dr. Sun Yat-sen, a chief mover of the 1911 revolt that overthrew the crumbling Manchu Dynasty, was peasant born, if Western educated.

From Mao's own account of his early life, his father was a hard-working, hot-tempered ex-soldier who gradually achieved modest wealth as a farmer and rice merchant; his mother was kind-hearted and a devout Buddhist.[22] He described himself as a rebellious youth, yet his father saw that he had a useful education—at that time, the study of Confucian classics and quotations. Mao and his classmates preferred adventure stories like *Shui Hu Chuan* (the Chinese version of *Robin Hood and His Merry Men*) and *Hsi-yu Chi* (a parody of an actual seventh-century Chinese monk's pilgrimage to India in search of Buddhist texts, with a supernatural monkey for its real hero). He lived through several famines and the peasant revolts and banditry that followed them. His family helped him achieve some advanced schooling, first marrying him to a local girl. He was fourteen at the time—he said—and the girl was twenty; he promptly deserted her. Thereafter, his life was a series of schools, in all of which he did very well, living on a tight budget and spending much of his money on newspapers, which he had just discovered, and books. He read translations of Adam Smith's *Wealth of Nations,* Charles Darwin's *Origin of Species,* and works by John S. Mill, Jean Rousseau, Montesquieu, and about everything else he could get his hands on. During the 1911 revolt he was briefly, and uselessly, a soldier. His political opinions drifted steadily leftward, from mild liberalism into anarchism. By late 1920 he had definitely become a Communist. He also married Yang K'ai-hui, who became a leading woman Communist, and was killed in or about 1930.

In the complicated power plays and double-crossings that followed the 1911 revolt, Dr. Sun Yat-sen had been crowded into a minor area around Canton. His Kuomintang (Nationalist) party had lost its cohesion, and the local warlords had abandoned him. When the Western powers ignored his pleas for assistance, he turned to Soviet Russia in 1923. Delighted with the opportunity to gain control of a major Chinese political party, the Russians provided weapons, money, and advisers. The small Chinese Communist party, formed in 1921, was told to cooperate. Under expert Russian guidance the Kuomintang was thoroughly reorganized into an aggressive revolutionary party. Chinese Communists joined it; Mao was for a while secretary of its propaganda department. The two parties formed an alliance to overthrow the regional warlords and force foreign "imperialist" nations to surrender their extraterritorial rights in China. Both had their own quiet plans as to what would happen afterward.

A new school, the Whampoa Military Academy, was set up to train officers who could create "a new revolutionary army for the salvation of China."[23] Command of the school was assigned Sun's chief military adviser, Chiang Kai-shek, with a mysterious Russian called General Galen to tell him what to do. (Galen was really Gen. Vassily Blucher, field marshal in 1936, purged by Stalin in 1938, rehabilitated as a Soviet hero in 1957.) The school's political commissar was French-educated Chou En-lai, later Mao's foreign minister and China's one indispensable man. Chiang proved inspiring and energetic. He also began building up a personal Mafia among instructors and students.

Chiang's career still has several blank pages. Born in or about 1886, the son of a middle-class salt merchant, he had chosen a military career and studied his profession both in China and Japan. He was involved in the 1911 uprising and subsequent civil brawls. He joined Sun Yat-sen sometime between 1907 and 1918 and apparently served him loyally. By Chinese warlord standards, Chiang was a formidable general, skilled at intrigue, the deft alternation of military and political pressure, the exact placement of "silver bullets" (bribes), the shaping of face-saving compromises, and finally the massing of overwhelming numbers of troops if combat proved unavoidable. To Western soldiers who served with him during World War II, he would become "Chancre Jack"—stubborn, perverse, and incompetent. That, however, was much in the future.

Sun Yat-sen died in 1925, leaving the Kuomintang split between right and left wings. Chiang still was not one of its principal figures, but he had carefully secured control of its armed forces. Hobbled by tight control from Moscow, which still intended to gain control of the Kuomintang from within, the Communists were unable to take advantage of its disorganization. Balancing one Kuomintang wing against the other, snubbing up overeager Communists, flattering the Russians, Chiang rapidly built up his personal authority. The two parties managed to cooperate long enough to assert their control over South China, Communists handling the mass agitation; Kuomintang, the armed forces. Then, a "March North" in 1926 conquered Central China. But in April 1927 Chiang broke with the then-dominant left wing of the Kuomintang and began destroying the Communists in the area under his immediate control. By luck or foresight, Mao had gone back to his home province to check on the peasant movement. Two months later the Kuomintang left-wing leaders saw a directive from Joseph V. Stalin, Secretary General of the Russian Communist party. It was a confused document, based on total ignorance of conditions in China, but it did direct the Communists to take over control of the Chinese armed forces and increase their strength in the Kuomintang. That sufficed; the Russian advisers were sent home, and—after complex shuffles and deals among the Kuomintang left, various major warlords, and Chiang—the northern warlords were brought to heel in 1928. China was nominally reunified under an internationally recognized Kuomintang government, with Chiang as its president. His power, however, was far from consolidated; warlords, though they might acknowledge the central government's primacy, still had direct control of more Chinese territory than the government could claim, and the Chinese Communists were yet to be dealt with.

That seemed simple. In August 1927 the Communists had sparked mutinies and revolts across China, but both Kuomintang and warlord forces had come down on them and whipped them around the compass. Mao had organized a revolt, poetically titled "Autumn Harvest Rising," among the traditionally unruly Hunan peasantry. It was a flash in the pan, ending quickly with Mao and a few survivors on the run for the backcountry tall timber. As a final buffet, the Chinese Communist party's confused Central Committee rebuked Mao for his "purely military viewpoint," whatever that may have meant. Other risings fared no better and often worse. The Communist party was reduced to its governing Central Committee, operating from a secret headquarters in Shanghai and a scattering of refugee bands hiding out in wilderness areas. They were little better than bandits, but the Kuomintang made the gross error of regarding them as such—something to be mopped up by local authorities.

Chinese Communist hagiology has it that Mao's organization of peasant uprisings was his own original idea, carried through in defiance of Moscow directives. In fact, Moscow had been urging since around 1923 that the peasants be organized as the main force behind the future revolution; other Communist leaders had begun this task well before Mao put his hand to it. It was Mao's good fortune to be joined in his hideout by another defeated Communist unit under Chu Teh, a former army officer who had been recruited into the Communist party by Chou En-lai while studying in Germany. Chu probably was the most competent of the Communist military leaders. He and Mao made a formidable team, Chu functioning as army commander, Mao as political commissar in charge of political and propaganda activities. Just which one deserves major credit for their overall strategy is uncertain. Once the wars were over, Chu's importance dwindled, while Mao appropriated credit for the whole show.

They promptly reorganized their forces into the Fourth Red Army and began disciplining and enlarging it. Like any other Chinese warlords, their first concern was to secure a base area large enough to feed their soldiers. Being weak, they could do this only if the local inhabitants were willing to support them—in Mao's concept, to be the "sea" in which the Red Army "fish" could swim and survive. Their only source of weapons and ammunition would have to be their enemies; consequently, their strategy and tactics must be designed to make large-scale captures possible. To be certain that their troops—a not-too-promising collection of deserters, fugitives, bandits, raw recruits, and Communist enthusiasts—conducted themselves properly, Mao/Chu enforced three basic rules for soldier behavior: "Obey all orders. Do not steal from the people. Turn in all captured and confiscated supplies."[24] To these were added eight "Remarks" that enjoined honesty, courtesy, cleanliness, and not bathing in the presence of women. Such proper behavior was rare in Chinese armies, especially among the usually rowdy warlord contingents, but it took some years and much effort to make it habitual Red Army doctrine.

To win over the peasantry, Mao/Chu confiscated land from "landlords" and "rich peasants" for distribution to the poor—land of his own being a Chinese peasant's major desire. Such government as they could establish was honest, if brutal toward Kuomintang supporters. This policy would have the added advantage of leading liberal-minded Westerners to view the Chinese Communists as beneficent "agrarian reformers"—which was about as accurate as considering Tamerlane an avant-garde architect. (Being a Communist, once he was in power, Mao took the peasants' land away from them and forcibly crammed them into cooperatives and communes.)

It must be kept in mind that this Mao/Chu Fourth Red Army was only one of the three major Communist groupings, at least one of which was sometimes stronger than the Fourth. With Chiang tied down by rebellious warlords and Kuomintang political rivals, all Communist forces were able to gain territory and recruits. By 1930 the Chinese Communist Central Committee, still underground in Shanghai and apparently not too aware as to the actual state of affairs back in the boondocks, decided that its forces had sufficient strength to begin capturing major cities and developing a proletarian (urban industrial worker) movement in the proper Russian style. Communist field commanders reluctantly obeyed orders and moved out of their base areas, only to be chased back into them with heavy losses.

Thus alerted to the seriousness of the Communist threat and finally able to give it some attention, Chiang began offensives against their base areas. He had comparatively few first-line divisions, carefully trained by German advisers, and dared not expend them since they were his only real source of power. Consequently, he turned his first two offensives (1930 and 1931) over to warlord troops. Most of these were of low quality, not much better armed than the Communists, and poorly commanded. They moved into the base areas in widely separated columns, usually out of mutual supporting distance. Once they were well entangled in the rough terrain, the Communist commanders—operating consciously or unconsciously after Sun Tzu's advice to be "many in collected mass . . . amidst his separated parts"—concentrated their forces to strike short, heavy blows at the head of one isolated column after another. Lightly equipped and thoroughly familiar with the terrain of their base area, the Communists could move far more rapidly than their opponents. They captured large quantities of weapons and thousands of prisoners, most of whom gladly joined the better-managed Red forces. A third offensive, commanded by Chiang himself and employing some Kuomintang regulars, in mid-1931 had initial success but also ended in failure. A fourth attempt had to be postponed when the Japanese suddenly seized Manchuria and attacked Shanghai. There was much pressure on Chiang to declare war on the Japanese, especially from strictly noncombatant Chinese intellectuals. The Chinese Communists seized the occasion to improve their public image by declaring war on Japan in April 1932—a completely phony gesture, since they nowhere were in contact with the Japanese.

Chiang, whatever his limitations, saw clearly that it would be suicide to risk war, especially with hostile Communists and dubious warlords ready to knife him in the back. Determined to first unify

China, he continued his anti-Communist operations, explaining that "the Japanese are a disease of the skin. The Communists are a disease of the heart."[25] His fourth offensive in 1932 broke up one major base area but failed against the Mao/Chu enclave in South China. It is unlikely that Mao and Chu were responsible for the successful defense; they were in the Central Committee's doghouse for some reason and had been relieved of their commands. Chiang came back in late 1933, but this time, instead of attempting to find and fight the more mobile Communists, he followed a plan apparently provided by his German advisers. The base areas were surrounded by fortified lines, roads were built to facilitate Kuomintang troop movements, and effective counterrevolutionary land-reform and population-control measures were introduced. Air and ground-force raids put increasing pressure on the beleaguered Communists, who could no longer secure essential supplies such as salt from outside their areas. As they weakened, the Kuomintang lines were tightened. Unable to halt this remorseless boa-constrictor strategy, the Communists put the best possible face on their troubles by announcing a "Northwards Anti-Japanese Expedition." They managed to break out of the trap in October 1934 and run for their lives in hope of reaching a small, recently established base area far to the north in Shensi Province.

This famous "Long March" was a true epic of courage and determination. It has passed into Chinese Communist legend, but certain basic facts can be deduced. It involved most of the surviving Communists from base areas in central and southern China—three armies following different routes and march schedules. The Mao/Chu group (now the "First Front") marched approximately 7,000 miles in 365 days, crossing eighteen mountain ranges and twenty-four rivers, waging constant battles with the scratch forces of minor local warlords. The rear guards they had left behind were practically annihilated. There are no reliable figures on losses suffered. The First Front had broken out with possibly 100,000 people; between 5,000 and 7,000 of these finally came through to momentary safety with several thousand recruits it had collected en route. They had been lucky as well as determined. Had the local warlord troops been better armed and officered, had the Kuomintang pursuit been pushed a little more energetically, the First Front would have been trapped and destroyed, and probably the other fronts as well. As it worked out, Mao was able to proclaim a Chinese People's Soviet Republic in their new base on January 1, 1937.

Mao's authority increased during and immediately after the Long March. Why this happened is not clear and possibly never will be. There was considerable squabbling among the Communist leaders; apparently one of Chinese Communist party's founders, Chang Kuo-t'ao, who had gotten his Fourth Front away in somewhat better order than the First, opposed Mao. Since Chang went over to the Kuomintang in 1938, it is obvious that Mao bested him. Physically tough, clearheaded, clever and glib, with an instinct for survival and no troublesome scruples, Mao outargued and outlasted his rivals. Behind his plump face with its ready smile, he also was a zealot and a tyrant in search of an empire. Sometime between 1935 and the early 1940s he became the Communist party's unchallenged chief.

Chiang had readied a final extermination campaign against the exhausted, isolated Communists. The most convenient troops for this purpose were the strong ''Manchurian'' army of ''Young Marshal'' Chang Hsueh-liang, which had been evicted from Manchuria by the Japanese five years earlier. These soldiers, however, wanted to recover their homeland and so were natural suckers for the Communists' plea that the Reds' one burning desire and purpose was to fight the Japanese, if only Chiang Kai-shek would let them. When Chiang came north to get the Young Marshal's offensive rolling, he was kidnapped by the Manchurians; after two weeks of obscure, convoluted dickering, he reemerged as the leader of a supposedly reunited China and vowed to lead it against the Japanese aggressors. The Chinese Communists pledged themselves to be loyal citizens and to cease all attempts to overthrow Chiang.

China was not ready for a war against Japan. Chiang had long bucked growing public opinion in his determination to avoid it until the Communists were destroyed and the Kuomintang forces strengthened and better trained and equipped. He apparently gave in now rather than chance being pushed aside by more bellicose leaders. He obviously did not believe the Communists' promises. Neither did the Communists: In ''very confidential'' instructions to his troops, Mao explained: ''Our present compromise is designed to weaken the Kuomintang and to overthrow the National Government.''[26] Having maneuvered Chiang into a war he could not win, the Communists would wait like buzzards to pick over the remains.

Their apparent reunion nevertheless alarmed the Japanese militarists who had been gradually edging down from Manchuria to seize sections of Northern China. With that amazing obliqueness Japanese sometimes use to explain their motives, they announced that it was time to achieve a basic adjustment of Chinese-Japanese relations and to establish peace in the Far East without delay. To this end, they urged "grave self-reflection" upon all Chinese leaders. On the night of July 7, 1937 a chance clash between Kuomintang and Japanese troops rapidly brought on a full-scale war.

Both Chiang and Mao realized that this would be a long-drawn-out war and not a Sun Tzu blitzkreig. Mao/Chu proposed to apply a strategy they had evolved in their struggle against Chiang, as embodied in the now-familiar creed:

> When the enemy advances, we retreat.
> When the enemy halts and encamps, we harass them.
> When the enemy seeks to avoid battle, we attack.
> When the enemy retreats, we pursue.[27]

Mao explained this in more detail in his book *The Protracted War*.[28] Facing a superior enemy with better arms and equipment, Mao/Chu would wage a mobile campaign, inflicting whatever casualties they could on the advancing enemy but not grudging any temporary loss of territory. This was the old nomad strategy: lure your opponent deep into the interior of your territory, entangle him in unfavorable terrain, and wear him down by partial engagements before closing in to destroy him.

To this, however, Mao/Chu had added a new factor in their organization and indoctrination of the "poor" peasants throughout Communist-held territory. This was a shrewd and imaginative program. Local governments were organized and honestly administered; taxes were reduced, debts canceled. The peasant masses were "educated" by traveling theatrical troupes, card games, agitators, and other methods to accept the Communist political viewpoint. Never before given any voice in their government, the peasants responded wholeheartedly in most areas. From them the Communists organized home-defense militia units and intelligence networks.

Mao/Chu planned to employ this militia, strengthened by detachments of Red Army troops, for guerrilla operations against the Japanese communications and supply bases. They realized the resulting irregular warfare would be ruthless and costly to both sides, but it would gradually weaken the Japanese and stall their advance, giving the regular Red Army units time to regroup. Once the Japanese were halted and disorganized, Mao/Chu would seize the initiative and launch a massive counteroffensive. The guerrillas, now armed with captured weapons, would join in this more conventional warfare. Every possible form of deception would be utilized; offensive action initially would be directed at weak or isolated enemy units, which would be encircled and overwhelmed by rapid concentrations of far stronger Communist forces. It was especially important to fight only when they had all the odds in their favor and were certain of winning; otherwise, they moved away and avoided action. Mao/Chu's major objectives were to annihilate the enemy and preserve their own troops. An ''ordinary'' victory, from which a defeated foe might extricate his troops to fight another day, was of no particular use. ''Injuring all of a man's ten fingers is not as effective as chopping off one, and routing ten enemy divisions is not as effective as annihilating one of them.''[29] This would be demoralizing to the enemy and would result in a harvest of captured weapons.

Such strategy and tactics depended much on China's vast size and poor communications; the first allowed the Communists to duck, run, and maneuver; the second slowed the advance of a conventional army with its artillery and supply trains. It also was based on the fact that (as of 1941) Japan's total population was approximately 70 million; China's, 450 million.

Actual combat is a rude test of theories. Mao/Chu strategy began well enough with the ambush and partial destruction of an overconfident Japanese division, but a new Japanese commander introduced a ''cage strategy.'' This divided and subdivided the area of operations by roads, barriers, and fortifications, cutting Red Army units off from the civilian population and enabling the Japanese to mop up one subarea at a time. Mao/Chu countered this in late 1940 by their ''Hundred Regiments Campaign,'' a large-scale surprise attack that did considerable damage to the ''cage'' system. The Japanese retaliated with a ''Three All'' (Kill all! Burn all! Destroy All!) strategy that removed large chunks of the Red Army and the population that supported it and left a very peaceful desolation where it passed. Continuing through 1942, this strategy seems to have cured Mao/Chu of any desire for further large-scale operations.

Chiang meanwhile had faced the major Japanese offensive. Trading space for time, he managed to save most of his troops as he fell back into the interior of China. His generals won at least one considerable victory, and some six hundred factories were dismantled and moved westward. The Japanese occupied as much of China as they could conveniently control, including the major seaports, and waited for the Kuomintang government to collapse.

The eventual outcome of this war was decided when Japan foolishly attacked the United States, the British Commonwealth, and Holland in December 1941. More and more divisions of the tough young veterans who had cowed Kuomintang and Communists alike had to be transferred to Burma and the Pacific; their replacements were middle-aged reservists with no training in counterguerrilla warfare and a live-and-let-live attitude. Since it was obvious that the Allies would eventually defeat the Japanese for them, both Mao and Chiang concentrated on preparing for something more important—a final civil war to decide which of them would rule China. Meanwhile, they indulged in minor mutual backstabbing whenever convenient.

Having lost the coastal areas that had been his greatest source of support and revenue, Chiang was largely dependent on the reactionary western warlords. His rule grew increasingly harsh and dictatorial, his administration increasingly corrupt and brutal. Chiang's early land-reform program was scrapped, and the peasantry was taxed unbearably. Inflation led to hoarding and speculation. Even much of the Kuomintang army was neglected and half starved. An American general compared its hospitals to Nazi extermination camps. Too many Chinese officers not only failed to look after their men but cheated them. (American officers training a U.S.-equipped Chinese tank battalion in Burma had trouble convincing the battalion supply officer that he was supposed to *issue* the rations and gasoline turned over to him instead of peddling them for his own benefit!)

Mao concentrated on organizing and indoctrinating the civilian population, extending his efforts through the Japanese lines. Methods varied from area to area: In some, their appeal was to patriotism and self-defense against the Japanese; in others, it was social reforms and/or redistribution of land holdings. The Communists might even work with patriotic landlords and local gentry, who were used for the time being and disposed of later. On the whole, Communist-controlled areas were orderly and efficiently governed; there were none of the appalling abuses so prevalent under Chiang. Red Army units were strictly disciplined and capably commanded; their morale was high. The contrast between these two parts of China was only too clear to foreign observers. The Communists waged occasional guerrilla war against the Japanese, usually in defense of base areas they had established behind the Japanese lines. This received considerable publicity, but whether it really troubled the Japanese appreciably is difficult to determine.

At the war's end the United States made a well-meaning, if naive, attempt to mediate between the two parties, putting pressure on Chiang by suspending military aid. Meanwhile, the Communists pushed 100,000 men into Manchuria, where the Russians (who had overrun that area) armed them with captured Japanese weapons. Hastily dispatched Kuomintang troops defeated them badly, but the American-imposed truce and asylum offered by the Russian forces saved them. Mao, all sweet reason, was willing to enter a coalition government, naturally intending to take over gradually from within. Chiang would not share power even with loyal subordinates or small groups of liberal intellectuals who wanted a reformed Kuomintang government. And so the war continued across a wrecked China.

The Mao/Chu strategy worked splendidly against the Kuomintang armies. In the Red Army—soon renamed the Chinese People's Liberation Army—command assignments and promotion had to be earned by proven merit and valor. Operations were carefully planned and tightly controlled; the supply system worked efficiently.

Initially harassing their enemies by intensive guerrilla warfare, Mao/Chu rapidly increased the strength of their field armies. Any sort of recruit was welcomed and quickly indoctrinated—Kuomintang deserters and prisoners of war, ex-soldiers of the puppet Chinese states the Japanese had set up to administer occupied territory, and veterans from the second-class Kuomintang divisions Chiang had demobilized. Large numbers of militiamen were summoned to active duty. With Russian help, Mao/Chu also organized efficient artillery units, enabling their forces to carry out major conventional military operations.

The Red Army's initial offensives were aimed at the smaller, isolated Kuomintang commands. Communists would suddenly mass "absolutely superior forces"—as many as five or six to one—around them and literally swamp them with unceasing "human sea" assaults in the old Mongol way. A series of such easy victories raised Communist morale and made Kuomintang troops apprehensive.

Chiang's insistence, despite the protests of his American advisers, on scattering some of his best divisions across Manchuria greatly simplified Mao/Chu's task; it proved impossible to supply them adequately, and they were destroyed piecemeal.

As their forces grew, Mao/Chu moved against the bigger Kuomintang commands, shrewdly selecting those whose generals were known to be fainthearted or incompetent. They seldom bothered to attack strongly garrisoned cities, preferring to flow around them and strike Chiang's field armies—those shattered, the cities soon surrendered. Determined leadership, light equipment, and the use of tens of thousands of peasants to transport their supplies gave them an unusual mobility—even good Kuomintang generals might be surprised by the speed of their advance. In short, the Chinese People's Liberation Army was a much more modern "Western" army than Chiang's miscellaneous forces.

Even so, a cynic might argue that Chiang Kai-shek was Mao's best general. Chiang had a number of excellent commanders, both Kuomintang and warlord, who repeatedly defeated Mao/Chu's best troops and commanders, yet he trusted only a small clique of his comrades from the Whampoa military academy days. These included a disproportionate number of incompetent stumblebums; the best were arrogant and uncooperative, refusing to accept orders from any superior but Chiang himself. Nonclique generals, especially if they were aggressive and successful, found themselves transferred to unimportant commands, or their armies would be left short of rations and ammunition.

Chiang had no apparent strategy, unless it was to hold the major cities. He interfered constantly with his generals' operations; his orders were always imperative, often unrealistic, and increasingly erratic. Kuomintang morale, never very high, sagged more and more under such treatment. Larger and larger commands tended to mutter the Chinese version of "Ah, the hell with it" and go over to the Communists. To make matters worse, Chiang not only did nothing to improve the efficiency of his Defense Ministry but allowed the Whampoa clique to shortcut its authority at will. Unable to adjust to a rapid war of maneuver, the Ministry lost track of various units for weeks and failed to keep them supplied.

By early 1949 Chiang had lost Manchuria, most of northern China, and at least half of his army; massive Communist forces were pushing toward the north bank of the Yangtze River, the last good defensive line left to the Kuomintang forces. Domestic and foreign criticism was intense. Chiang therefore resigned the office of President of China and ostensibly went into retirement, but—continuing to operate through his Whampoa clique—deliberately sabotaged his vice-president's determined effort to hold the Yangtze. Some 150,000 picked troops, most of China's gold reserve, much of the weapons and equipment furnished by the United States, warships, and aircraft were moved to Formosa (Taiwan), which Chiang already had selected as his personal refuge. Once the Communists attacked, Chiang's meddling killed any chance of a successful defense. Safe on Formosa, Chiang proclaimed himself once more President of all China, insisting he soon would return to the mainland. He continued this claim (but limited his martial feats to periodic grand reviews) and ruled despotically until his death in 1975.

Chiang had many admirers and supporters in the United States (the so-called China Lobby), but very few Americans appreciated how limited and tenuous his actual power was. His second wife, the beautiful, imperious, and cultured Soong Mei-ling, daughter of a Christian, American-educated family, was an effective lobbyist on his behalf, but he was the most difficult, devious, and demanding of allies. In his own way and mind he was a great Chinese patriot and leader, but his constant tendency was to think first of his own personal interests and later—if at all—of the Chinese nation and people. However, whatever his motives, he can be honored for his long resistance to the Japanese.

Mao took over China and spent the rest of his life reworking its stubborn people and society to his ideas, subjecting them to periodic renewals of revolutionary purges and violence. His goal may have been a "pure communism," with individual Chinese little more than interchangeable parts within an omnipotent state machine. One of his chief lieutenants in this was his fourth wife, Chiang Ch'ing, the "white-boned demon," former movie actress and a she-hellion of rare force, now considered a public enemy. (Wife number three left little trace.) Mao achieved semidivine status and such adulation as no previous Chinese ruler experienced. He died in 1976, leaving an utterly changed China, possibly a stable one, and a worldwide tradition of revolutionary violence and guerrilla warfare.

The application of Mao/Chu strategy outside China has been indecisive or unsuccessful, in part, possibly, because of the rather primitive equipment of Chinese troops. Chinese intervention in the Korean War in 1950 began brilliantly with the movement of 300,000 soldiers into North Korea, undetected by the U.S. Air Force. The Chinese, far more mobile in that rough terrain than the road-bound UN forces, won a series of startling victories—though sometimes at exorbitant cost, as when

> Those slant-eyed Chink soldiers
> Struck Hagaru-ri
> And now know the meaning of USMC.

But Americans and their allies quickly comprehended the Mao/Chu strategy and tactics; the Chinese won no more big battles but lost terribly. In Vietnam, the North Vietnamese and their Viet Cong auxiliaries attempted a version of the Mao/Chu strategy. U.S. forces won most of the small fights and all of the big ones.

Asiatic strategy has always depended heavily on deception and propaganda. Conventional warfare will go on amid constant guerrilla actions which, however, can be stamped out, as they were in Korea. There is little or no concern for human life; Asiatic commanders usually have plenty of manpower and expend it freely.

Deception involves not only the customary employment of evasive maneuvers, surprise, ambush, and night attacks but extends into sometimes more effective employment of psychological warfare and diplomatic maneuverings. Sun Tzu's teachings include two especially pertinent warnings: "Placatory words and increased preparations are signs that the enemy is about to advance" (naturally, it is up to us to spot the preparations), and "Peace proposals unaccompanied by a sworn covenant indicate a plot."[30]

In Korea, midsummer 1951 saw a major Chinese Communist/North Korean offensive shattered in what Americans called the May Massacre, and a UN counterattack shouldering relentlessly northward. Badly hurt, the Communists called for negotiations. However, they insisted that these be opened at Kaesong, a few miles inside the Communist lines, one of the few points they still held below the 38th Parallel. In a mood that can be described only as one of innocent simplemindedness, the UN command agreed. They got the works: To get into Kaesong, they had to pass through the enemy lines with white flags on their vehicles; the senior UN negotiator found that the legs of his chair had been sawed off so that he must look up at the shorter senior Communist. The UN team was composed of military men ready to work out an armistice. The enemy team was of that type of uniformed specialist peculiar to Communist nations—men who handle purges, espionage, subversion, civil disturbances, and similar chores; they wear uniforms and many medals, have military titles, and are not any sort of soldier. (One of them is presently head warden in Poland.) The angered Americans got the meetings shifted to Panmunjom, in the no-man's-land between the armies, but the world already had seen motion pictures of the UN representatives coming, apparently as humble suppliants for peace, with their white flags (and, it must be confessed, some apprehensive looks), before haughtily condescending Communist officers. Having gained their primary objective of halting the UN advance and thus having won time to rebuild their armies and fortify their lines in depth, the Communists proceeded to dicker, delay, and probe in hope of further military or propaganda success. With suspensions and disagreements, the truce negotiations did not end until July 1953. After the cease-fire was established, the Communists immediately began violating its provisions.

The war in Vietnam showed the same skilled trickery. The famous North Vietnamese Tet offensive (Tet was the major annual Vietnamese holy and holiday week) was launched during a truce the North Vietnamese themselves had proclaimed, which they had sanctimoniously begged the American/South Vietnamese forces to honor, and which they had hinted might be extended as a prelude to peace negotiations. And when Pres. Richard M. Nixon, launching the war's first intelligently planned air and naval operations, finally forced them to sign a peace treaty, they immediately set about preparing to violate it.

Obviously, the major lesson from Asiatic strategy is to expect deception and to be prepared to counter it. War against an opponent utilizing Mao/Chu teachings will involve ruse and stratagem on the battlefield but also propaganda campaigns directed at the American public and world opinion—such as the faked-up charge that we were employing germ warfare in Korea. Every effort will be made to stir up guerrilla warfare and sabotage behind our armed forces. Assassination may well be utilized; Sun Tzu recommended it.[31] Patience is essential. We are known to be an impatient people; Mao/Chu strategists believe that we can be defeated simply by denying us a clear-cut victory until the American public becomes bored, that "given even a feeble but sustained effort over a long enough period of time, success in reaching any goal, no matter how remote, is certain."[32]

Sun Tzu said:

> If you know the enemy and know yourself, you need not fear the result of a hundred battles. . . . Hence the saying: The enlightened ruler lays his plans well ahead; the good general cultivates his resources. . . . If it is to your advantage to make a forward move, make a forward move; if not, stay where you are . . . the clever combatant imposes his will on the enemy, but does not allow the enemy's will to be imposed on him.[33]

X

CHAIRBORNE WARRIORS; BATTLE CAPTAINS

> Anyone can do the little job of directing operations in war. The task of the Commander-in-Chief is to educate the Army to be a National Socialist [Army]. I do not know any Army general who can do this as I want it done. I have therefore decided to take over command of the Army myself.
>
> HITLER[1]

> When President Roosevelt began waving his cigarette holder, you never knew where you were going.
>
> GEN. GEORGE C. MARSHALL[2]

> Do your own reconnaissance. See for yourself and then get down to the job without delay.
>
> ERICH VON MANSTEIN[3]

For all its intensity and extent, World War II did not produce a comparable crop of great captains and superstrategists. In fact, there were times and places when even ordinary strategists seemed in short supply.

In part this was because some national leaders tended to keep a far tighter control over their military commanders than in previous wars, insisting on shaping the details of their current military operations to secure the long-range objectives of their national strategy. If the national strategy were realistic, this might not be a serious clog to their commanders' initiative. Most of these national strategies, however, contained more wishful thinking than common sense and so wasted their commanders' best skills and their soldiers' lives.

Hitler and Stalin rode their generals with whip, spur, and curb bit. Churchill had his fingers in all British operations. One of his generals observed that he undoubtedly was the world's greatest strategist—nobody else would even dream of attempting to use one division simultaneously in three different parts of the world. By contrast, in Japan the armed forces had effective control of the government and established their own strategy. Only in the United States, once Roosevelt and Churchill had agreed that Germany was the more dangerous of their enemies, were the military leaders allowed a relatively free hand. However, beyond a tenebrous indication that the Russians were valuable allies and should be so cherished, they were not given clear-cut guidance as to the United States' objectives for the postwar period. In the Pacific, MacArthur (see page 134) and the U.S. Navy used this freedom to cripple Japan in a series of brilliant campaigns, ending the war in control of most of the strategic points of that area. In Europe, Gen. Dwight D. Eisenhower concentrated largely on military objectives, leaving "what-ifs" such as Berlin and Prague to clutter our history.

The United States, Great Britain, and Russia shared one paramount wartime objective—to defeat Germany. With considerable wrangling and backbiting, they managed to coordinate their military strategies to that purpose. Their postwar objectives were as different as the personalities of their leaders, and those leaders were an ill-matched trio indeed.

Franklin D. Roosevelt (1882–1945), president of the world's most powerful nation, was a politician's politician, dexterous, devious, and opportunistic, but gifted with a certain patrician charm that camouflaged those unctuous qualities. He evoked either loyalty or strong dislike, but it is probable that neither friend nor enemy ever fully understood him. Crippled by polio at thirty-nine, he had the inner toughness to make his disability of small account. Ebullient and endlessly self-confident, he considered himself equal to any problem and delighted in the opportunity to set the world to rights, according to his somewhat simplistic conception of it. But he was possessed by an honestly idealistic concept of a new world order of peace and democracy, guided by an organization of "United Nations," that must follow the war, and he strove for it with all his politician's arts of compromise, appeasement, charm, and under-the-table dealings. Unfortunately, neither his upbringing, education, nor experience had prepared him to comprehend Stalin. Moreover, Roosevelt's efforts were sapped by his progressively failing health and complicated by a tendency to run his war extemporaneously. "Having a thousand and one things on his mind . . . He forgot that he ever had made some detailed agreements and usually there was no record of them in any regular government department."[4] Also, he "never was much of a stickler for language"[5] and so was careless concerning the wording of international agreements. Whatever Russian leaders may have thought of him, Russian combat soldiers admired him. After his death they would greet Americans with "We condole with you for the loss of your great leader. He gave us everything we wanted." (Responses they got from some iconoclastic Americans gave them a dreadful shock.)

Winston Churchill (1874–1965) had been a staunch defender of the British Empire, and he proposed to preserve it so far as fate allowed—and that despite Roosevelt's not-too-subtle hints that it should be abolished. At the same time, his dependence on American aid left him little choice but to support Roosevelt's grand design for a United Nations, though his greater experience with the world's nations and their particular cussednesses left him doubtful of its ability to reform international relations. He was an awkward politician, a gifted orator and writer, an amateur artist, and a competent bricklayer. Alone of the three leaders he had been a combat soldier, against Afridi hillmen along India's northwest frontier, dervish fanatics in the Sudan, Boers in South Africa, and Germans in Flanders. He had strategic talent but was apt to attempt more than his forces could handle and to have too many pet schemes—some potentially disastrous—on his mind at any one time. But, beyond all, he possessed an absolute courage that made him a natural leader in a desperate hour. Russian leaders respected him as "far sighted and dangerous,"[6] which is to say they knew he regarded them without illusions.

The third leader was out of quite another world.

"Joseph Stalin," Winston Churchill mused forebodingly in 1943, "is an unnatural man. There will be grave troubles."[7]

Joseph Stalin was indeed unnatural, a bloody tyrant without ruth even for his own begotten children. God alone knows how many millions of inoffensive people he sent to death because they were inconvenient. He also was World War II's most successful national strategist.

Born Iosif Dzhugashvili in the Caucasus Mountains state of Georgia in 1879, he began by studying for the priesthood, as his widowed mother wished, but drifted into the revolutionary movements that troubled the last decades of czarist Russia. His real activities are hazed by Communist mythology, but he slowly won a reputation as an intelligent planner and organizer. He was still a minor figure during the Russian revolutions of 1917 but while other Bolshevik leaders made speeches and disputed over policy, Stalin methodically gained control of the Communist bureaucracy, concealing his ambition within a carefully contrived image of a kindly, devoted party functionary. After Nikolai Lenin's death (which Stalin may have arranged) in 1924, he became more and more the absolute master of Russia. He improved its industries, collectivized its farms, increased its armed forces, established a vast system of penal labor camps, and scourged the land with purges, famine, and the terror of his secret police, the NKVD. With World War II, he played jackal to Hitler, savaging the small nations on his frontiers and waiting for the time when Hitler would become too thoroughly engaged with England to oppose further Russian expansion. Then Hitler outscoundreled him by hitting first, and Stalin suddenly found himself, with Winston Churchill and Franklin D. Roosevelt, an alleged champion of democracy, the "four freedoms," and the rights of man. Stalin had his own strategic goals: the achievement of traditional Russian territorial ambitions and the imposition of Communist rule across as much of the world as he could manage. The latter, at least, would not please his new allies, but he desperately needed their wholehearted help, both in supplies and in active military operations to draw German divisions out of Russia, to accomplish it. Accordingly, he set about using his allies' own resources to win a postwar empire that would confront their interests and their ideals alike.

No accomplished gold digger ever worked her boyfriends more expertly. Stalin could shift unexpectedly from warm affection to shrill abuse; dicker like a back-alley old-clothes dealer; importune like an underemployed streetwalker; rise to the noble indignation of a betrayed virgin; or outpathos MacArthur in posing as a neglected champion, left unsupported to wage humanity's battle against impossible odds. Cat-smoothly courteous, sensitive, and charming when the occasion required, he could casually chop off the translation of a proposal by Pres. Harry Truman with an abrupt "Nyet!"[8] Being untroubled by elections, public opinion, a free press, or political parties, he could negotiate with a free hand and Oriental patience. When Roosevelt showed occasional unhappiness over the possible reaction of Polish-American voters to the various brutalities Stalin was heaping upon Poland, Stalin nimbly distracted him by bringing up the more seductive subject of Russian participation in the future United Nations. Probably his cutest ploy was his periodic posing as friendly, liberal "Uncle Joe," who longed to please his allies but was restrained by a clique of hard-line Communists led by his foreign minister, Vyacheslav Molotov. (Molotov, whom Churchill defined as the closest possible imitation of a robot, could look and act that sinister part but actually lived in servile terror of his master.) Back of all this was a remarkable intelligence, memory, and tenacity—"better informed than Roosevelt, more realistic than Churchill."[9]

Stalin kept his allies guessing, always a little worried that he might suddenly switch sides again and, if they did not court him ardently enough, make a separate peace with Hitler. They slathered him with gifts. (Some 12,755 tanks, 22,206 aircraft, and 375,883 trucks were only part of them.) He always demanded more. They acquiesced, when they did not cooperate, in his annexation of the Baltic states and portions of Finland, Poland, Germany, Czechoslovakia, Romania, and Japan. He generally coordinated his military operations with theirs, but he kept no promises as to the treatment of Balkan countries he overran.

By early 1945 the dying Roosevelt at last realized that Stalin could not be charmed or trusted into honorable behavior and had no sense of noblesse oblige. But it was too late to save much in Eastern Europe. Russian armies were in possession of all of it except Greece, and Communist governments were being imposed willy-nilly on non-Communist populations. (Churchill had unilaterally decided to establish a Western foothold in Greece: The American mass media was denouncing his "imperialism.")

In matters of military strategy, Stalin maintained a steely control of the entire Russian war effort. After his death his successor, Nikita Khrushchev, would claim that Hitler's sudden attack in 1941 had put

Stalin into a panic. That is quite possible—there is nothing in Stalin's known history to suggest personal bravery. If he did panic, however, he recovered, remaining in Moscow through the worst of the German offensive. (The actual risk was slight, and it was an effective bit of propaganda.) Aided by a small but usually competent "High Command" staff, Stalin raised armies and fed them into the battle, evacuated industrial plants from western Russia to havens east of the Ural Mountains, and launched a masterful propaganda campaign to convince his subjects that they were fighting for the "Mother Russia" of old tradition and not for communism. He was briefed three times a day on the current military situation and took part in the planning for all major operations. The real extent of his own strategic talent will never be known, but the general level of Russian strategy was not impressive. It certainly was quick to take advantage of Hitler's tendency to shove his head into a sack; its greatest victories—Stalingrad (1942) and Kursk (1944)—were won by taking advantage of such errors. Its other victories, however, were largely a matter of *mass*—throwing in hordes of men, tanks, and guns until something cracked, or the attackers were used up. As much as possible, this was coupled with *surprise*. Stalin stressed the *offensive*, hounding his generals forward: One general who complained he needed more armored units was told "my grandmother could have waged war with the help of tanks!"[10] Always he insisted, "Don't spare the men!"[11]

Stalin's command system was highly personal. No senior general was left in the same assignment long enough to gain the personal loyalty of his troops and thus become a possible rival. All army headquarters were haunted by representatives of the High Command and the Communist party—and the NKVD watched all of them. For Stalin trusted no one; having come to absolute power by intrigue and the deaths of many, many opponents, he lived in fear himself and knew the uses of fear better than any man alive. "When we were called to his office," Khrushchev said later, "we never knew whether we were going to see our families again."[12]

Nothing of any importance could be done without his express approval. One bitter American, awaiting Russian reaction during the Poltava disaster (June 21, 1944), with the Luftwaffe methodically demolishing a fleet of grounded American bombers which had landed at that base after a "shuttle bombing" mission, put it in earthy words: Russian officers "can't even take a piss" without Stalin's OK.[13]

The load of such detailed, incessant responsibilities must have been crushing, but Stalin proved himself equal to it. Whatever his contribution to Russian military strategy, he absolutely controlled Russian national strategy. Once the war was turning against the Germans, Stalin chose his *objectives* for their after-the-war political results. He would advance almost to Warsaw, call on the pro-Western Polish underground to rise in open revolt—and then halt in place while the Germans demolished it, leaving his Polish Communist stooges a free hand to reorganize Poland. He could suspend a promising offensive toward Berlin to launch an offensive that overran the Balkans. With the war in its last stages, he agreed with Eisenhower that Berlin was of no real value as a military objective and, once Eisenhower had turned his attention to largely nonexistent German forces on his flanks, immediately put everything into a climactic assault on Berlin. Finally, he bluffed Eisenhower into halting Patton's advance to rescue a Czech revolt in Prague so that Russian troops might gain the political leverage and glory of liberating that capital.

Stalin's national strategy thus achieved practically all its objectives. (Some of his final attempts to hornswoggle his allies into further subsidizing him would fail. They would not put pressure on Turkey to grant him effective control of the Dardanelles nor on Norway to grant him bases. The United States refused him a share in the occupation of Japan, and his self-sacrificing offer to fortify and defend Italy's former North African colony of Libya was considered a mite too expansive.) But he had his new Eastern European empire and did with it as he wished, willingly forfeiting his allies' future good will.

To most Russians he was a slightly modernized version of an Asiatic god-emperor, aloof and isolated, demanding constant adulation from his subjects. Where Churchill lived on the official British food ration, Stalin had every available luxury for himself and his entourage. Yet this terrifying potentate was a small, slouchy man with bad teeth, thin hair, and abnormally long arms. His "lively and impish avid yellow eyes" impressed the Yugoslav guerrilla-intellectual Milovan Djilas, who felt Stalin had "a special intuitive power to see into men."[14] He was nervous, constantly doodling or fussing with his pipe, alert to everything about him. His sense of humor ran from the crude to the subtle, but he never joked about himself.

Tamerlane would have comprehended Stalin. One of the few Americans who did grasp something of his essential character was

Gen. George C. Marshall: "blunt as an old battle-axe that has seen many campaigns."[15] Amazed at the general American ignorance of Stalin's history, Marshall declared, "I always thought they made a mistake in treating Stalin like [a Western statesman]. He was a rough SOB who made his way by murder and everything else and should be talked to that way."[16]

The leaders of the Axis nations were a weaker lot. Mussolini possessed no military qualities whatever; except that men were killed and maimed in them, his own campaigns in Greece and Africa were studies in low-comedy frustration. Japanese leaders had deadly skills, but they were as much samurai warriors as modern military officers. Their surprise offensive of December 7–8, 1941, striking from Hawaii westward to Malaya, is an unequaled masterpiece of planning and execution. But it was launched with full knowledge that the United States had, by their own estimate, "seven to eight times" the war potential of Japan and that Japan had no hope of "directly vanquishing" the United States.[17] Their choice, as they saw it, was either to attack the Americans, English, and Dutch in the Far East or to give up their great dreams of "contributing to world peace by establishing the Great East Asia Sphere of Co-Prosperity."[18] Clausewitz would have told them to iron out the creases in their brains and realize that their plan was dangerous self-deception. But Japanese minds had their own peculiar logic; like the foredoomed samurai of their legends, they drew their swords and went against the world. "Rather than await extinction, it were better to face death."[19]

The Allies' major opponent, Adolf Hitler (1899–1945), had great repute as a strategist and war leader. His rise to power had been an amazing rags-to-riches success story. Orphaned before he was twenty, an unsuccessful candidate for training as an artist, sometimes an odd-jobs worker, Hitler had left his native Austria and enlisted in the German Army in World War I. Though he served with zeal and valor, winning the Iron Cross, first class (then seldom awarded enlisted men), he still was a corporal at the war's end,[20] which found him temporarily blinded by a British gas attack. In the chaos of revolt and mutiny that shook Germany after the war, Hitler remained loyal. Then, gradually, he turned from soldiering to politics, gradually discovered his terrible abilities as a demagogue and politician. His honest patriotism twisted into a drive for personal power. Once relaxed and friendly in his private life, he became a withdrawn megalomaniac with fewer and fewer close human relationships. Failing health (and probably a quack doctor he trusted) steadily sapped his energies. In Poland in 1939 he still was the courageous "front soldier": Erwin Rommel, who then commanded Hitler's personal bodyguard, wrote his wife that "he was always wanting to be right up with the forward troops. He seemed to enjoy being under fire."[21] Four years later he would refuse to speak to chance-met soldiers returning from the Russian front or to take the slight risk of visiting rear-area headquarters in Normandy. He finished as an exhausted, untidy, prematurely old man, almost passively awaiting his end in his Berlin command bunker.

Becoming chancellor (prime minister) of Germany in 1933, he quickly made himself its absolute ruler and began the crash rebuilding of its armed forces. Up to 1939 he showed diabolical skill at combining aggressive diplomacy and the threat of military action to bring Austria and Czechoslovakia under German rule. His military leaders had urged a more gradual approach, fearing French and British reaction while the German armed forces still were half organized. But, at every successive crisis, the French and English—still sore from their losses in World War I—flinched from the prospect of an armed confrontation and stood aside. These proofs of his superior political acumen, daring, and strength of will gave Hitler a moral ascendancy over his military leaders. There were generals ready to oppose him openly at the time of the Munich crisis (September 1938) had England and France taken a hard stand. But when British Prime Minister Neville Chamberlain chose instead to babble of "Peace in our time," they could only retire from active service or acquiesce in Hitler's future adventures.

Hitler's first wars—Poland in 1939; Denmark, Norway, France, Holland, and Belgium in 1940; Greece and Yugoslavia in early 1941—were lightning-swift applications of *mass* and *surprise*, notable for their daring use of panzer troops across difficult terrain and for their expert coordination of air, ground, and naval forces. Their planning and execution were almost entirely the work of German officers: Hitler contributed modestly to the plan for the invasion of the Low Countries and France and seldom interfered in day-to-day operations. (The major exception was his two-day halt of panzer columns closing in on the Anglo-French forces at Dunkirk, and this apparently had the concurrence of some of his senior officers.) Hitler's inner staff, however, learned that their "Leader" tended to lose his nerve at the crisis of a campaign. Being professionals of the old Great General Staff breed, they upheld and steadied him. Unfortunately for everyone, the victories they won for him left Hitler convinced that he truly *was* a great captain. From the beginning of his Russian campaign in 1941, he interfered more and more frequently and always more or less disastrously (see page 293). He was strengthened in his conviction of his own genius by the Russian 1941–1942 winter counteroffensive. Rather than allow his commanders to wage a mobile defense that would involve some withdrawals, he ordered his troops to hold in place and fight for every inch of ground. *That* they managed, though at great cost, the hurriedly regrouped Russian armies being short of artillery. Thereafter, "hold in place" became his pet strategy. Though German armies excelled in a mobile, flexible war of maneuver, he constantly tied them down to static defense of untenable positions—Stalingrad, Tunisia, and many others—where the enemy's superior numbers and fire power eventually would overwhelm them. His message to Rommel, who was desperately attempting to get his overmatched army away from El Alamein (November 1942), was typical of those received by too many German commanders:

> In the situation in which you find yourself there can be no other thought but to stand fast, yield not a yard of ground and throw every man and gun into the battle. . . . It would not be the first time in history that a strong will has triumphed over the bigger battalions. As to your troops, you can show them no other road than that to victory or death.[22]

Increasingly, Hitler invoked will power and morale, in a manner reminiscent of Du Picq and Foch, as a substitute for his dwindling material strength in manpower and weapons. He also increasingly meddled with the tactical conduct of battles already in progress, and practically never for the better. Isolated in his command bunker, Hitler lost all personal contact with the battlefronts and all concept of the conditions under which his troops were fighting. He wasted them wholesale; a report of heavy junior-officer casualties brought only the remark "But that's what the young people are there for."[23]

In addition to his strategic and tactical blunders, Hitler failed to manage his war intelligently. His civil government of Germany was a hash of conflicting and overlapping jurisdictions wherein the established governmental organizations and civil service had to function in competition with more or less parallel organs of Hitler's Nazi (National Socialist German Workers) party. Local party leaders (*Gauleiters*) frequently ruled like powerful feudal lords; they often were inefficient or corrupt, but many of them had been Hitler's earliest followers, and no one else could discipline them. Both dreaded the SS (*Schutzstaffel*), originally a Nazi party security element but now, under Heinrich Himmler, rapidly becoming a state within a state. The German armed forces became similarly riven. The SS developed a combat arm, the *Waffen SS;* Hermann Goering, the number-two Nazi leader, controlled (and damnably mismanaged) the Luftwaffe (air force) but also several Luftwaffe Field Divisions—one of them a panzer unit—formed from excess Luftwaffe personnel. These troops normally served under Army command, but if unhappy could always appeal to Himmler or Goering. Moreover, these private armies were better armed and equipped and got a better class of recruits than veteran Army divisions. But, being officered, respectively, by fanatical Nazis and Luftwaffe officers, they often suffered unnecessary losses because of insufficient training and hamhanded leadership.

Similarly, responsibility for German war production was split between different governmental agencies. Production of military aircraft actually declined during late 1940; unbelievably, army weapons production fell off by more than a third during the six months after the invasion of Russia. At the same time civilian consumer goods were turned out at peacetime levels. It was 1942 before Hitler thought to bring some order into the business. His young architect, Albert Speer, made "Minister of Armaments and Munitions," amazed the world (and probably himself) by tripling production, and yet Hitler eventually allowed Speer's efforts to be crippled by intrigues among Nazi party leaders who either envied his authority or resented his demand for an all-out mobilization of Germany's resources. He would not permit the employment of German women in factories in place of dubious foreign "slave labor," and he interfered in weapons development as he did in military operations (see page 324). Had Speer been given a freer hand, the German Army might have numbered up to three million men more; its panzer divisions would have had their regulation number of tanks. (During 1940–41 Hitler doubled the number of his panzer divisions for his attack on Russia, only to learn that he lacked the industrial plant to produce the necessary additional equipment in time. He solved the problem by reducing the number of tanks in each division.) In that case, any modern book on strategy would have considerably different contents.

Hitler undoubtedly was a man of considerable intelligence, plus much low animal cunning. He was an avid, if uncritical, student of military history and theory. Frederick the Great was his hero, but he also studied such modern works as Charles de Gaulle's book on armored warfare. At first he showed a real comprehension of strategic problems and a rare ability to isolate the key points of conflicting proposals and deduce a solution from them. His willingness to take risks few professional soldiers would consider made him a tricky, dangerous opponent, and his furious will extracted their utmost efforts from his subordinates.

His mind, however, worked more by intuition than disciplined reason. Also, he had a seemingly inbred dislike of regular routine, clear-cut divisions of authority, and firm decisions. Though he had a computer's memory for facts and figures, his knowledge of the world at large was superficial and flawed with prejudice. He had not learned from his readings in history; like "Kaiser Bill," he had built up a futile surface navy and neglected his submarine force. Worse, he involved Germany in a two-front war against the British Commonwealth, Russia, and the United States. His megalomania blinded him to his great opportunity to go into Russia as a liberator and to strengthen his armies with hordes of volunteers from the Ukraine and other minority areas oppressed by "Great Russian" rule.

His character was an unguessably tangled skein of arrogance, common sense, superstition, brutality, romanticism, and a hundred other contradictory qualities. He loved beautiful landscapes and "Wild West" stories (by a German author), could have been a successful architect, and did his best to wipe out Europe's Jews and Gypsies. Beyond all that, he possessed a strange hypnotic quality that could make intelligent, strong-minded men his willing servants, and burned out his closest associates' ability to think and act independently. Some of this remained until his death. Introduced to him in February 1945, a young officer saw an apparently "senile" man but remembered his "strangely penetrating look . . . an indescribable, flickering glow in his eyes, creating a fearsome and wholly unnatural effect."[24]

As the war went more and more against him, he sought to control personally even small-scale operations, but the task was too grinding; long, regular working hours and his crumbling health finally broke him. As he weakened, the diabolical element of his character increasingly controlled his actions. He wished to destroy all Germany to make himself a fitting funeral pyre. And yet he thought at the last to marry the quiet, devoted girl who had been his mistress of many years before they went down into the dark together.

There were some extremely able men among the military commanders who served these national leaders—born battle captains like Patton and Erwin Rommel; able administrator/managers such as George C. Marshall and Dwight Eisenhower; great strategists like MacArthur and Erich von Manstein. However, since World War I, command had developed into a more collective process, to match the increasing complexities of modern war. U. S. Grant had pondered over his crude maps amid a cloud of cigar smoke, consulted briefly with his quartermaster general as to supplies and transportation, sent a few staff officers out to check on various details, discussed his roughed-out plan with his corps commanders, then moved out to find and fight his enemy. The World War II commander might state a strategic objective, but this had to be thoroughly "staffed," both to determine its feasibility and to obtain the coordination of all the forces to be involved. His weapons were far more powerful than those Grant possessed, but they required immensely greater varieties and amounts of ammunition and more ordnance specialists to keep them in order. Motor vehicles and aircraft gave him greater speed and flexibility but added motor fuels to the bulk supplies he must carry with him. Grant's men, horses, and mules could live on little, scrape supplies from the countryside, or even press on empty-bellied, but no method of persuading a motor vehicle with an empty gas tank to keep moving has yet been devised. Armored divisions could not filter silently across country and down back roads like Forrest's cavalry. Radio gave those commanders—barring "technical difficulties"—almost instant communication with their subordinates but put their words up for grabs by enemy radio intercept stations: Staff conferences to tie in air and naval cooperation sometimes resembled crisis negotiations with unfriendly foreign powers—and all these were only samples of the commander's tribulations. Consequently, any major strategic plan normally embodied the work of many unnoticed majors and lieutenant colonels, some of them highly specialized technicians on esoteric subjects such as weather forecasting and radar jamming. Only a strong, intelligent commander could imbue it with his own thought and personality or grasp the opportunities offered by unexpected developments in the midst of a battle fought over hundreds of miles of front. And the supreme military commander who must control a war waged all around the world on widely separated fronts, each of which required its own strategy, must be a commander indeed—trusted, respected, and obeyed.

Without doubt, the greatest of World War II's nonfighting soldiers was George C. Marshall (1880–1959), Chief of Staff of the U.S. Army. Born in Pennsylvania, a graduate of the Virginia Military Institute, Marshall secured a Regular Army commission and saw service as an infantryman during the Philippine Insurrection. From the first he was marked as a brilliant officer; most of his assignments were with schools and staffs, culminating with his outstanding work under Col. Fox Connor in handling intricate troop movements during the 1918 Saint-Mihiel and Meuse-Argonne offensives—some 600,000 troops, 3,000 guns, and 40,000 tons of ammunition shifted by night without lights into positions some 220,000 French troops were evacuating. Thereafter, Pershing championed Marshall's military career, which otherwise might have withered into retirement before World War II. (Marshall's perfectionism and drive apparently lacked a certain common touch, making him less successful as a troop commander than as a staff officer.) But Black Jack Pershing, still austerely grandfathering the Army from his apartment at Walter Reed Army Hospital, eventually went directly to President Roosevelt to secure the post of Army Chief of Staff for Marshall.

It was a most fortuitous appointment. Marshall took over in September 1939 an army that had been increased from a 1927 low of some 118,750 to some 190,000 in response to the worsening European situation. It had no combat-ready divisions; its weapons and equipment were mostly outdated. By May 1945 that Army (including the Army Air Force) numbered 8,291,336 men and women; though some combat infantrymen might have registered disbelief, it was the best-cared-for army in American history—and Marshall had inspired and guided practically every detail of its growth and employment. Industrial mobilization, selective service, global strategy and logistics, troop training and morale, rearming the French, the special problems of women and black soldiers, sticky questions of commissions for politicians and Roosevelt's sons, and dozens of other problems came to him for solution and were solved.

Experience, intelligence, integrity, and sheer force of character made him the unquestioned leader of the U.S. Joint Chiefs of Staff and its spokesman in the Anglo-American Combined Chiefs of Staff who were responsible for planning the strategic conduct of the war.[25] He worked harmoniously with Roosevelt and could head off or handle that president's occasional ebulitions—though usually relying on the advice of his senior military officers, Roosevelt did have interludes when he could outfantasize Churchill. Congress found him a cooperative, reliable guide on defense questions. Like Halleck, he was "all for the service," without political interests or any desire except to serve well. His one specific ambition was to command the invasion of France, which he had especially devoted himself to preparing. How successful he might have been as a field commander is, of course, impossible to guess. At least, being firmer and more awe-inspiring than Eisenhower, he probably would have squelched the bickering between British and American senior commanders that haunted Eisenhower's campaign. Probably he would have waged a bolder, more aggressive war. But he had made himself indispensable in Washington, and Roosevelt felt unable to let him go.

A demanding, exacting superior, Marshall won the loyalty of his staff, though he could be a veritable Schlieffen in getting work out of them. "He'd give you a six-inch stack of papers, and tell you he wanted a report on their contents in thirty minutes. You'd sprint back to your desk, and find your phone ringing—it would be Marshall wanting to know why you weren't already finished!"[26] At the same time, he was a man of endless quiet kindnesses and thoughtful acts, courteous and gentle, always concerned with the welfare of the American soldier.

From his Washington desk, Marshall coordinated the operations of American armies all around the world and these operations with those of the U.S. Navy and the forces of our allies. Yet he was far more than a "chairborne" manager of American military resources.

Whenever he could, sometimes as part of his travels to various high-level international conferences, he visited the armies in the field. There he made a special point of talking with enlisted men and junior officers without the presence of their superiors. He also used trusted liaison officers, proved veterans who "knew how to get soldiers out of the rain,"[27] to prowl the combat zones, talk to soldiers, and pick up combat lessons. By such measures he kept intellectual and emotional contact with the fighting fronts. His subordinate commanders found him understanding and quick to support them against outside interference but strict in demanding their best efforts.

Marshall saw war in the same stark terms as U. S. Grant: Get at your enemy and smash him! His strategy was to mass the maximum possible weight of men, materiel, and fire power against the enemy where he considered success would bring the quickest, most decisive results. Hence his insistence on an all-out invasion of France at the earliest possible date (he would have preferred to launch it in 1943 instead of a year later) and his opposition to many British-proposed minor operations around the Mediterranean.

To him, the armed forces' one responsibility was total victory, to be won as quickly as possible. The shape of the postwar world would be the responsibility of the nation's civilian leaders. Unless so ordered by the president, he would not risk his soldiers' lives for the capture of objectives that had little military value, regardless of their political or psychological importance. It was a narrow, powerful conviction based on the American tradition of the supremacy of civil authority over the military. The tragedy was that the gradually dying Roosevelt failed to provide effective political guidance during the last crucial year of the war, and also kept his new Vice President Truman in complete ignorance of his international dealings. Marshall continued to shape the general strategy and provide the means to win the war. By his own creed, whatever came, he could do no less—and no more.

By traditional standards, Marshall hardly can be classed as a superstrategist. He never personally, in Iago's words, "set a squadron in the field" or matched wits with an enemy commander. Yet in his own way he was one of the mightiest soldiers in all history, truly—as Churchill hailed him in 1945—the "Organizer of Victory."

Marshall also selected the American commanders who led its armies, corps, and even divisions. Early in his career he had begun studying his fellow officers, noting their strengths and weaknesses. On the whole, he proved an excellent judge of character. Practically all of his selections would do at least an acceptable job; very few were failures. Two of his most hopeful selections were Dwight D. Eisenhower and Omar N. Bradley.

It is one of military history's wryer facts that the mightiest expeditionary force the United States ever put into battle—by March 1945, over three million strong, to which were added some forty-one British, Canadian, French, and Polish divisions—was commanded by a general who never had heard that proverbial "shot fired in anger" that makes the field soldier out of a mere man in uniform. Dwight D. Eisenhower began as an infantryman, transferring to the Tank Corps in World War I. He did not get overseas but was one of the few field-grade officers to receive the Distinguished Service Medal for duty in the United States. Most of his subsequent service was with staffs and schools; clearminded and industrious, he made an excellent record everywhere. At the same time, he inwardly remained (as service talk occasionally put it) the tough kid from the wrong side of the tracks in Abilene, Kansas; the West Point cadet who was a middling good student, a promising football player (until he injured a knee); and not infrequently in trouble over his tendency to dodge or bend regulations. His temper was a potent concealed weapon; when crossed or thwarted, he could release a fury "that shriveled everyone in the room."[28]

His appointment in 1942 to command all American forces in Europe, and the next year to command the Allied Expeditionary Forces that would invade Europe in 1944, surprised many, but he soon exhibited a positive genius for getting officers of different nations to work willingly and efficiently together, and in picking his way through the coils of conflicting French political/military factions in North Africa. The detailed planning for the invasions of North Africa, Sicily, Italy, and Normandy was done for him by Anglo-American staffs, but Eisenhower was responsible for approving their work and especially in 1944 for making the decision to launch the invasion of Normandy through a brief "window" in a prolonged period of bad weather.

Ashore in Normandy, Eisenhower began a careful, conservative war, with many hesitations. He could keep his wits in an emergency, as during Hitler's surprise December 1944 Ardennes counteroffensive (Battle of the Bulge), and made no gross errors. His most consistently held strategy—that of attacking on a broad front, from the English Channel south to Switzerland—did bring him a major victory in the Rhineland (February–March 1945) but largely because of Hitler's senseless insistence that the riddled German armies stand and fight west of the Rhine instead of withdrawing behind it. Completely overmatched, they fought and were shattered. Too few got away to defend the east bank against subsequent Anglo-American crossings.

Seemingly, Eisenhower never acquired a sensing of the actual fighting or the combat soldier, remaining very much of a rear-area operator, sometimes not too effectively in touch with the front. Even in the last weeks of the war, with the German armies obviously collapsing, he tended to take counsel of his fears. In North Africa and Sicily (1942–43) he had been given Gen. Harold R. L. Alexander, a much-decorated North Ireland aristocrat, who had been waging various wars since 1914, as his ground forces commander to handle the actual fighting. Though much senior to Eisenhower, Alexander discharged his mission with imperturbable aplomb, impressing even Patton. Eisenhower had hoped to have Alexander with him again in France, but Churchill chose to leave Alexander to direct the sideshow Italian campaign, replacing him with Bernard L. Montgomery, of whom more later. It was not a happy choice.

Generous toward his favorite commanders, Eisenhower could be picky with others, such as the capable Gen. Jacob M. Devers, who commanded the U.S. 6th Army Group that held his right flank. Envisioning himself as a crusader against evil, he was curt, if not contemptuous, toward surrendering German generals and supported harsh occupation policies for a conquered Germany. Lacking definite political guidance as to America's postwar objectives in Europe, he kept loyally to what he understood to be Roosevelt's grand design, doing his utmost to maintain excellent relations with the Russians. Years before, between the great wars, Fox Connor had put Eisenhower to reading Clausewitz, three times in all. Possibly Connor should have ordered a fourth perusal. Eisenhower comprehended the fact that wars are waged for political objectives; he made certain that his advancing armies sealed off the land approaches to Denmark from the Russians, and occupied as much of Austria as possible. But when the leaders of a Czech revolt in Prague appealed to him for help, he would not cross the Russians by advancing out of the thin strip of western Czechoslovakia they were willing to allow him. Accordingly, the Russians got the glory and credit for liberating the Czech capital (*and* numberless watches) somewhat later, and the American zone was promptly plastered with posters showing a beautiful blonde Russian tank commander, red banner in hand, urging the Czechs to be properly grateful to the heroic Red Army. (From close observation of that army, the author is convinced that the blonde was a figment of the artist's imagination.) Abruptly halted by this decision, Patton dourly observed that it might be well to let the Russians begin worrying about maintaining good relations with the United States. However, such an opinion then was downright subversive.

This placating attitude toward the Russians was reinforced by Eisenhower's naiveness concerning the nature and functioning of the Russian state and of the Communist party organization that ruled it. He had expected to deal directly with his apparent Russian opposite number, Marshal Grigori K. Zhukov, and it took time and frustration before he comprehended that Zhukov, too, could not blow his own nose without clearance from Stalin.

Eisenhower's exact military qualities are unexpectedly illusive. He was more of a chairman of a council of war and a military manager than a Great Captain. His strategic concepts were pedestrian at best; possibly they represented his real abilities less than his desire to run no unnecessary chances. The American public and mass media have seldom been kind to unsuccessful generals.

In the end, he accomplished his assigned mission (see page 323). Considering his overwhelming superiority in air power, artillery, armor, and logistic support, he hardly could have done otherwise. But in the flush of victory most people remembered only his normal geniality and contagious self-confidence. He went on to be a competent two-term president of the United States, popular at home and across most of the world.

Omar N. Bradley, sometimes called ''Omar the tentmaker,'' was a West Point classmate of Eisenhower's. (He stood forty-fourth in the class of 1915 to Eisenhower's sixty-first place.) Like Eisenhower, he began as an infantry shavetail and did not get overseas during World War I. Soon noted as a reliable, hardworking officer, he had risen to brigadier general and commandant of the Infantry School by the beginning of World War II. In early 1943, Gen.

George C. Marshall sent Bradley to North Africa to study the far-from-satisfactory performance of American troops. He functioned efficiently as a corps commander in Tunisia and Sicily; Eisenhower made him commander of the U.S. First Army for the invasion of Normandy, then raised him to command of the U.S. 12th Army Group, the largest field command ever led by an American general. Bradley was to Eisenhower as Sherman had been to Grant. Eisenhower reported him as having "brains, a fine capacity for leadership, and a thorough understanding of the requirements of modern battle. He has never caused me one moment of worry."[29]

Homely, bespectacled, plain and matter-of-fact in dress and conduct, speaking with a soft Missouri drawl, Bradley reflected quiet competence and homespun virtue. War correspondents christened him "the G.I. General"; his staff was devoted to him. The inner man was sensitive, ambitious, quickly irritated. He would come to regard Montgomery as practically a personal enemy; his first book, *A Soldier's Story,* dealt unkindly with him. (His last book, *A General's Life,* is an unfortunate blackguarding of other, sometimes better, soldiers, now mostly dead.[30])

Possibly promotion to army group command stretched Bradley's capabilities. He rendered much capable service and continued to please Eisenhower, but a string of missed opportunities and blunders followed him. Bradley flinched from attempting to close the Falaise-Argentan gap, to bag German units escaping from the Normandy front. He fed his troops into the "death factory" of the Hurtgen Forest instead of bypassing it and was surprised off-balance by Hitler's Ardennes counteroffensive. After the passage of the Rhine, he devoted too many troops to mopping up miscellaneous Germans trapped in the "Ruhr pocket," thereby seriously reducing the number of divisions available for continuing the advance into Germany, Austria, and Czechoslovakia. He also was violently opposed to any Anglo-American drive on Berlin, possibly because he was convinced that large German forces would make a last-ditch stand in a "National Redoubt" in the mountains of southern Germany/western Austria. This Redoubt proved a myth, after Bradley flooded the mountains with troops looking for it, but there was no advance on Berlin. In one of his many disgusted moments, Patton complained that every time Eisenhower and Bradley got together they turned timid.

Bradley continued to serve the United States to the best of his abilities, not in the least during 1945–47 when he took over and rebuilt the politician-ridden Veterans Administration. Eventually he drifted away into a hospitalized invalid's existence.

Only one of the senior American generals in Europe led, like MacArthur, in Old Army style. George S. Patton never quite reached independent command, but he was a legend among fighting men. The last of the great American cavalry generals, born—unfortunately for him, luckily for the United States—a century or so too late, he came out of the older, tougher horse-soldier Army that Eisenhower and Bradley never knew, where a first sergeant's fist embodied the Articles of War. Tall, a dead shot, and an internationally famous swordsman, he loved danger, thought war the highest adventure, and believed utterly that an American citizen had no prouder privilege than to serve his country in arms.

Acknowledging that Patton had "extraordinary and ruthless driving power," Eisenhower insisted that he was "unbalanced" and possessed by a "love of showmanship and histrionics"[31] and so allowed him no higher command than that of the U.S. Third Army, a part of Bradley's 12th Army Group. Obviously Patton must have grated on Eisenhower, probably in more ways than the furors he caused among nice people back stateside by slapping apparent malingerers in Sicily (an action much approved by his combat soldiers) or allegedly failing to pay suitable tribute to Russia in a short talk to a ladies' group in England.

"Georgeous Georgie" Patton (also the "Green Hornet" and "Old Blood and Guts") came from a wealthy California family; his ancestors had followed "Bonnie Prince Charlie" Stuart and Robert E. Lee. He went naturally to VMI and then West Point; a certain infelicity with mathematics kept him there an extra year until 1909. He made a love marriage with a gallant gentlewoman who rode beautifully and would willingly ruin a dress showing nervous generals how easily an early model of the Christie tank could be driven across country. As Black Jack Pershing's aide-de-camp in Mexico, he killed one of Pancho Villa's "generals" in a man-to-man shootout; in World War I he commanded the first U.S. tank brigade, winning the Distinguished Service Cross and a "hole in his hip about the size of a teacup" in the Meuse-Argonne.[32]

Between the wars he was a fox hunter, polo player, sometimes a sore trial to various commanding officers. He also was both a devout Episcopalian and one of the Army's most gifted speakers of "Profane." His big personal library dealt with great captains such as Hannibal; Scipio Africanus, who defeated Hannibal; Caesar; Napoleon; and Stonewall Jackson. Reading himself into bygone wars, he stored their military geography—roads, rivers, hills, and the key chokepoints they formed—in a near-photographic memory. Especially he pondered the problems of morale and leadership. "To be a successful soldier," he later advised his soldier son,

> you must know history. Read it objectively. . . . What you must know is how man reacts. Weapons change, but man who uses them changes not at all. To win battles, you do not beat weapons—you beat the soul of man, the enemy man. . . . You must read biography and especially autobiography. If you will do it you will find that war is simple. Decide what will hurt the enemy most within the limits of your capabilities to harm him and then do it.[33]

Like Pershing, he was the strictest of trainers, disciplinarians, and commanders. He made no compromises with perfection, seldom was satisfied that his victories were as complete as they might have been, and led from the front by example, applying blistering invective or quiet sympathy as might be needed to keep his offensives rolling. His men wore neckties, saluted smartly, kept their hair cut and their weapons clean, and came to feel they were the world's finest army—as they may well have become. Patton saw to hot meals, dry socks, regular mail, and quick and skilled medical attention. He worked with a quiet staff that proved itself one of the best in American history. His most dashing operations were carefully planned and organized. Older, far more experienced than Eisenhower or Bradley, he had a better sense of strategy than either of them. Repeatedly he foresaw and won an opening for a quicker victory, only to be checked because his superiors could see only risks where he sensed opportunities. His decisive part in the Battle of the Bulge (see page 327) got him small thanks from Eisenhower.

Possibly his self-created image (he was an excellent actor, known to practice martial poses and facial expressions before his mirror) of a flamboyant, profane, hell-for-leather martinet was actually too impressive, hiding the imaginative, devoted soldier he truly was. But he *was* overeager for glory and men's applause, too impulsive in word and action. He understood soldiers, but never politicians (in or out of uniform), public opinion, or the carnivorous species of war correspondent. Come what might, he would speak his mind, without regard for time, place, or audience. And his exacting standards of dress and military courtesy, his open enjoyment of the pomp, pride, and occasional comforts of high command outraged certain sensitive observers who confused democracy with slack discipline and dirty necks and rifles. Moreover, unlike Eisenhower (and very much like Montgomery), once the war was over Patton instinctively treated a brave enemy in captivity as a fellow soldier. Also, he regarded the Russians with open suspicion. Neither attitude was popular in an America still seething with unexpended crusading zeal. When Patton compounded such unfashionable views by speaking bluntly on the asininity of American occupation policies in Germany, he was overwhelmed by what old Winfield Scott once called "the fury of the non-combatants."[34] A few weeks later he was fatally injured in an automobile accident and died slowly, gallantly through twelve days.

Since Eisenhower refused to trust Patton as an army group commander, we shall never know if he could have adapted his highly personal style of leadership to that higher command level. The Germans had few doubts: To them, he was another Guderian, the one American general they feared.

Among the "bomber barons" only Gen. Carl A. ("Tooey") Spaatz had solid distinction as a strategist. A methodical Pennsylvanian (his name originally was "Spatz"), he graduated midway in the West Point class of 1914, went into the Infantry, but soon transferred to the Army's tiny Air Service (then part—since it was conceived as a courier service—of the Signal Corps). Assigned to Pershing's 1916 Punitive Expedition into Mexico, Spaatz flew flimsy, underpowered aircraft that were apt to fold up anywhere over the wastes of northern Mexico, leaving the pilot to limp home across a frequently hostile countryside. World War I brought him repute both as an instructor and a fighter pilot. Afterward, he had a major hand in developing the Air Corps theories of the employment of masses of heavy bombers in precision raids on an enemy's strategic industries. At the same time, his habit of personally checking his commands and their functioning to the last detail had his subordinates jesting about always seeing "Spaatz before their eyes." In 1944, when he took command of the U.S. Strategic Air Forces in Europe and, of necessity, became another desk general, he could issue an order involving 1,500 airplanes with no more show or ado than "a groceryman ordering another five cases of canned peas."[35] He also opposed aerial operations that might involve heavy civilian casualties. His "oil strategy" (see chapter VIII) was the only truly decisive World War II air campaign in Europe.

The ablest of British strategists and probably the best Anglo-American strategist in Europe was Gen. Harold Leofric Rupert Alexander, later Field Marshal the Earl Alexander of Tunis. A second lieutenant of the elite Irish Guards in 1911, he came out of World War I an acting lieutenant colonel, much beloved by his "Micks" for his efficiency, courage, fairness, and care of them—and also for a jaunty humor that found comedy in the most desperate affairs. After the war, on detached service in Latvia with a British mission, he commanded local Baltic German militia against the Bolsheviks with great success, then went back to his Irish Guards for duty at Constantinople, holding Greeks and Turks from each others' throats. Advanced schooling and brief staff duty were followed by command of a brigade of British and Indian troops and two victorious campaigns on India's Northwest Frontier.

A division commander in 1940, he led the British rear guard at Dunkirk and was the last Englishman off that battered beach. His next assignment was Burma, where the sudden Japanese offensive was overrunning the weak British forces. Unable to send substantial reinforcements in time, Churchill dispatched Alexander. There was no hope of saving Burma, but Alexander got his troops out in decent order. His next command was the Middle East, with Montgomery under him as commander of the British Eighth Army. Montgomery actually fought and won the battle of El Alamein—and naturally claimed credit for everything connected with it—but the overall campaign was Alexander's. Thereafter, Alexander (under Eisenhower's general command) would bag the Axis forces in Tunisia and conquer Sicily. Through 1943–45 he slowly mastered the German forces in Italy. This Italian campaign, waged chiefly to pull in and use up German forces that otherwise would be committed in France or Russia, had elements of heartbreak. Alexander's best divisions, English,

French, or American, were steadily drawn off to reinforce Eisenhower; in replacement he got a Brazilian division, a Jewish brigade, reorganized Italian troops, the famous 442d Regimental Combat Team of Japanese Americans, and the unfortunate U.S. 92d Infantry Division of black soldiers. There was always one more jagged line of mountains, one more flooding river to force; the Germans fought stubbornly and shrewdly. But on May 2, 1945 the German forces in Italy surrendered—the first of the unconditional capitulations that ended the war in Europe.

Alexander, in even the captious Bradley's judgment, was "the outstanding general's general of the European war," with the "reasonableness, patience, and modesty of a great soldier . . . self-effacing and punctilious."[36] After a brief period of mistrust of American troops and commanders, he learned to work with them as effectively as he had worked with other nationalities and rivaled Eisenhower in his handling of multinational forces. As a general he was highly competent rather than brilliant. Based on his long experience, he evolved his own unique "two-handed" strategy of confusing and wearing down an enemy with alternating attacks from different directions until he was off-balance and ready for a knockout punch, which might be delivered by either "hand." He disliked Montgomery's "set-piece" battles, with their rigid planning and scheduling, preferring to give his subordinates general missions, then modifying these as the ebb and surge of battle revealed the enemy's vulnerabilities—much like Napoleon's system of engaging all across an enemy's front until he could determine the battle's shape. Alexander also had the knack of shifting his forces suddenly and secretly, to give unexpected weight to a sudden punch by one of his "hands," and was unsparing in following up a beaten enemy.

In some respects, he was more of a leader than a commander. Out of an old North Ireland family with traditions of service and loyalty to their King, a soldier, an officer, and a gentleman by breeding, choice, and education from boyhood, he took his orders and did his duty—an attitude he expected from his subordinates, preferring, like Robert E. Lee, to lead by example and persuasion. This method usually succeeded, but not against Montgomery's abrasive self-righteousness or Mark W. Clark's vanity. (Clark, commanding the U.S. Fifth Army under Alexander, was able, energetic, and courageous, but cankered by lust for publicity. He would turn aside from the destruction of the German Tenth Army, which probably would have broken German resistance in Italy, for the fleeting newspaper fame of entering Rome.) It was agreed that Alexander's campaigns went best when he had a sharp, no-nonsense chief of staff.

He was always the same, unmoved by either disaster or victory, faultlessly uniformed (though usually with an individual nonregulation touch), perfectly at ease with soldiers of any nation, setting an example of cheerful calm and common sense. His headquarters always were well forward, and he did his own reconnoitering before any major attack, coming forward quietly and almost unaccompanied. He wasted neither time nor men.

Alexander is practically unknown to most Americans, but most Americans have heard of Field Marshal Bernard L. Montgomery. Like Patton, whom he greatly resembled in a puritanical British way, he never quite achieved the independent command that would give him the opportunity to prove himself as a strategist. Also, like Patton, he was an iron disciplinarian, an inspired and inspiring trainer, a soldier out of instinctive patriotism and a love and pride of soldiering. He also sought praise and glory at least as avidly as Patton. The two were natural rivals. Having endured the bloody years of 1914–18, Montgomery had no intention of wasting his men in futile, ill-conceived attacks. England was a small nation, with definitely limited manpower. He preferred a careful ''set-piece'' battle, carefully planned and prepared beforehand and employing all available forces. El Alamein had been such a battle; the British landings on D day in Normandy were another, better handled than Bradley's, in part because of Montgomery's judicious use of Hobart's ''funnies'' (see page 214). His subsequent conduct of the Normandy battle, launching limited attacks to draw the German panzer reinforcements against his British troops while Bradley's Americans wore their way through the German lines farther west, was well done.

Also, Montgomery was not without—as is so often charged—dash and vision. In fact, his plan for Operation MARKET GARDEN (a combined three-and-a-half-division airborne drop and massive armored thrust in September 1944 to seize a crossing over the Rhine at Arnhem) was the most audacious Anglo-American operation of the war. It failed because of bad weather, insufficient air transport, the snowballing of many minor tactical errors, and also Montgomery's own failure to accept the proffered help of the Dutch resistance movement, and to listen to an overworked intelligence officer's warning that there were German tanks near Arnhem.

Small, brisk, cocksure, and finicky, a nonsmoker and teetotaler, Montgomery was always his own man and his own worst enemy. His quick, incisive mind picked out the flaws and pretensions in other officers' plans, and his frequently acid tongue dissected them without courtesy. Mentally, some English soldier remarked, Montgomery was quick as a ferret—and about as lovable. Somehow his scheme of things acknowledged neither superiors nor equals; few of his actual superiors could control him and those only with much difficulty. Supremely certain of his own capabilities, he expected to be given whatever he needed to carry out his own plans, without interference, and could be utterly indifferent to everything else. He never learned when to drop a lost argument and apparently was proud of his ability to infuriate his fellow officers. Eisenhower treated him with vast patience, which Montgomery too often repaid with snubs and arrogance. Eventually, concluding that Montgomery was a psychopathic egocentric, Eisenhower would come close to relieving him.

Montgomery's major complaint was quite simple: Eisenhower was a nice person and an excellent supreme commander, but he knew nothing about handling troops in the presence of the enemy. During preparations for and the first months of the Normandy invasion, Montgomery had, like Alexander in Tunisia, acted as ground forces commander for Eisenhower and did so with considerable tact and understanding. As the number of American divisions in action increased, however, such an arrangement began to raise unfavorable publicity in the United States. Marshall ordered Eisenhower to take over the ground forces command himself; Montgomery reverted to command of his 21st Army Group only, on a par with Bradley's 12th Army Group. It was a distinct comedown, and Montgomery greeted it with outrage and no understanding whatever. He would spend the rest of the war attempting to recover that lost authority, sometimes with help from Churchill but much to the embarrassment of his own staff and other British officers. This continual bickering was an emotional wound to the efficiency of the Anglo-American command, but Montgomery persisted. His belief that Eisenhower lacked practical experience and was overloaded as both supreme commander and ground forces commander was not unjustified. Montgomery himself, however, was not the man to take over and put things straight; he had made himself detested by most senior American officers and was unpopular (if sometimes for trivial reasons) with the American press. Also, he had been showing a lack of the final quality of a great strategist. He would close with his enemy and defeat him by able application of *mass* and *maneuver,* but he never managed that enemy's destruction. At El Alamein, Normandy, and Antwerp, hard-core German units fought their way out of impending disaster, to fight again.

Yet there was another Montgomery, the idol of the British public and of his soldiers. Like the Duke of Wellington, to whom he often was compared, he exacted the strictest discipline but combined that with an easy laxity as to dress and unofficial modifications to increase the comfort of army motor vehicles. Some of his units on the march resembled Gypsy caravans; his soldiers might or might not wear their steel helmets. Constantly visible to his men, he would pause to talk in clear-cut terms concerning the existing situation and his intentions. An amazingly effective speaker, he could put the most complex subject into simple terms, talking without notes or strain, effortlessly getting and keeping his listeners' attention.[37]

The son of a bishop in far-off Tasmania, a constant rebel against a strict mother, a sharp boy athlete, obnoxious cadet at Sandhurst, and deadly serious officer, Montgomery had been almost forty when he fell in love with and married a gentle girl who disliked things military. Her death ten years later left him in desolation. And for all his thorny, insensitive pride, he treasured good soldiers and brave men. After El Alamein he shared his meal with a captured *Afrika Korps* general while discussing their recent battle. And in December 1944, with Hitler's counteroffensive cresting through the Ardennes and all confusion loose, Eisenhower gave him command of Bradley's units north of the ''bulge'' the panzers had driven into the American lines. At the vital crossroads town of St. Vith, the U.S. 7th Armored Division had waged a splendid delaying action but finally was being knocked to pieces by converging panzers and volksgrenadiers. Montgomery, organizing a new line behind it, overruled an American general who wanted them to hang on in what would have been a reenactment of Custer's last stand: ''They can come back with all honor. . . . They put up a wonderful show.''[38]

Because of Stalin's system of high command—and also the traditional Russian belief that history should be used to indoctrinate, not educate—it is impossible to estimate the true strategic skills of the Russian marshals. For example, Grigori K. Zhukov was much publicized, being credited for halting the last 1941 German drive on Moscow and planning the Stalingrad counteroffensive, yet one high-ranking Russian defector reported him notable only for stupidity and ferocity, a sort of military front man whom Stalin could use and then deposit in some one-jackass backcountry garrison when finished with him.

The British have a tendency to ''adopt'' worthy foemen. Napoleon was ''Boney'' to English soldiers and sailors who fought him; not a few of them regarded his return to France and eviction of the Bourbons in early 1815 with a curious approval. It was in this tradition that Winston Churchill, somberly reporting fresh defeats in North Africa to the House of Commons in January 1942, ended with a sincere salute from one warrior to another: ''We have a very daring and skillful opponent against us, and, may I say across the havoc of war, a great general.''[39] Not all Englishmen, especially the bloodthirsty-by-proxy types, approved, but British soldiers in North Africa consciously agreed. As Gen. Sir Claude Auchinleck, one of his ablest opponents, complained, ''There exists a real danger that our friend Rommel is becoming a kind of magician or bogey-man to our troops, who are talking far too much about him.''[40]

Erwin Johannes Eugen Rommel (1891–1945) began as a small, quiet, towheaded child in a little town in the once-independent state of Württemberg in southwest Germany. By tradition, Württembergers are thrifty, shrewd, and realistic, with their own wry humor and ''Swabian'' dialect—excellent fighting men, if somewhat short of parade-ground graces. Rommel's father and grandfather were schoolmasters and mathematicians; there was no family military tradition, but Rommel went into the Imperial German Army at eighteen as an officer candidate, like Clausewitz, serving in the ranks and qualifying as corporal and sergeant before he could become a second lieutenant in 1912.

From the start he was an excellent officer, totally dedicated to his profession except that he fell in love early with a slender, sweet-faced girl, whom he married while on leave from the front in 1916. She had courage, understanding, and humor, could ski, canoe, ride, and climb with him; when they were apart, he wrote his ''Dearest Lu'' almost every day, even from the midst of desperate battles.

Rommel's service in World War I was a succession of hairsbreadth feats, especially with Württemberg mountain troops in Romania and Italy. Gifted with incredible hardihood and physical toughness, he could move across country like death in the wind and rain, suddenly surfacing in the enemy's rear to dislocate their front. Repeatedly wounded, he earned the Iron Cross, both second and first class, and the rarely bestowed order *Pour le Mérite* (established by Frederick the Great in 1740) and became a legend while still only a captain. Soldiers swore he possessed a mysterious *Fingerspitzengefuhl*—a "finger sense" or combat intuition that warned him of dangers and opportunities. He could be relied on absolutely; his hunches and tactical skill kept his casualties low.

Between the world wars, while an instructor at the Infantry School, he wrote a short, vivid book, *Infanterie Greift An,* on infantry tactics.[41] Hitler read it from the viewpoint of a frontline infantryman; liked it, and gave Rommel the command of his personal escort battalion. After the Polish campaign he offered Rommel his choice of assignments: Rommel asked for an armored division. He got the 7th Panzer Division, which he led across France in 1940 with the same skill and daring he had shown with his Württemberg mountaineers twenty-odd years before, his unexpected in-and-out tactics earning it the nickname "The Ghost Division."

Thereafter, the Italians having managed to get themselves thoroughly routed in North Africa, Hitler dispatched Rommel with a small "blocking force" (at its largest, just before El Alamein, four German divisions) in early 1941 to stiffen his allies. Rommel would be under Italian command and largely dependent on Italy for supplies; his mission was to prevent further English advances. Such were his orders, but Rommel quickly saw the strategic truths of his position. If he stood on the defensive, he eventually would be overwhelmed as the British built up their forces in Egypt. His only chance lay in keeping the *offensive* through *surprise* and *maneuver* to knock the English off-balance. If successful, he might break into Egypt and even cut the Suez Canal, thereby completely dislocating the British position in the Middle East. Acting on his own initiative, he attacked. Through months of swirling fighting, he broke one English army after another as the battles shifted back and forth across the Western Desert (the coastal area between the Libyan port of Tripoli and the Egyptian seaport of Alexandria). He had better weapons and better-trained soldiers, and he knew how to use them. Cautious and calculating before launching his offensives, reckless and driving once they were launched, he dominated the battlefields and his opponents' minds. Even his Italian auxiliaries, badly armed and neglected by Mussolini and his Duke of Plaza Torro generals, could sometimes act the ''lions'' their mountebank *Duce* exhorted them to be. On occasion Rommel bit off more than he could chew, but—like Caesar—he was a master at improvising sudden shifts that always got his troops out, if they did not abruptly reverse the battle. Constantly in the middle of the fighting, he could react swiftly; he was as dangerous retreating as advancing; apparently defeated, he would be back on top of his opponents before they could regroup. He set the personal example of toughness, endurance, and drive, and he fought a gentleman's war. Hitler made him a field marshal for the capture of Tobruk in June 1942; Rommel wrote ''Dearest Lu'' that night that he would rather have had another division. Had Hitler done so, Rommel probably could have smashed through the partially organized El Alamein position where his pursuit finally ran down from exhaustion, sixty-five miles short of Alexandria and the canal.

Because of the complete inability of the Italian Navy and Air Force to protect convoys during the approximately four-hundred-mile voyage from Italy to North Africa from British submarines and aircraft, and Hitler's preoccupation with his Russian campaign, Rommel's logistic support dropped from little to almost nothing. Unable to maneuver extensively for lack of gasoline, forbidden to withdraw, Rommel could only dig in and watch the newly arrived team of Alexander and Montgomery build up their forces and the Royal Air Force gain control of the skies. In late September, Rommel, too ill to command effectively, had to go home on sick leave. Montgomery attacked on the night of October 23, throwing masses of infantry and new American tanks forward behind a thunderous artillery barrage. He had inspired and trained his army for this battle; a large-scale deception plan involving hundreds of dummy vehicles and installations had persuaded the Germans that it would be a month yet before the British were ready to attack. It was a hammer blow, but daybreak found the British still entangled in Rommel's ''devil's gardens'' of minefields and barbed wire that covered the Italo-German front. Leaving his hospital bed, Rommel flew back to Egypt to take over a battle already three-fourths lost. Still, taking a high price in Montgomery's tanks and men for every position he had to yield, Rommel had hopes of extricating most of his troops—only to have his tautly controlled withdrawal halted by Hitler's order (see page 266) for a last-ditch fight to the death. It took Rommel one day to decide to disobey Hitler, but that delay cost him heavy casualties.

In the end, aided by unexpected rainstorms and Montgomery's inability to get reorganized for a prompt pursuit, Rommel brought the survivors of his best units away. He fell back across Libya into Tunisia—over fourteen hundred miles in three months—baffling Montgomery's efforts to head him off.

Meanwhile, the situation all across North Africa had changed drastically. Through November 8–10, Anglo-American forces had landed in Morocco and Algeria, brought the French forces there into a somewhat grudging alliance, and begun to edge into Tunisia from the west. Hitler was setting up a new North African base in Tunisia, pouring in troops that, sent six months earlier, would have won Egypt for him. Rommel won his last victory when he gave the Americans a spanking at Kasserine Pass, but his heart was not really in the business. He was very sick, apparently with jaundice and various complications thereof. (Anyone ever afflicted with that disease can only marvel how he managed to keep on his feet, let alone wage a canny war.) It was plain to him that the Italo-German forces in North Africa were in the position of a small nut dropped into the jaws of a heavy-duty nutcracker. He flew to Germany to plead with Hitler to withdraw them while there still was a chance. Hitler refused, and sent him back to the hospital. Eisenhower would loftily write that "Rommel himself escaped . . . earnestly desiring to save his own skin,"[42] which, coming from the most-unshot-at of American generals, *does* seem unnecessarily sanctimonious!

Once released from the hospital, Rommel held several commands but finally was sent to inspect Hitler's Atlantic Wall (the line of coastal fortifications Hitler was building from Denmark to Spain against a foreseen Anglo-American invasion) and to take any measures necessary to perfect it. This created a typical Hitler command muddle. Most of this area was under the command of the very senior Field Marshal Gerd von Rundstedt, himself an excellent, if conventional, strategist. Rundstedt requested the situation be clarified. Hitler then made Rommel an army group commander under Rundstedt, with the mission of defending the coastline of Holland, Belgium, and northern France—some thirteen hundred miles in extent and the area against which the Anglo-American invasion was certain to strike.

Rommel's war in North Africa had been largely a matter of tactics—or to use an old term, grand tactics—but he had grasped the essential strategy by which that war might have been won. Confronted with his new mission, he promptly showed a vital quality of a great strategist—the ability to learn by experience and to apply his new knowledge. Egypt and Tunisia had taught him that, as he expressed it, "Anyone who has to fight . . . against an enemy with complete air superiority, fights like a savage against modern European troops . . . with the same chances of success."[43] Captured American equipment he had inspected after Kasserine had convinced him of the "overwhelming industrial capacity of the United States."[44] At the same time, El Alamein had shown how long a properly prepared defensive system *could* hold out against the attack of superior enemy forces, even though that enemy had complete command of the air.

When he took over his command in January 1944, Rommel found the Atlantic Wall was more fiction than fact—something out of Hitler's cloud-cuckoo-land. The major ports had been fortified and some work done in the Calais area where the English Channel was narrowest, but most of the Normandy coastline was practically open. Many of the troops available were second- and third-line units, filled up with foreign volunteers and short of transportation. Officers had become lax through years of inactivity in that quiet sector. Also, Rundstedt considered coastal defenses of little value, preferring to hold his panzer units in a central reserve and counterattack the enemy after they had landed. Most of the panzer officers who had served on the Russian front agreed with him. (The Red Air Force was not particularly efficient even when it had air superiority.)

Rommel read them his lesson: Any central reserve will be under constant air attack as it moves up; bridges will be bombed out, movement will be at a crawl, losses heavy. Give the Anglo-Americans a toehold on the French coast and their vast superiority will enable them to consolidate their beachhead before your delayed and weakened counterattack can be launched. The enemy must be defeated on the beach as he tries to come ashore. The whole coastline must be organized for defense in depth, the panzer divisions held under cover well forward so as to be able to intervene immediately.

Rommel wrote his fifteen-year-old son, Manfred, just called up for antiaircraft duty, that he was "raising plenty of dust wherever I go."[45] Defenses were strengthened, rows of improvised obstacles suddenly sprouted along the beaches, possible drop zones for paratroopers and gliders were studded with posts and mines. When no land mines were available, deadly substitutes were contrived from artillery shells. (Rommel had a knack for such mechanical improvisations.) The *Luftwaffe* (which controlled the antiaircraft artillery) and the Navy (which had the heavy coast defense artillery and the offshore mines) would not help, but Rommel was everywhere and into everything, and troop morale lifted as the work went on. Both he and Hitler became convinced the Allies would attempt to land in Normandy, and Rommel intensified his preparations there. In London, Montgomery looked at aerial photos of that work and gave it his professional approval: ''Rommel is an energetic and determined commander; he has made a world of difference since he took over.''[46]

It was a splendid effort, but Rommel had neither time nor manpower enough, and Hitler insisted on holding most of the panzer divisions too far back. When the invasion came, backed by naval gunfire and hordes of warplanes, Rommel held doggedly to his half-ready position, repeatedly blocking Montgomery's and Bradley's drives. On July 27 he was desperately wounded when a strafing Allied fighter plane wrecked his car. He then baffled his doctors by surviving.

The whole business is as hard to nail down as a handful of fog, but it is definite that Rommel had joined the anti-Hitler underground sometime after his return from Africa. His reasons probably were as simple and straightforward as his character. He had been away from Germany for two years and so could see Germany's changed position with an outsider's clarity. The Hitler he had left had been a brave soldier and apparently a dedicated, successful national leader. Now he was timorous; some of Rommel's interviews ended with a semihysterical Führer snarling his contempt and hatred for humanity in general and his own soldiers in particular. Worse, Hitler had first neglected and then deliberately sacrificed his troops in North Africa. His blunderings had brought disaster in Russia; Germany was on the defensive everywhere, with no hope of final victory.

From Normandy, Rommel repeatedly warned Hitler that no soldiers could long endure the punishment his were taking and urged him to, somehow, make peace. Rundstedt backed his warnings. Hitler told his marshals to mind their own military business, implying they were disloyal and cowardly. Rommel, however, did not approve of the underground's periodic attempts to assassinate Hitler, preferring to rally enough support in the Army to arrest him. (Such support was building; apparently even some of the SS commanders in Normandy would have followed Rommel.) Now his wounding robbed the underground of their most popular member; on July 20 badly maimed young Col. Count Claus Schenk von Stauffenberg failed in an attempt to kill Hitler with a time bomb, and an attempted underground coup d'etat in Berlin failed. Hitler's vengeance, extending into the families of the plotters, was horrible. The questing Gestapo established Rommel's underground connections, but even Hitler hesitated at a public arrest and trial of one of his most famous soldiers. Rommel was given the choice of standing trial, with the plainly implied risk to his wife and son, or taking poison. He chose the poison. Hitler gave him a stately funeral.

We have here a man who was a devoted soldier. Beyond his Lu and Manfred, he had no other interest than his profession; probably he never read a book on any other subject. He wrote on it in clear, effective language, extracting the important lessons from every operation in which he took part. He lived simply, did not smoke, drank only a little wine. His pleasures were simple things—dogs, horses when there was opportunity to ride, an occasional day to hunt or fish. Of middle-class origin, he showed nothing of the traditional General Staff attitude or manners. Courageous beyond the usual standards of human bravery, demanding the last wheezing gasp of effort from his men, always sharing dangers and hardships, he brought an intensely personal style of command into the complexities of swift-moving armored warfare. His officers found him exigent, sometimes rough in speech, but willing to admit his occasional errors and always quick to praise good work. His soldiers worshipped "that bastard Rommel"; he liked to talk with them, sensed their problems, and took care of them. As unsparing a fighting man as ever went into battle, ruthlessly determined on his enemy's destruction, he yet waged war with rigid observance of its traditional laws and courtesies and a sometimes whimsical mingling of chivalry and common sense. Like Patton, he could be a headlong, sometimes rash, tactician, but was a longheaded planner before action, with a far greater grasp of strategy than he ever was given the opportunity to exercise.

And there remains that unique aspect of his career. A few World

War II generals—again like Patton—were respected by their enemies, but only Rommel won their admiration and liking. Let Auchinleck express it: "I say, now that he is gone, that I salute him as a soldier and a man and deplore the shameful manner of his death."[47]

There were many able German commanders on the Russian front, as well as a fair number who either proved to be natural fumblers or wore out prematurely. Their average efficiency, however, was very high. At the 1943 Teheran Conference, Stalin paid them the honor of proposing that 50,000 or more of them be liquidated once the war was won to ensure the future peace of Europe. (Churchill instantly thundered that idea into oblivion as a piece of senseless savagery.) Among these officers—and many more—Erich von Manstein was considered Germany's outstanding general and strategist.

In many ways he was the epitome of the old German Imperial Army. Both his father, General of the Artillery Eduard von Lewinski, and his mother were from long-established families of military aristocrats. Since Erich was their tenth child, it probably was agreeable enough to allow the childless von Mansteins (Frau Manstein was his mother's younger sister) to adopt him. Young Manstein thereafter was educated in the Royal Prussian Cadet Corps, commissioned in the 3d Regiment of Foot Guards, selected for General Staff training. Lean, tall, and impressive, he had a quick, analytical mind and a cool reserve no emergencies could ruffle. A lieutenant and adjutant of a Guards reserve regiment when World War I began, he at once showed himself an unusually skilled tactician and leader of men. After a bad wound, he served largely on staff duties; his skills there made him one of the officers retained in the 100,000-man army allowed a defeated Germany by the 1919 Treaty of Versailles. General von Seeckt (see page 182) assigned him high-level staff functions, among them emergency planning against a possible Polish invasion of East Prussia and making certain that every German soldier would be capable of handling a job at least one grade above his present one. Promotions came to him regularly; in late 1939 he contributed considerably to the final plan for the 1940 invasion of France, Holland, and Belgium. During that invasion he commanded an infantry corps that he kept up close behind Rommel's "Ghost Division," in one burst marching seventy-five miles in less than five days. Pleased with his intelligence and aggressiveness, Hitler assigned him a panzer corps for the 1941 invasion of Russia; with it, he proved very much another Rommel, moving with his leading units and always appearing at the critical point. He was just and correct in dealings with his subordinates, trusted and admired by his troops. Though he lacked the common touch of Rommel and Guderian, he looked after his men and had a rather rare humanity toward prisoners of war and the overrun civilian population; like Guderian, he refused to obey Hitler's orders for merciless warfare.

Nevertheless, in September 1941 Hitler gave him command of the small Eleventh Army, with the mission of clearing the Crimean Peninsula and capturing Sevastopol, its strongly fortified seaport. Manstein had the peninsula in hand when the great Russian winter counteroffensive began. Amphibious landings poured hordes of Russians ashore at the western tip of the Crimea and reinforced Sevastopol. By desperate fighting, he broke the new Russian army, bagging over 150,000 prisoners; then, using massive artillery and air support, he battered Sevastopol's defenders into surrender on July 1, 1942. Hitler promoted him to field marshal.

But Hitler also was engaged in a major military gamble of which Manstein's offensive had been only a minor part. He had sent his armies (still far from rebuilt and reequipped after the past 1941–42 winter's hard campaign) into an offensive across southern Russia and had thrown their initially successful advance into confusion by his continual changing orders and counterorders. His Army Group A was moving south into the Caucasus Mountains in a drive for the oil fields beyond them; Army Group B was headed east across the Don River toward the Volga and Stalingrad. Both, especially A, were outrunning their supplies. In this deepening crisis, Hitler was moved to order the Eleventh Army, his only sizable reserve in the south, to shift northward some twelve hundred miles to complete the capture of the city of Leningrad!

By early November, A was stalled deep in the Caucasus; B was engaged in a costly "rats' war" around Stalingrad. On November 19 a great Russian counteroffensive shattered B's front, trapping the German Sixth Army at Stalingrad. Lacking other reserves, Hitler recalled Manstein to restore the situation. (The Eleventh Army was occupied mopping up an unsuccessful Russian offensive south of Leningrad.) The situation was desperate and rapidly getting worse. Manstein found himself in command of little more than a vast gap in the German lines; much of his front was held only by little battle groups rallied by undaunted officers. (A typical one, led by young Col. Helmuth von Pannwitz, consisted of eighteen German tanks; 1,000 Cossack volunteer cavalrymen; a battery of Romanian field artillery; a few German service troops; and one small liaison airplane. It ate Russian divisions for breakfast.) Army Group B was crumbling away to the north as the Russian offensive overwhelmed the poorly armed and equipped Romanian, Hungarian, and Italian units that formed its left flank. Worse, the expanding Russian drive to the west and southwest from the Stalingrad area was threatening to cut Army Group A's communications and pin it against the Black Sea.

Of these crises, Sixth Army's position seemed the most crucial; it was short of supplies, and attempts to resupply it by air were proving unsuccessful. Collecting three weary panzer divisions, Manstein attempted to break through to Stalingrad.

The Sixth Army was commanded by painfully conscientious Gen. Friedrich Paulus, a military version of Shakespeare's Hamlet who had every gift except that ultimate self-confidence that can make a quick decision when the world is falling apart and the best decision may be bad indeed. Paulus knew he should break out, and quickly; being overly housebroken, however, he asked for orders. Hitler dithered over the question, finally ordering him to hold where he was. Manstein's relief expedition fought its way against heavy odds to within thirty miles of Stalingrad, but Paulus quibbled against attacking to link up with it without Hitler's permission. Hitler, against the advice of most of his staff, still demurred. During this bickering the rest of Group B's front came apart; the relief column had to fall back; Group A was under attack from all directions, with Russian tanks within forty-five miles of its one supply line and the bulk of its forces still four hundred miles off to the southeast. Hitler's initial contribution had been to fire its commander, replacing him with *Generaloberst* Ewald von Kleist, an old-breed Prussian aristocrat cavalryman who, recalled from retirement in 1939, had proved himself a gifted panzer general. Now, he ordered Kleist also to stand and fight—an order Kleist felt was Group A's death sentence.

Then something peculiar happened: Hitler suddenly told Kleist to get out as quickly as possible and reinforce Manstein. Moreover, except for hunting up reinforcements for him, he finally left Manstein to run his own battle. Paulus had to make a hopeless last stand to pin down as many Russians as possible while Manstein and Kleist regrouped; the last pocket of his starving, frostbitten men held out until February 2, 1943. (Hitler was furious with Paulus for surrendering instead of dying gloriously.) Yet for all the Sixth Army's sacrificial valor, it seemed the jig was up for the remaining German forces in southern Russia. Masses of Russian armor, cavalry, and motorized infantry swept southwestward, cutting in behind the scattered forces that composed Manstein's so-called Army Group Don (later renamed "South"). The troops that had been besieging Stalingrad followed them. By February 20 leading Russian units had broken the one railroad that supplied Manstein and were within sight of his headquarters. But, two days earlier, Manstein had opened his own counteroffensive. It was deep winter across the steppes; in places, the odds were as much as seven to one against him; his own troops were almost exhausted. But the Russians too were outrunning their supplies and had allowed their forces to become scattered in the excitement of their pursuit. Concentrating his troops, Manstein hit hard and swiftly, pocketing and destroying one Russian unit after another. Only the spring thaw of mid-March halted his advance, which had recovered almost all the territory the Germans had held in southern Russia at the beginning of their 1942 campaign.

This masterpiece of mobile warfare should have shown Hitler the most effective strategy for the open countryside of southern Russia. Hitler, however, reverted to his old meddling self, ignoring Manstein's advice to lure the Russians into attacking and then crush them in a campaign of maneuver. He wasted most of his restored panzer strength in an attack on a thoroughly prepared Russian salient around Kursk in July 1943, leaving it too weak to halt a prompt Russian counteroffensive. Thrown on the defensive, he insisted on holding useless territory, whatever the cost, instead of shortening his front so that some troops could be pulled out of line to build up a reserve. At best, Manstein could only fight a wearing delaying action, struggling to keep his army group intact, lashing out when opportunity offered to destroy Russian spearheads, disputing Hitler's orders to hold untenable positions and sometimes bluntly ignoring them. His losses were heavy, but he brought his troops back still in fighting spirit. That, however, counted for little with Hitler; on March 30, 1944 both Manstein and Kleist were decorated and sent home.

Living quietly in retirement, Manstein was captured by British troops at the end of the war. Accused of alleged war crimes, he was brought before the Allied Nuremberg tribunal and sentenced to eighteen years' imprisonment but was released in 1953.

It is notable that his final year of rear guard actions did not detract from the esteem in which his fellow soldiers held Manstein. To them, he remained a true *Feldherr*—"master of the field." Though he had a free hand as a strategist for only a few months, his accomplishments, in the teeth of all disaster, were startling enough to qualify him as a superstrategist. It is no exaggeration to say that—had Hitler given him a free hand earlier and longer—he probably would have changed the whole course of the war on the eastern front.

XI

PANDORA'S LITTLE BOX

But I ha' dreamed a dreary dream.
Beyond the western sky,
I saw a dead man win a fight
—And I think the man was I.[1]

To declaim against war . . . is to beat the air with vain sounds, for ambitious, unjust, or powerful rulers will certainly not be restrained by such means. But what . . . must necessarily result, is to extinguish little by little the military spirit, and some day to deliver up one's own nation . . . to the yoke of warlike nations which may be less civilized but which have more judgment and prudence. GUIBERT[2]

The most important leg of any three-legged stool is the one that's missing. FINAGAL'S BASIC LAW[3]

One evening every week in our quiet New York village the going-home traffic passes a knot of people holding improvised signs: "Nuclear Freeze Now"; "Out of Central America"; "Disarm Before It's Too Late." Mostly they are a next-door neighbor sort of people. Their sincerity, through all winds and weathers, is obvious. Improbable as it may seem, they unwittingly also are part of the abrasive edge of Russian national strategy.

The Russians have a name for them: "Useful idiots."

World War II ended in the apparent triumph of righteousness and chastisement of the wicked. Italy, Germany, and Japan had been conquered. Russia had attained its stated territorial objectives. The British Empire remained intact, free to dismantle itself in appropriately dignified fashion. President Roosevelt's vision of a United Nations Organization wherein all nations might be reconciled was soon realized.

The nations, however, chose not to be reconciled. The apparent victory, and the world peace it momentarily had won, became, in an ancient French soldier's phrase, "a halt in a swamp."

A multitude of new strategic problems erupted. In Africa and Asia the breakup of the prewar colonial empires loosed a mob of raw, new nations and semi-nations, mostly impoverished and unprepared for self-government but full of grandiose ambitions. Strategic choke points and base areas that England had held for generations—the Suez Canal, Malta, Aden, and Singapore—passed into the custody of weak minor states and became tribulations to the nations of the West.

At the same time a far more deadly strategic problem emerged. American development of nuclear weapons and their employment against Hiroshima and Nagasaki had forced Japan's surrender and—ironically—undoubtedly had saved millions of American and Japanese lives (see page 331). In 1945 the United States held a monopoly—how fleeting a monopoly few Americans realized—on the world's most powerful weapon. Once perfected and used, it inevitably became an important factor in the national strategy of the United States. The whole world waited to see what use we might make of it.

U.S. nuclear strategy has gone through a series of modifications and changes, but it has been shaped by two often-conflicting factors. The first, accepted as a basic tenet of our national strategy since World War II, is expressed in professional jargon as "second strike, but possible first use."[4] Translated, this means that the United States will not go to war unless attacked. The enemy must strike us first. Consequently, our enemies need not fear that we shall launch a sudden preemptive nuclear strike against them, however threatening their attitude and preparations may become. However, if attacked, we are prepared to be the first to use nuclear weapons if we think the situation requires it—for example, should the Russians attack Western Europe with overwhelming conventional[5] forces beyond the ability of the NATO armies to handle. (It is interesting to note that

Russian Premier Konstantin Chernenko's October 16, 1984 list of the "practical deeds" he wanted from the United States, to prove the sincerity of its desire for negotiations, included a promise, to be given *before* such negotiations began, that the United States would never be the first to employ nuclear weapons.)

Since any attack by Russia against the United States itself necessarily would include a bombardment by intercontinental nuclear missiles, our "second strike" strategy implies that we must accept and endure that destruction before launching our own nuclear weapons in retaliation—assuming that enough of our nuclear forces have survived to deliver an attack of any consequence. This allowing the enemy the first punch is a damnably risky business. It grows steadily riskier as the Russian buildup continues, while assorted pressure groups here and in our allied NATO nations agitate for a "nuclear freeze" that would prevent any compensatory weapons developments by our side. However, given our established American political and social creeds and systems, we undoubtedly are committed to a second-strike strategy. Practically all of our wars have caught us with our trousers at half-mast. Next time, we will be lucky to get them up again, let alone buttoned and buckled.

The second factor determining U.S. nuclear strategy is that it evolves in the midst of a six-ring circus of politicians in search of publicity and office; pressure groups that want the money spent on something else; peace movements of all types that can visualize nuclear war only as an utter holocaust; Army, Navy, and Air Force competition for larger slices of the defense budget; intellectuals and educators (the terms are *not* synonymous) of all calibers and subcalibers; and scared local groups who insist they are patriotic Americans, but don't want nuclear installations anywhere in their home state. Also, nuclear weapons are expensive. Consequently, there is the constant temptation to postpone or stretch out the acquisition of new models needed to replace such overaged weapons as the B-52 heavy bomber and the Titan II intercontinental missile. When possible replacements—the B-1 bomber and the MX missile, for example—are presented, they are opposed as too expensive, not sufficiently advanced, or "destabilizing" (see page 304). Politicians and "experts" (usually self-anointed) assure us we should wait a few years more, either for the coming of allegedly stupendous new weapons such as the "Stealth" bomber, or in hope of some diplomatic miracle that will make nuclear weapons unnecessary. All this misses the brutal facts that we are right now in a time of crisis and that no blueprint for a future perfect weapon ever frightened an enemy or won a battle. It also results in our nuclear weapons systems and strategy being generally considered item by item rather than as a coordinated whole. Therefore, whatever we have in the way of a nuclear strategy today tends to be somewhat disheveled. It has been shaped as much by temporary political expediency, blind hope, and timorous emotion as by practical military factors. Such, however, always has been the usual American method of dealing with serious long-term problems.

In all honesty, the basic purpose of all the various nuclear strategies the United States has considered since 1945 has not been conquest but *deterrence*—a search for some method of persuading Russia to cease its aggressions and live in peace with the rest of the world. Even in crises like the Berlin Blockade (1948–49), the Korean War (1950–53), and the Russian conquest of Hungary (1956) when the United States had clear superiority in nuclear weapons, it did not threaten their use.

Nuclear weapons are, in fact, explosive charges: They can be "delivered" to the target by aerial bomb, missile, artillery projectile, or even by hand. In the years just after World War II, only the first of these methods was feasible. Consequently, our nuclear strategy was simplicity itself: If Russia began hostilities, the U.S. Air Force would use the few atomic bombs then available as Hiroshima-style "city busters" against the major centers of the Russian armament industry. It was hoped that this implicit threat would be sufficient to support the then-current U.S. national Strategy of Containment, which sought to curb any further Russian expansion, while avoiding confrontations and seeking some basis for peaceful coexistence. As an incentive for such long-term cooperation, in 1946 the United States offered the so-called Baruch Plan, which proposed a veto-free international agency, under the general supervision of the United Nations, to control the development of atomic energy. America would contribute all its hard-won knowledge and technical skills to that authority; once it was functioning, the United States would cease manufacturing nuclear weapons and dismantle its existing stock of them. Russia, then busily tooling up to produce its own bombs, refused to participate. Only two actions were required to guarantee world peace, Moscow insisted: The United States must destroy all its nuclear weapons immediately; all other nations must pledge never to produce any.

Meanwhile America cut back its national defense appropriations drastically. The Air Force, now viewed as the nation's first line of defense, got most of what money there was. The Navy came next.

Sad to relate, the Air Force reacted to its new mission by rapidly growing too big for its britches. In the best Billy Mitchell style, it soon was suggesting that the Army and Navy had become unnecessary vestiges of ancient history, of no more use in modern warfare than the vermiform appendix is to human digestion. Should Russia be so foolish as to risk a war, the starkly trained U.S. Strategic Air Command alone would suffice to promptly slap those Muscovites back into the Early Stone Age. Air Force speakers at staff conferences and service schools expansively explained how they not only would win any future war single-handed but would afterward "occupy Russia from the air." If its patrols noticed any sign of reviving industry, a messenger would be dropped by parachute with the ultimatum "Desist, or be bombed again!" (At Carlisle Barracks, a lanky infantry lieutenant, a survivor of the Benevolent and Protective Brotherhood of Them What Has Been Shot At,[6] brought down the house with the seemingly innocent question "Will the man you send down by parachute be Air Force too, sir?") In its diplomatic intercourse with other nations the United States never rattled its nuclear saber. At worst, some loudmouth politicians and Air Force types stated our Strategy of Containment simplistically as "Draw a line on the map and tell the Russians we'll clobber them if they cross it."[7] But, however expressed, containment didn't work. The Russians tested their first atomic bomb in 1949. Their armed forces crossed no lines, but in 1950 the Russians first recognized and then helped to arm Ho Chi Minh's Communist revolt in French Indochina, then turned their North Korean satellite loose to invade South Korea. There was no fitting military target for nuclear weapons in North Korea; the United States had to send in its neglected, understrength Army, equipped with malfunctioning World War II hand-me-downs. The Air Force gave all-out support, but it had lost many of its World War II skills; in late 1950 it even failed to detect the coming of some 300,000 Communist Chinese troops to rescue the defeated North Koreans.

This experience, the increasing numbers and variety of American nuclear weapons, and the growing Russian nuclear strength caused a restudy of U.S. nuclear strategy. Gen. Curtis Le May, SAC's exigent commander, and other officers urged that greater priority be given to the destruction of the Russian Air Force to protect the United States from nuclear attack. Russian conventional forces and command and communications systems would be given second priority; Russian cities would be struck only in retaliation for attacks on American cities. The intentions behind this new strategy were excellent, but its application might have been messy. Reliable information on the locations of Russian air bases was scant and difficult to come by. The Air Force therefore proposed to throw nuclear bombs around like rice at a wedding to be certain of taking out all suspected base sites, without much regard for friendly surface forces or friendly, neutral, or enemy civilians.

Beginning with the late 1950s, nuclear strategy became more and more complex. Russia tested its first thermonuclear (hydrogen or H) bomb in 1953; the United States duplicated this feat during 1952–54. New means of delivery proliferated—ballistic-missile-launching submarines, intercontinental ballistic missiles, and a variety of short- and intermediate-range rockets and missiles on mobile mounts. Miniature nuclear missiles that could be fired in conventional artillery shells were developed, and the first cruise missiles (pilotless aircraft) were tested. At the same time, by the use of free-floating, high-altitude, camera-carrying balloons and the U-2 "spy" plane, Americans began learning more concerning Russia's interior. Naturally there was much interservice squabbling over powers and responsibilities: The Navy suggested that SAC had become obsolete; the Air Force wanted operational control of all U.S. nuclear forces, including submarines. This, however, was a beginning of combined planning to establish a flexible nuclear strategy, designed to cover contingencies ranging from limited local squabbles to total war. This general national strategy, sometimes termed "Massive Retaliation," visualized the United States maintaining such a decisive nuclear superiority over Russia that even a Russian surprise attack would leave us with sufficient second-strike weapons to carry that retaliation through. If Russia *did* begin hostilities, our first priority would be the obliteration of Russian nuclear forces; the second, other military targets; the third, cities and industries—which would not be struck unless Russia attacked American cities or refused to make an acceptable peace.

When the Cuban missile crisis of 1962 broke, the United States still possessed such superiority. That, plus the apparent willingness of the United States to take any necessary military action, forced Russian Premier Nikita Khrushchev to back water. At that, Khrushchev scored a victory which was overlooked amid the Kennedy administration's celebration of its decidedly overdue firmness: the United States pledged not to invade Cuba, even though a frustrated Fidel Castro snarled down an agreed UN inspection of the Russian withdrawal. In addition, the Russians were quietly promised that the United States would soon withdraw its missiles from the territory of its NATO ally Turkey. The excuse presented for this voluntary appeasement was that those missiles were "antiquated and useless."[8]

It is another of history's ironies that this apparent American victory was the point at which America's nuclear strategy started downhill. First cause was that Russian leaders, like Philip II after the destruction of his Armada, studied the reasons for their failure and concluded they had lacked the naval power to challenge the U.S. "quarantine" of Cuba and enough nuclear weapons to impress the Americans. They went to work on those problems and are still cranking out warships of all types, improved aircraft, and missiles. Worse, their weapons technology, much aided by industrial espionage and purchases of sensitive equipment from Western nations, threatens to equal that of the United States.

This growth of Russian first-strike nuclear strength appalled Robert S. McNamara, Secretary of Defense for both the Kennedy and Johnson administrations. A banker with no patience for military men's opinions and little understanding of warfare, he faced a number of honest problems. The U.S. involvement in Vietnam (1954–73) consumed too much of this nation's time and energy and ended by making the military unpopular. The need to build up the Army, Navy, and Air Force's capabilities for conventional warfare reduced the money available for nuclear armaments. And the increasing size and variety of Russia's nuclear forces made American chances for a decisive second strike less and less without a proportional—and *very* expensive—strengthening of ours. Eventually, McNamara began to argue that the real need was for *parity* between American and Russian nuclear forces so that each would possess the capability for "assured destruction'' of the other. At that happy point, he felt Russia would have no incentive to continue building up its forces; both nations could sit down and negotiate the end of the arms race. Probably it never occurred to him that Russians do not necessarily calculate like respectable, cost-conscious bankers, and might not be interested in mere parity. In addition to limiting the size of the American nuclear forces, he cut back American antiaircraft defenses and discouraged work on antimissile defense systems. Heavy missile warheads, designed to crack ''hardened'' Russian missile silos, were not built.[9] While planning for attacks on Russian nuclear forces did not cease, there was more emphasis on the destruction of Russia's critical industries and political centers, with McNamara urging that an American ability to wreck Russian industry beyond easy reconstruction and to kill off a quarter of the Russian population would be a sufficient deterrent to a Russian attack. McNamara's ''assured destruction'' has become known as ''Mutual Assured Destruction,'' or by its appropriate acronym ''MAD.'' At best, it is a strategy of despair—reliance on a balance of terror, under which both nations' civilian populations are held hostage. To supporters of this policy it is vital that any nuclear war be a mutual national suicide, inflicting uncountable civilian casualties and prolonged misery; otherwise, there will be no deterrence. To them, any measure, active or passive, to protect the United States against a Russian nuclear strike therefore would be ''destabilizing'' because it might reduce casualties and so make war seem acceptable. They have not grasped the brutal fact that our second-strike strategy, plus our dwindling capacity for a successful second strike, makes it increasingly unlikely that the ''assured destruction'' and suicide really would be ''mutual.''

Pres. Richard M. Nixon attempted to achieve an agreement on nuclear parity, adopting the general principle that the United States should maintain forces sufficient to defeat any Russian attack, but that their makeup and disposition should pose no direct threat that might inspire a Russian nuclear attack during international crises. His 1972 Strategic Arms Limitation Agreement (SALT I) with Russia achieved mixed results. Both sides agreed to limit themselves to two antiballistic missile sites (reduced to one apiece in 1974), which reinforced the MAD concept. Unfortunately, the suave incompetence of Henry Kissinger, Nixon's national security adviser and leading negotiator, allowed the Russians to claim an unjustifiable superiority in missile-launching submarines. Nixon did attempt to modernize the American nuclear forces, but Congress strangely saw fit to abolish the remaining American antimissile site in 1975. (The Russians kept theirs and are industriously expanding it.) No coherent strategic doctrine emerged from the confused administrations of Gerald R. Ford and James Earl Carter. Carter tried to carry on the SALT negotiations, but a renewed buildup of Russian combat troops in Cuba and the Russian invasion of Afghanistan fortunately killed prospects of the United States accepting a SALT II agreement.

Through these years American offensive nuclear strategy has evolved into something very much like Finagal's three-legged stool—a "triad" of specialized, mutually supporting weapons systems: SAC's heavy bombers; intercontinental ballistic missiles launched from fixed silos; and missile-launching submarines. This mixture provides a variety of possible responses to hostile action and is a guarantee that our whole nuclear force would neither be entirely knocked out by a surprise attack nor rendered obsolete by some technological breakthrough in weapons development.

At present, none of these three legs to our strategic stool is missing, but one—SAC's bombers—is badly in need of replacement; another—our intercontinental missiles—needs repair. The submarine leg is being strengthened by the introduction of new *Trident*-class ships. Smaller, more accurate land-based missiles are under development; the basic triad is now being supplemented by land- or sea-launched cruise missiles.[10] Nevertheless, our ability to launch a successful second strike is presently dubious because of increasingly superior Russian nuclear power. A modified strategy—"launch on confirmed attack" (or "launch on warning")—which would allow fixed-site ballistic missiles to be launched once a Russian nuclear attack was known to be on the way, has been considered. However, the difficulty—within the six to ten minutes available—of confirming the attack, securing orders from the proper authority, and going through the complex safety procedures built into our weapons systems makes it dubious that we could get many missiles away before the Russian bombardment burst upon us, even if this strategy were adopted.

This dreary state of affairs inspired Pres. Ronald Reagan to propose the development of an antiballistic missile defense system, initially for the protection of our missile silo sites, eventually to be extended into space to protect the entire United States by intercepting hostile missiles in flight. As might be expected, this proposal—titled "High Frontier" by its advocates, "Star Wars" by its opponents—has roused much emotional opposition. MAD supporters label it "destabilizing"; a major accusation is that it would introduce nuclear weapons into space. In fact, the system now envisaged would employ conventional missiles. Russia's loudly proclaimed anxiety to strangle High Frontier, so to speak, in its cradle should be evidence enough of its strategic value.

Nuclear weapons are far deadlier than other weapons yet developed, but they *are* simply weapons. Russians regard them as such. Their Communist party leaders and military staffs have proceeded methodically with the development of their nuclear and conventional forces alike, untroubled by public opinion or domestic peace movements. They devote whatever money, scientific talent, and manpower they consider necessary out of their available resources for that purpose. Their nuclear strategy seems based largely on land-based intercontinental ballistic missiles (heavier than their American counterparts and increasingly supplemented by mobile, intermediate-range SS-20 missiles) and missile-launching submarines. There is less emphasis on bombers, though Russia possesses some excellent modern aircraft; development of cruise missiles is being pushed. The Russian civil defense program appears to be thoroughly organized; at worst, it is far better than our faint excuse for one, which now is concerned only with natural disasters. In addition, Russia already has established its own frontier in space, with a permanent militarized space station and "satellite killer" weapons designed for the destruction of our "spy" satellites. They want no competition there; with Russian logic they can declare that the proposed American High Frontier would be a "transfer of the arms race to outer space."[11]

The obvious deadliness of even limited nuclear war has spurred efforts to find alternate military strategies since the day that the first atomic bomb gutted Hiroshima. This search produced one unintended semicomic interlude. Fuller, Liddell Hart, and other theorists were touting a return to the strategy of the eighteenth century—a period, so they claimed, of limited wars waged without rancor by strictly disciplined armies under the command of punctilious gentlemen who treated their opponents with bounteous courtesy and shielded civilians from all possible discomfort. American commanders who wished to be proper guardians of Western civilization, they inferred, would find some means of returning to those good old days.

This thesis was mostly nonsense, begotten out of those theorists' ignorance of real history. The factors limiting eighteenth-century wars had been the armies' fumbling command and logistic systems, plus Europe's miserable roads. But the U.S. Army, hopefully helped by our State Department, eventually committed a sizable staff study, designing a new strategy that would rule thickly inhabited areas off-limits to military operations and otherwise stress the humanities. Apparently nobody bothered to do any deep research on the Russian version of eighteenth-century military gentility—according to citizens of that era who experienced it, a very rare and transitory commodity. But the difficulty of getting the Russians to agree to such rules soon became overwhelmingly obvious; in fact, our own forces could hardly have observed them! The study went into "File 13."

Other alternative strategies have no humorous aspect whatever. Biological Warfare (BW)—the use of bacteria, viruses, fungi, protozoa, rickettsiae, and their toxins as weapons—is older than history. However, its actual employment has been small-scale and sporadic. Chucking catapult loads of dead animals or filth into besieged cities in hope of starting a pestilence was an occasional feature of classical or medieval warfare; polluting waterholes always has been an effective tactic in dry country. One outstanding example was self-inflicted: in 1757 following the surrender of Fort William Henry, the Indian allies of the French slaughtered the British sick in the fort's hospital and dug up and scalped corpses of soldiers who recently had died of smallpox. From them and from prisoners they carried off there promptly developed a smallpox epidemic that ate out tribes as far west as the Mississippi. "News," noted French officer Louis Antoine de Bougainville,

> from St. Joseph River and Detroit, dated January 4 [1758]. The Indians who came to the army during the last campaign have lost many people from smallpox. Their custom in such a case is to say that the nation which had called upon them [for military service] has given them bad medicine. The commanders of the [French] posts must dry their eyes and cover the dead. . . . It is a mourning which will cost the King dearly.[12]

News of this epidemic undoubtedly inspired Sir Jeffrey Amherst, that "soldier of the King" who commanded the British forces in North America through 1759–63, to attempt a similar affliction of the Indian tribes who rose in 1763 against British rule under the Ottawa war chief Pontiac. "Could it not be contrived to Send the Small Pox among these disaffected Tribes of Indians. We must . . . use every Stratagem in our power to reduce them." It seems one attempt was made, using two blankets and two handkerchiefs "got from People in the Hospital," but apparently with little result.[13] (Amherst had the character of an honorable man, but the Indians had inflicted several nasty little defeats on his forces—frequently by what professional soldiers would consider treachery—and had done it messily. He was especially angered by the case of a Lieutenant Gordon whom the Indians had spent several happy nights roasting alive.)

Biological warfare reappeared during World War I; German agents were accused of spreading glanders (an equine disease) in Romania and the United States. During the 1930s Japan, Germany, and Russia, and possibly other nations, conducted BW research; the Japanese apparently tested it on a small scale against the Chinese. The North Koreans and their Chinese Communist allies accused the United States of employing biological agents against them during the Korean War (1950–53). These charges were publicized and amplified by Communists and Communist-front organizations throughout the world. (Today, they undoubtedly would be seriously considered by some American TV producers.) The physical "evidence" displayed was obviously faked; the most impressive portion of it was "confessions" wrung from American prisoners by deliberate torture. When the United States challenged the Communists to submit their charges to the International Red Cross for a complete investigation, the Communists hastily refused. Echoes of that accusation, however, linger on—and there still are helpful idiots who will reecho it.

Because of intensive Russian biological warfare research and training, however, the United States quietly began a parallel program for its possible deterrent effect. It was not favorably viewed by either the American public or much of the armed forces. In November 1969 Pres. Richard M. Nixon unilaterally renounced American use of biological warfare. All existing stocks of BW agents and weapons were destroyed; BW research was confined to defensive measures such as immunization. Two international treaties, which the United States ratified in early 1975,[14] forbid the production, stockpiling, or employment of biological weapons. Unfortunately, they include no provision for international inspection of possible violations. An example was the accident in April 1979 at a military installation in the "closed city" of Sverdlovsk in south-central Russia that caused at least three hundred deaths from anthrax. When queried, the Russian government admitted a number of anthrax cases in that area but asserted that they were caused by the consumption of meat from anthrax-infected cattle—and accused NATO of planning to launch biological warfare against Russia and its satellites.[15] Since then there have been increasing reports of Russian use of BW agents in Southeast Asia and Afghanistan.

Biological agents may become the most potent and selective weapons in history. They can be used to attack, as desired, human beings, their domestic animals, or their food crops. They can be programmed to kill or merely disable. (One projected American BW tactic was to have sloshed bypassed enemy units with an agent that produced quick and violent diarrhea, to keep them harmlessly occupied until they could be gathered in.) Secret, silent, and invisible, BW "weapons" are extremely difficult to detect. Normally they require an "incubation time" to become active and so would be more effective against civilian populations and rear-area installations than combat units. An enemy could subject the United States to a crippling covert biological offensive months before a planned overt attack. (One method would be the release on a favorable wind of aerosol clouds of BW agents from merchant ships off our coast.) Even if the resulting epidemic finally were identified as a planned hostile action, the problem of determining the attacker's identity still would remain. (We must keep in mind that even minor nations *could* be capable of developing a dangerous BW arsenal.) BW agents also could be employed for small-scale sabotage against key installations and personnel, both military and civilian, and would make ideal terrorist weapons. Since biological warfare does not destroy industrial plants and natural resources, it conceivably offers a means of taking over a nation practically intact. By using an agent such as dengue (breakbone) fever, which has a low "kill" rate, the aggressor could spare most of that nation's population, the potentially useful members of which could be screened out to serve the conquerors' various requirements. Or, should the aggressor want more room for his own surplus population, he might employ bubonic plague, which can kill approximately half the people affected. The major danger would be an aggressor nation's development of a mutant strain of some lethal microorganism *and* an effective immunization for its essential personnel. A nation attacked with such an agent would lack medical defenses against it and, quite possibly, the time to develop them; the aggressor could follow up a covert attack by such an agent with an invasion under the disguise of a massive rescue and relief effort.

So much is theory. Whether such attacks would succeed against a major power like the United States is difficult to estimate. But it *is* a danger and one that probably will increase. Security against it requires both an efficient intelligence system, alert to any indications of preparations for BW operations, and effective medical/public health systems capable of quickly detecting and treating any unusual type or incidence of disease.

There are other weapons, such as the deadly "nerve" gases, that were available but not used during World War II and older types like those revived recently by Iraq to deal with massed Iranian attacks. Also, there are new incapacitating chemical agents that induce fear, disorientation, or the equivalent of complete drunkenness. Research continues into the use of lasers and, probably, charged-particle beams[16] as weapons. And various nasty surprises undoubtedly are being tested in laboratories around the world.

None of these, however, whether nuclear, biological, or chemical, troubles the more perceptive high-level strategists of the non-Communist nations so urgently as Russia's continual drive to weaken and break them by pressures short of a major war—pressures applied within their own frontiers as well as from beyond their borders. This is a protean, opportunistic, unceasing offensive, variously dubbed the Strategy of Disruption or Destabilization. It shifts from sizable open hostilities (waged, as in Korea, by Russian satellite states) to solitary assassinations, sometimes disguised to appear as death from natural causes. With these come the subversion and seizure of weak neighboring states like Afghanistan; the continuing acquisition of naval and air bases along the major sea lanes; the worldwide support of terrorist societies and traffic in illegal narcotics; sabotage of NATO installations; forgery, planted news items, and propaganda of all sorts; and the sponsorship and support of an endless, always-shifting variety of front organizations, ostensibly devoted to peace, disarmament, and brotherhood.

Combined with the current strengthening of its nuclear and conventional forces, intensive military and industrial espionage, and militarized space program, these activities compose Russia's strategy for world dominance, to be achieved, if possible, without a major war but designed, in case the Western nations refused to be nibbled and cowed into submission, to give Russia every possible military advantage before the fighting begins. No nuclear freeze will cancel it.

We probably still are unaware of the full extent of this strategy. A good many Americans reject, sometimes violently, the idea that Russia could have conceived so grandiose a project. But, whether ruled by czar or commissar, Russia always has been an aggressive, expanding nation. Its growth has been not only—like ours—the occupation of a thinly peopled wilderness but also the conquest of older, more civilized nations. (Dubious readers should get a good historical atlas; then compare the territory ruled by Ivan IV, first Czar of Russia, rightly named "the Terrible," at the end of his reign in 1584 with the present spread of Russian territory and vassal states, three centuries later.)

Much of this strategy is nothing particularly new. Subversion and propaganda are ancient techniques of statecraft. As a single example, in 1918 Pres. Woodrow Wilson's Fourteen Points[17] had at least as great a share as Foch's military leadership in sapping the German will to continue the war. What *is* new is the extent of the Russian effort and the variety of agencies it utilizes. A complete study of this strategy would require several fat books, but certain features of it deserve quick mention here.

Russia's relationship with international terrorism is a known fact but one generally not mentioned in polite society, America and its allies not having been able to devise an effective method of dealing with the terrorist menace. While directly controlling some terrorist organizations, the KGB[18] has preferred to work through the political police/intelligence systems (modeled, naturally, on the KGB) of Russia's vassal states or of its peculiar ally Libya. Cuba is its essential, tightly controlled agent in the Western Hemisphere; Bulgaria does dirty jobs in Europe—such as the attempted murder of Pope John Paul II. (The latter deed was covered by a flow of disinformation to establish the would-be assassin as a member of a Turkish right-wing society who had been trained by German neo-Nazis.) But the present worldwide stew of terrorist groups of all types and doctrines is far too seething for any effective central control, however ruthless. Russia offers weapons, training, sanctuaries, staff work, encouragement, sometimes money. Whatever the terrorists do with it—even, as in the Middle East, when they fight each other—contributes to the crumbling away of established order and the strength and prestige of the Western nations. The weapons (which include late-model missiles and remote-control demolition equipment) and the training have vastly increased the devastation even small gangs can inflict.

Terrorists are frustrating enemies, especially to Western nations with their traditional codes of human rights and the rule of law. They are exceedingly difficult to detect; if you catch them red-handed, they must have a fair trial, which their lawyers try to convert into a propaganda circus, at the taxpayers' expense. Punitive actions against foreign terrorists such as those who blew up our installations in Lebanon are blocked by the near impossibility of identifying the guilty group among the dozens of shifting gangs in the area and the danger of hitting innocent people. To Russians and terrorists alike, there are no innocent bystanders.

It is notable that no terrorist actions are reported from Russia and its vassal states and few indeed directed at their citizens abroad. Of course, Russia has competitors in this field, such as Col. Muammar Qaddafi, ruler of Libya, quite possibly unbalanced and dedicated to international terrorism. And there is the reverend, rabidly anti-American Ayatollah Khomeini, master of Iran. Both men, however, are decidedly small bore compared to the KGB. Qaddafi depends on Russia for weapons; Khomeini may be having trouble at home with Iranian Communist guerrillas.

Recent Russian territorial expansion has multiple objectives: establishment of strategically located air and naval bases, development of vassal regimes in minor states possessing valuable raw materials, and the denial of such territory and resources to the Western nations. One of Russia's first triumphs was the gradual absorption of the former British Aden Protectorate at the southern tip of the Arabian peninsula. Its final conversion into the People's Democratic Republic of Yemen came after June 1978, when Russian air and naval forces attacked the holdout port of Aden and inconvenient officials were purged. (That whole business seems to have been resolutely ignored by most of the Western media.) Yemen is a mostly barren area but provides a possible base for operations against both the small, oil-rich Arab states of the Persian Gulf area and traffic through the Red Sea. It also has furnished an excellent secluded site for terrorist training camps. Its seizure was a rehearsal, if an insufficient one, for Afghanistan.

Russian adventures in Africa met mixed results, Russian diplomacy frequently proving awkward and most of the new African states being highly unstable. But, by employing large forces of Cuban troops, Russia has succeeded in establishing sufficient armlocks on Angola and Ethiopia. (Reportedly, there now are over thirty thousand Cubans in Angola; several thousand more uphold Ethiopia's Marxist government.) Russia's contribution to the relief of the Ethiopian (and general African) famine has been measly at best, but meanwhile Russians are driving ahead with the completion of a base in Ethiopia's Dahlak Islands in the throat of the Red Sea, excellently sited to block Suez Canal traffic. A similar base is growing on an island cluster just off the shores of the minor west-coast African state of Guinea. North Korean troops serve as instructors and praetorian guards in Zimbabwe (formerly Rhodesia) and the Seychelles Islands in the Indian Ocean, where Russian penetration is not so far advanced, and East Germans, Bulgarians, and other vassals make themselves handy in a variety of places and jobs. In Vietnam, the Russians have free use of the former great American base at Cam Ranh Bay. In the Mediterranean, the Marxist government of the strategic island of Malta has allowed Russian ships refueling privileges at its former British naval base. More important, Russia now has a functioning base in Cuba and undoubtedly is seeking one in Nicaragua, closer to the Panama Canal. This burgeoning base system, coupled with the rapid growth of the Russian Navy in general, and its submarine force in particular, is a grim challenge to the U.S. Navy's ability to protect essential sea traffic, or even our own coastline.

All these facets of Russia's overall national strategy are genuine threats and problems, yet Russia's most effective thrusts at the United States have been in a half-intangible war of words and ideas. Many of the wounds we have taken in this exchange have been self-inflicted or self-aggravated. By what must be one of history's greatest ironies, we have an infinitely more decent life than Russian communism will ever offer, yet we are lax in its defense. By tradition and heritage, we are many individuals with many ideas, largely busy about our own lives; Russia speaks with one voice, utilizing the largest, most elaborate propaganda machine in history in a strategy ranging from bloodcurdling threats of instant nuclear obliteration to tours by troupes of folk dancers. It trains propagandists and espionage agents as it does its soldiers—intensely and in great numbers.

A principal force in this Russian offensive is the KGB. Its functions include "disinformation"—the development and planting of false information. One such project involved the expert forgery of a "US Army Field Manual 30-31B," dealing with covert operations, which stated that the United States might employ extreme left-wing groups to defend its interests during foreign political crises. (There was no such American publication.) When Red Brigade terrorists kidnapped and murdered former Italian Prime Minister Aldo Moro, Communist propaganda claimed that they had acted under orders from the CIA; shortly thereafter, a copy of the fake manual surfaced through obscure channels to "prove" American guilt. Recently, African and Asian athletes received threatening letters from the Ku Klux Klan—soon proved to be forgeries, also, but clumsy ones.

Most disinformation is planted in foreign newspapers, magazines, and TV and radio programs through active Communist agents or sympathizers on their staffs, bribery, or some reporter who is teased with the opportunity to make a sensational scoop and lacks the intelligence or knowledge to check its accuracy. The material is published—and thereafter republished in Russia as originating abroad: see what wicked things those capitalist imperialists openly admit they are doing! A recent example was articles that appeared in American newspapers following the Russian destruction of Korean Air Lines flight 007. These stories stated categorically, on the authority of alleged intelligence specialists and a science magazine, that 007 had been on an American intelligence-gathering mission. When a few hard-nosed individuals checked those sources, both specialists and magazine proved surprisingly elusive—but meanwhile the stories had been published.

As a captured North Vietnamese officer explained in 1966, "Americans are naive . . . and do not know what is happening to them . . . newspapers, radios, magazines, and television are all instruments of war.''[19] Russians are not so ignorant. In 1975 a statement by a senior Communist party official outlined their strategy succinctly:

> By all overt and covert means, we shall manipulate public opinion of Western countries as we like and drown out criticism of our military buildup. We have the resources to create dozens of new organizations in the West and to reinforce existing frontline organizations. Our glorious intelligence services will seize all the new opportunities to operate on a much higher and wider scale. . . . We shall turn public opinion . . . against the USA. Everywhere we shall plant seeds of mistrust against the main enemy.[20]

For this mission, the Russians can deploy major propaganda/espionage forces, including the KGB and their Academy of Sciences with its busy Institute for the Study of the U.S.A. and Canada. These may operate directly or through Communist-controlled international organizations such as the World Peace Council, the World Federation of Trade Unions, or the International Association of Democratic Lawyers, all of which have branches or associated groups in the United States. Other work is done through small, secret ''cells'' of American Communists, usually consisting of professional people of some influence in their communities. The nature of American society makes it as easy for Russian agents (even known KGB officers) and money to move freely into and around the United States as it is impossible for Americans to do likewise in Russian territory.

Communist-controlled or sympathetic organizations within the United States sponsor other groups that appear, merge, vanish and reappear, or throw out new organizations as the situation may require. When some important Russian propaganda objective develops—at present, the nuclear freeze movement, which would leave Russia in a position of nuclear superiority—all sorts of new organizations with noble-sounding names appear out of nowhere, each designed to appeal to particular groups of Americans. These groups can draw on a number of frankly left-wing ''think tanks''—for example, the Center for Defense Information—for books, data, and motion pictures to support their activities. The relationship between all of these organizations is often tenuous, their financial support sometimes obscure, their memberships overlapping, their actual scope and influence undetermined, especially since the Congressional committees dealing with internal security were closed down during the 1970s. Also, the FBI's authority to investigate domestic subversive organizations was severely curtailed during the Carter administration. In both cases, the inhibiting action was taken by congressmen sympathizing with or pressed and beguiled by these leftist groups, to the immense profit of Russian covert activities.

One particular question of this moment is the degree to which these Russian-controlled/influenced groups may have penetrated the American mass media. While this possibility has been furiously denied by the media, the openly hostile attitude toward U.S. policies shown by some newspapers and TV personalities is cause enough for interest, especially since infiltration of the mass media has been a prime Communist objective in other countries.

This is not to say that Russians control the nuclear freeze and other disarmament movements in this country. They do not need to, any more than they must control international terrorism. Americans of all political views do their work for them; Russia needs only to provide occasional deft guidance to the upper levels of the movement, money when necessary, possibly professional advice on staging mass protests. When the cause of the moment can be packaged as a movement for peace, disarmament, jobs, and the brotherhood of man, good Americans will take it up out of innocent zeal. They will stand on windy street corners with homemade signs, hold candlelight vigils, picket, parade, and demonstrate. Some can be coaxed into more active roles; as this chapter was written several groups were hoping to disrupt U.S. cruise missile tests in Canada. Russian leaders, whose subjects would never dare such behavior, may consider their sincerity and gullibility idiotic, but they use them happily as extensions of the Russian propaganda/disinformation offensive. And they are effective; pressure by antinuclear, pacifist organizations in New Zealand and Australia threatens to destroy the defensive ANZUS alliance between those nations and the United States.

Our losses in this conflict of propaganda and ideas have been continuous and sometimes severe. Our failure in Vietnam began with inaccurate media reports of the North Vietnamese Tet offensive (see page 255), which echoed the enemy claims of victory. Also, there was the case of the "neutron bomb"—actually an economical nuclear warhead for artillery shells and tactical (battlefield) missiles, designed to produce little heat or blast but an intense burst of neutron radiation that could pierce tank armor. Intended as an "equalizer" in Eastern Europe where Russian tanks outnumbered NATO's at least two to one, its introduction in 1975 produced a worldwide Russian propaganda assault that hardly would be equaled until the recent appearance of the High Frontier/Star Wars concept. Cleverly, the Russians depicted it as intended to kill "people"—by implication, innocent civilians instead of Russian tank crews—while sparing property. There were demonstrations in front of the White House. The Russian Orthodox Church, which operates under KGB control, appealed to American churches to halt production of such an "inhuman" weapon. Though our NATO allies had agreed to the deployment of this neutron weapon, President Carter wilted and postponed its production. An even more grievous defeat was the long campaign of defamation by which the CIA, like the FBI, was drastically hamstrung during the 1970s. The Reagan administration has reversed this losing tide, but some of the damage done may be beyond repair, and the sniping against the CIA continues.

For these and many other reasons the problem of designing an effective national/military strategy to govern our relationship with Russia is one of the most difficult to confront the United States throughout its history. For one thing, it is a long-abiding problem that has been with us in one form or another since Communists seized control of Russia in 1917 and promises to be with us for a long, long time—and we are an impatient people.

Russians are interesting, often likable. They put their pants on one leg at a time and have to boil water to make tea, just like us. Few of them have any appetite for a war, nuclear or otherwise. Hitler, as we are reminded plaintively on all possible occasions, killed some twenty million of them and devastated much of western Russia. That Stalin killed off another twenty million or so—and still remains something of a national hero—is *not* mentioned.

They differ from us in being, through all their history, a herded, frightened people. Today, in an interesting variant of apartheid, the Soviet Union's roughly 274 million inhabitants are ruled by—and largely for the benefit of—the 18 million members of the Russian Communist party. It is with the elite of that mere 18 million that we must deal, most of them aging and in uncertain health. They are products of Lenin's teachings and Stalin's system; many were Stalin's willing errand boys. They rule the Russian people much as Stalin did. Russians are little freer today than in 1945, when this author watched veteran Russian combat officers with rows of medals cringe before a bumptious MVD[21] lieutenant, who obviously had seen no combat but did embody the authority of the Russian Communist party.

Negotiations may accomplish some relaxation of international tension, but it will be a long, tricky business, even with reasonably good intentions on both sides. The Russian language is rich and flexible, framed in its own thirty-two character Cyrillic alphabet and capable of deep and sinuous subtleties. Too few Americans really have mastered it. Add that the Russian, mentally and emotionally, is (as Rudyard Kipling once expressed it) the most Western of the Asiatics and not the most Eastern of the Europeans. He may study Clausewitz, but Sun Tzu is in his instincts.

Unfortunately, we cannot count on Russian good intentions. They are an expanding power; their national strategy is, in general, operating successfully. The Western powers remain almost entirely on the defensive. It would be thoroughly stupid to expect them to give up any part of their expanding superiority in both nuclear and conventional forces and their existing military presence in space without exacting a maximum *quid pro quo*—which well could be more than we could risk giving. Also, no agreement on disarmament will end the Russian propaganda war or its support of terrorist groups. Optimistic Americans might recall that Russia's Foreign Minister Andrei Gromyko, recently applauded for agreeing to disarmament negotiations, made a special visit to President Kennedy in October 1962 to assure him that there were no Russian missiles in Cuba and that Russia would never consider such a naughty thing.

However, so long as there is an opportunity for a verifiable reduction in armaments and tension, we must negotiate—carefully. So far, too many of our agreements with Russia have resembled that reached between a missionary's nubile daughter and a marine sergeant over a bowl of spiked punch. We might take as our guide the doctrine stated by those incredible gentleman adventurers of Queen Elizabeth's Muscovy Company who braved the North Cape's fog and storms, the uncharted White Sea, and unpredictable Ivan the Terrible to dare southward across his realm and out of it to the Caspian Sea and Persia. Russians, they soon found, were kittle folk to deal with: "you have need to take heed how you have to do with him . . . and to make your bargains plain, and to set them down in writing. For they be subtle people, and do not always speak the truth, and think other men to be like themselves."[22]

So far as we can judge from a study of their publications and their senior officers' speeches, the Russian armed forces believe they could win a nuclear war without prohibitive damage to Russia. They know, however much they may pretend to believe otherwise, that we have surrendered the initiative (see page 298) to them by our adoption of a second-strike strategy; that if they struck heavily and shrewdly enough, their first blow might shatter America's defenses; that, America defeated, there would be no refuge from their mastery for free men, anywhere.

But, whatever we do know concerning Russia's capabilities and intentions, we do not know enough. Russia always has been a nation of secrets, to its own citizens as well as to foreigners. Our space satellites listen to Russian radio and telephone messages and snap photographs so detailed that we can read the automobile license plates shown in them. But we cannot be certain whether the messages and signals they intercept are real communications or transmissions doctored for our misinformation, and it does us little good to read a truck's license plate when we cannot tell what cargo the truck is hauling.

The average American citizen, confused by portentous rumblings out of Moscow, further echoed and amplified by American groups that vary from hard-core Communist fronts through useful idiots to traditional pacifist sects is being conditioned to think of nuclear warfare only as an utter holocaust in which all combatants blaze away with every nuclear weapon in their arsenals, leaving this earth thoroughly charred and then frozen. Exactly to the contrary, even the most gung-ho SAC officer never had any such suicidal intent—and it is most doubtful that the most brutal commissar has, either. America's whole nuclear strategy has been to cripple the enemy as quickly and efficiently as possible without risking major damage to our country or inflicting unnecessary harm in Russia. Moreover, it is obvious to both sides that victory would be pointless if all you win is a radioactive ash heap and a nasty relief problem. The latter might not particularly bother the commissar, but we can be certain that he intends to survive as comfortably as possible.

The really frightening possibility is the chance—however small—

that a fanatic like Qaddafi or Khomeini might obtain or develop nuclear weapons.

Following World War II, strategic studies became more and more public property. Academicians and other civilians of the type classified generally (and sometimes charitably) as ''intellectuals'' assumed an increasing role in strategic planning. In part, they met new, essential needs—the rapidly escalating development of nuclear weapons, electronic equipment, computer technology, and intricate weapons systems that combined all of these, required skills in research and analysis taught by neither military schools nor combat. Other influential civilian participants were the experts in games theory—a variety of disciplines, mostly mathematically or psychologically based, for testing strategic concepts and proposed international policies under controlled conditions. As a class, these civilians formed a hagiocracy of highly educated specialists who knew nothing much of war and less of fighting men but were seldom troubled by that lack. With Georges Clemenceau, they were convinced ''war is too serious a matter to leave to soldiers.''[23]

Their influence probably peaked during the Kennedy-McNamara period. It was largely the result of ruminations by dilettante civilian authorities on guerrilla warfare that our forces in Vietnam were shackled by a Strategy of Graduated Response. The North Vietnamese were allowed the initiative; when they did something to us, we were to do something a little worse to them. At some point in this process, North Vietnam was supposed to see the light and call the war off. The North Vietnamese leaders were not impressed, and any potential superstrategist among the American commanders never had a chance to show what he could do.

The superstrategist of today and tomorrow must be a master of warfare unimaginable to most of his predecessors. He must be prepared to employ weapons running deep under the oceans and ranging the black void of outer space—weapons that can span the Atlantic in a quarter of an hour and carry unprecedented destruction. He must have schooled himself to decide and act in a blinding crisis when two minutes' hesitation can mean abject defeat. At the same time he must be instant-ready to plan and carry through minor police actions such as the recent liberation of Grenada, or to deal effectively with a dragging, on-and-off terrorist menace. If he is an American, he must have the talent to operate effectively and cheerfully amid politicians, the mass media, and skittery allies—and, come trouble, he cannot expect a free hand until the situation is (in an old Far West phrase) "damn near shot to hell."

It may well be that we shall not require a superstrategist in our lifetimes. If we do, God grant there is one among us!

APPENDIX: THE PRINCIPLES OF WAR

The Principles of War are a concentrated distillation of centuries of military experience—an essential checklist that planners and commanders must keep in a handy corner of their consciousness.

Throughout history nations have sought the secret of how wars were won, whether for conquest or in self-defense. Many of them developed a considerable literature on the subject—histories, treatises on the martial qualities commanders should possess, general discourses on the art of war in which high strategy was intermixed with discussions of shoes, rations, and petty ruses. Much of this has vanished into history's lost corners; those works that survive show an amazing agreement on military principles, though produced in all parts of the world, during different periods, by men who sometimes knew little or nothing of one another's existence.

The modern-day specific listings of Principles of War began with Lloyd, Jomini, and Clausewitz during the late eighteenth–early nineteenth centuries. Thereafter, their number and character were widely debated in European service schools. Between 1912 and 1924 J. F. C. Fuller worked out a set of eight: *objective, offensive action, surprise, concentration, economy of force, security, mobility, cooperation.* These, with the addition of *simplicity,* appeared in American training regulations in 1921 as part of United States military doctrine. *Concentration* was renamed *mass; mobility* changed to *movement,* and later to *maneuver. Cooperation* became the more emphatic *unity of command.*

Objective

The purpose of military strategy is to achieve a decisive, obtainable result. This is the controlling Principle of War: Without a clearly comprehended objective there is no sound basis for the application of the other principles.

Once hostilities have begun, the national strategy should be to win the war and to make certain that a just peace follows it. In achieving these twinned final objectives, the contributing objective of a nation's armed forces is the destruction of the enemy's forces and of his will to fight. (''Destruction'' in this case means that the enemy has surrendered or has been rendered incapable of further resistance—not annihilation.)

Just as the armed forces' objective must support the achievement of the overall national objective, so must the objectives of subordinate military forces contribute directly to attainment of the major military objective. In World War II the Anglo-American political leaders decided to defeat Germany first and then to concentrate against Japan. Gen. Dwight D. Eisenhower, Supreme Commander of the Allied invasion of Western Europe, was given the mission: ''You will enter the continent of Europe and, in conjunction with the other Allied Nations, undertake operations aimed at the heart of Germany and the destruction of her armed forces.''[1]

To achieve this, orders, based on plans developed by Eisenhower's staff, were passed down through army, air force, and naval commands. The army commands went to Gen. Sir Bernard Montgomery's 21st Army Group; from it to Lt. Gen. Omar Bradley's First U.S. Army and Lt. Gen. M. C. Dempsey's Second British Army; from them to their component corps; from those corps to their divisions; and from the divisions to their subordinate units, each of which was assigned its own mission, all designed to contribute to the success of the whole.

Failure to select a decisive objective results in ineffective, scattered efforts. In 1941 the German offensive into Russia had its main attack aimed at Moscow. Moscow was the capital of both Russia and world communism; it also was the major center of the Russian railroad system and of much of the Russian armament industry. Consequently, German generals were certain that Stalin would commit the pick of his army to Moscow's defense, enabling the Germans to destroy it in one sustained offensive. Their opening drive was amazingly successful, but Adolf Hitler wanted to skin the Russian bear before he had killed it. He diverted much of his strength off to the north, against Leningrad, and to the south, to seize the Ukraine, Donets Basin, and Crimea with their natural and industrial resources. These new offensives also were successful, but they gave the Russians six weeks to improve their defenses before Moscow and wore down the German panzer divisions. Afterward, the German drive on Moscow was renewed with great initial victories. Then the autumn rains came, turning roads to mud wallows, and the Russian winter followed.

The Offensive

As Confederate Gen. Nathan B. Forrest put it, "Forward, and *mix* with them!" "Come a-running and fetch all you've got!"[2]

Seize, keep, and exploit the initiative. If the enemy strikes first, without warning, it may be necessary to stand on the defense long enough to block his thrust and build up your own forces, but such defensive operations can only prevent your defeat. Victory requires offensive action, carrying the war to the enemy and into his own territories. An army on the offensive has the initiative and therefore freedom of action; it can strike where and when it chooses. If its operations are skillfully planned and sustained, the enemy will be too busy trying to stay alive to manage a counteroffensive. (Ulysses S. Grant's merciless pressure on Robert E. Lee during 1864–65 is an excellent example.) Soldiers on the offensive commonly show higher morale and a more aggressive spirit than those on the defensive.

A nation engaged in offensive operations in one decisive area may go on the defensive on less important fronts in order to concentrate maximum combat power for that offensive (see *mass* and *economy of force*). But even when thrown on the defensive for any reason, a wise commander avoids a passive defense and attempts to keep his immediate opponent off-balance by limited spoiling attacks and raids.

An offensive launched with insufficient means seldom achieves much and may lead to disaster. Napoleon's invasion of Russia in 1812 and Hitler's in 1941 were amazing military achievements that dealt summarily with the Russian armed forces. However, they were not prepared to deal with Russia itself—bad roads, forests, marshes, violent shifts of weather, and seemingly endless distances. Supply systems faltered, marches slowed, and eventually winter came.

Reflecting that "Pacifists would do well to study the Siegfried and Maginot Lines, remembering that those defenses were forced; that Troy fell; that the walls of Hadrian succumbed; that the Great Wall of China was futile," Gen. George S. Patton, Jr. concluded that "In war, the only sure defense is offense and the efficiency of offense depends on the warlike souls of those conducting it."[3]

Unity of Command

As a young general in Italy, Napoleon (then simply General Bonaparte) learned that the semblance of a French government currently sitting in Paris thought best to have him share his command with steady old Gen. François Kellermann. At once he offered his resignation; though acknowledging Kellermann's greater experience, he felt that "it would be better to have one incompetent general than two good ones."[4] U. S. Grant echoed him sixty-seven years later: "Two commanders on the same field are always one too many." This is merely a military version of the ancient folk adage "Too many cooks spoil the broth."

The forces engaged in any military operation must be under one responsible commander who has full authority over his subordinate commanders and their forces. (MacArthur naturally would expand this to claim full authority over all military, political, and economic activities in the area he commanded.)

Military history is filled with sad old tales of disaster resulting from a divided command. Frequently, this has been the case in amphibious operations where admirals and generals cannot agree. In 1809, with Napoleon deep in Austrian territory, England launched a strong expedition to capture Antwerp, one of his major naval bases. Disputes between land and sea commanders resulted in delays; seasonal fevers ate through the invasion force, and the survivors crept home, inspiring a doggerel ballad:

> The Earl of Chatham, with his saber drawn,
> Stood waiting for Sir Richard Strachan;
> Sir Richard, longing to be at 'em,
> Stood waiting for the Earl of Chatham.[5]

A similar misfortune befell the United States at Pearl Harbor. The Army and Navy commanders and their staffs had cordial surface relations, but had not bothered to coordinate their intelligence, reconnaissance, and security measures. And nobody knew who, if anyone, had the ultimate command responsibility in case of an enemy attack on the Hawaiian Islands. As an extra complication, control and employment of the Army's new radar Aircraft Warning Service was in dispute between the Signal Corps and the Air Corps; the system therefore was not fully operational on December 7, 1941.

Similarly, American operations throughout the Pacific were bedeviled by two competing commanders for the rest of the war. Gen. Douglas MacArthur, committed to an advance northward from Australia to liberate the Philippines, ruled the Southwest Pacific. The rest of that ocean fell under Adm. Ernest J. King, Chief of Naval Operations, who was determined on a drive westward from Hawaii. These two competed furiously for all available supplies and reinforcements and still demanded more. Gen. George C. Marshall, Army Chief of Staff, who managed to keep the affair just within the bounds of sanity, wryly remarked that the King-MacArthur dispute became a "very vicious war" of personalities, with MacArthur the more culpable.[6] Considering the powerful interests (MacArthur's backers included the Australian government and potent American politicians) and the umbrageous personalities involved, the situation was left to work itself out. Fortunately, the Pacific is an immense area, and the Japanese leaders were even more at loggerheads among themselves.

Mass

The fundamental law of military strategy is: Be stronger than your enemy at the decisive time and place. Centuries ago it was sung in the medieval ballad *Robin Hood and Guy of Gisborne:*

> And it is said, when men be met,
> Six can do more than three.[7]

Mass has been measured in files of armored pikemen, or cannon emplaced wheel-to-wheel and row-behind-row along a front. It may be measured in armored divisions or megatons of nuclear explosive in a rocket's warhead. But it does not refer to numbers alone but rather to the overall combat effectiveness of a force, including numbers, weapons, discipline, training, morale, leadership, inherent fighting ability, and its backup of reinforcements and supplies. One trained, armed terrorist in a street or plane full of peaceful people can be sufficient mass for that occasion.

The principle of *mass* is easily misinterpreted as requiring the piling up of all available forces—a process, hard to conceal, that may strip other commands of essential resources. History is full of examples whereby a smaller army, utilizing the principles of *maneuver* and *surprise,* has achieved sufficient *mass* at the critical point to defeat a larger one. Gen. Thomas J. "Stonewall" Jackson explained that method (in almost the same words that Napoleon had used before him): "Never fight against heavy odds if, by any possible maneuvering, you can hurl your whole force on a part, and that the weakest part, of your enemy and crush it."[8] Rough-edged Nathan B. Forrest put it even more simply: "Get there first with the most men."[9]

Economy of Force

This principle balances that of *mass,* because it is the method whereby we are able to concentrate the necessary combat power for decisive offensives. Areas of secondary importance should be assigned only the essential minimum force, which must do its best to hold, delay, or confuse the enemy facing it while our main effort is made elsewhere. Thus in 1809, caught half prepared when Austria (thinking him fully committed in Spain) invaded the territory of his German allies without any formal declaration of war, Napoleon checked the main Austrian thrust with one corps (47,000 men) while massing 105,000 to smash in and around the Austrians' left flank in a rapid series of *maneuvers* and battles. This offensive (loosely termed the Ratisbon-Abensberg-Eggmuhl campaign) was a quick success; Napoleon would remember it as the "most brilliant and most able" of his operations.[10]

One of this principle's most daring practitioners during the Civil War was

Maj. Gen. Henry W. Halleck. In early 1862 he sent Brig. Gen. Ulysses S. Grant and Comdr. Andrew H. Foote up the Tennessee River in an offensive other Union commanders considered risky. To keep it moving, through big successes and short-lived checks, Halleck fed in men and supplies as rapidly as he could, in spite of troubles in Missouri. Accepting the chance that resurgent Confederates might recover much of that state, he unhesitatingly sent Grant a large part of his forces there. If successful, Grant's advance would dislocate the Confederate position in the West; Missouri could be reconquered later.

In one respect, *economy of force* can be troublesome to apply. Few commanders ever feel that they have sufficient combat power; most want reinforcements. Consequently, most object to giving up any portion of their forces to another command and may require thoroughly unambiguous orders. Army legend tells of Patton's reaction when ordered to transfer one of his divisions to another army. Patton ordered it committed to action immediately: "I'll get it in so deep they'll have to send me three more to get it out!"[11]

Maneuver

Maneuver is the art of moving combat power by land, water, air and/or space. By itself, it produces no decisive results, yet it is the means of attaining *mass,* the *offensive, surprise,* and *economy of force.* Normally, it should be rapid and conducted with proper *security* so that the enemy is unaware of it. *Maneuver* also may be used to confuse the enemy, distracting him from our concentration of combat power elsewhere. One of the most successful maneuvers in American history followed the defeat of Gen. William S. Rosecrans at Chickamauga in September 1863 by Gen. Braxton Bragg's Confederate army. To restore the situation, the federals brought two corps (25,000) from Virginia—1,200 miles over a series of railroads, the leading elements arriving in six days, the last closing in eleven and a half. At the same time, more troops came in from the West. The resulting concentration enabled Grant to rout Bragg at Chattanooga on November 23–25. A more recent example occurred during the Battle of the Bulge in December 1944. Secretly massing troops, Hitler struck suddenly in the Ardennes, shattering the American lines. Patton's U.S. Third Army was then attacking German positions in the Saar area, roughly sixty miles to the south. Alone among the Allied commanders, Patton and his staff had foreseen such an emergency and prepared for it. Through freezing weather, along ice-slicked roads, Patton extricated a large part of his army from combat, turned it ninety degrees to the north, and drove in against the southern flank of the German offensive—one of World War II's outstanding feats.

Surprise

Surprise is the result of striking the enemy at a time or place, or in a manner, that he does not expect. It is exemplified by an irreverent old jingle that embodies the basic fact of international relations:

Twice-armed is he who hath a cause that's just,
—But thrice-armed he who gets his blow in fust![12]

The purpose of *surprise* is to win our *objective* before our enemy can react effectively. It is psychologically powerful, stampeding enemy troops and numbing their commander's ability to think clearly and react quickly, and therefore is an excellent means of gaining combat superiority, even over larger forces. Forrest, whose career as a cavalry raider depended greatly on achieving it, would rather have "fifteen minutes of the bulge than three days of tactics."[13]

To get maximum results from *surprise,* it must be done with sufficient *mass* to exploit and snowball the resulting confusion. Otherwise, the enemy may rally in time to swamp their attackers—a development not entirely unknown even to Nathan B. Forrest. The German 1944 counteroffensive in the Ardennes failed in large part because of insufficient logistic support—specifically, the lack of enough gasoline to keep the offensive rolling while the American forces still were disorganized.

Surprise does not need to be the absolute "open mouth" type where, as at Pearl Harbor, the enemy knows nothing of the planned attack until it strikes him. It suffices if he does not know that it has been launched until it is too late for him to react effectively. The North Vietnamese offensive launched on January 30, 1968 during their proclaimed Tet truce was generally of this type. American intelligence had indications that something *might* happen in the near future but had no idea of precisely what, when, or where. When the major offensive broke loose all across South Vietnam, there was a frantic scramble to meet unexpected emergencies in many areas. They were met successfully, but the psychological shock to the American Presidency, Walter Cronkite, much of the mass media—and thus to the American public—was devastating.

Surprise can be attained by rapid *maneuver,* secrecy and deception, *security* which denies information to the enemy, and the introduction of new weapons or tactics. Some modern technological developments have decreased the chance of obtaining surprise, for example, electronic sensors that can be air-dropped in the enemy rear areas to broadcast sounds of movement, improved aerial observation and photography, "spy" satellites, and over-the-horizon radar. Other developments—swifter ships, ground vehicles, and aircraft and long-range missiles—make it easier. The possibility of another Pearl Harbor, featuring intercontinental missiles with thermonuclear warheads, is always with us.

Security

Security is the complement of *surprise,* just as *economy of force* is that of *mass.* It is the total of the measures a commander takes to prevent the *surprise* of his own forces and to keep the enemy in ignorance of his own plans and actions. *Offensive* action normally increases *security,* since it keeps the enemy on the defensive and so limits his freedom of action; other supporting factors are efficient intelligence and reconnaissance and general readiness for action. Napoleon stated it in simple terms:

> An army should be ready every day, every night, and every hour, ready to offer all the resistance of which it is capable. . . . A general should say to himself many times a day: If a hostile army were to make its appearance in front, on my right, or on my left, what should I do? And if he is embarrassed, his arrangements are bad; there is something wrong; he must rectify his mistake.[14]

The same observations apply to a nation.

Of all these Principles of War, *security* is the one as applicable in peace as during hostilities. It involves both intelligence (the collection and evaluation of information concerning other nations, especially potential enemies) and counterintelligence (the thwarting of enemy intelligence) and constitutes a continuous shadow war, waged to prevent a thermonuclear Pearl Harbor.

Security can be overdone. Before the Franco-Prussian War of 1870–71, Emperor Napoleon III had pushed the development of a secret weapon, the *mitrailleuse* (an early machine gun, resembling the Gatling gun), using his own money when his Chamber of Deputies refused to vote funds for rearmament against the obviously growing danger from the east. This *mitrailleuse* promised to be an effective antipersonnel weapon. Knowledge of it, however, was tightly restricted; none were sent to the field armies until hostilities were imminent. Sometime, amid the antic confusion that beset France's 1870 mobilization, *mitrailleuses* arrived without the specialists trained to handle them. Too many generals, seeing they were mounted on wheeled carriages, employed them as if they were field guns. The efficient Prussian artillery outranged them and quickly knocked them out.

To our continual discredit, the United States always has handled its security problems with eleven thumbs, two left feet, and no perceptible intelligence. Among our oldest national traditions is that of allowing crises to catch us with our trousers at half-mast. The recent massacre of U.S. Marines in Beruit is only the latest (and by this time possibly *not* the latest) of a series of disgraceful and unnecessary surprises that run back through Korea, the Battle of the Bulge, Pearl Harbor, the *Maine,* the River Raisin, Paoli, and a hundred frontier fights. Something in the American character opposes the establishment of an effective intelligence/counterintelligence/security system and inhibits the working of whatever system we may have—an apparent combination of "That violates our constitutional rights"; "But they're friendly people, just like us"; "Don't make waves"; and "They wouldn't *dare!*"

Simplicity

Simplicity is essential in plans and orders. Orders must be concise and clear; plans must be perfectly comprehensible and direct. Carl von Clausewitz explained the reason:

> If one has never personally experienced war, one cannot un-

> derstand . . . the difficulties. . . . Action in war is like movement in a resistant element. Just as the simplest and most natural of movements, walking, cannot easily be performed in water, so in war it is difficult for normal efforts to achieve even moderate results.[15]

In war, anything that can go wrong probably will; any order that can be misunderstood certainly will be. Elaborate plans that require the coordination of a number of different forces have a way of coming unglued. The North Vietnamese Tet offensive of 1968 was supposed to break with simultaneous attacks throughout South Vietnam. Had this taken place, it is just possible that it might have succeeded. But it was too complex a plan: Some units did not get the word, others moved into position too slowly, some were detected moving up, some officers flinched at the missions assigned them. This friction produced enough indications that something big was brewing somewhere, and the American command built up a reserve to deal with a possible emergency. Instead of one big effort, the offensive dribbled out in piecemeal attacks over several days, none with enough *mass* to be successful. The result was a savage and costly defeat for North Vietnam.

Simplicity however, must not prohibit the use of new weapons, equipment, and techniques. (''Simplicity, yes,'' some English staff officer supposedly observed. ''But *not* the simplicity of an ape with a paintbrush!'')

The British have two additional Principles of War: *maintenance of morale* and *administration*. The first gives official recognition to the fact, known since wars began, that troops with high morale are far more effective in combat than troops without interest in why they fight and/or without confidence in their leaders. *Administration* stresses the even better known fact that an army must be efficiently supplied and administered, the last including such matters as medical care, pay, mail, and personal records. Frederick the Great reduced this to its bedrock sense with his famous ''Understand that the foundation of an army is the belly. It is necessary to procure nourishment for the soldier. . . . This is the primary duty of a general.''[16] These two principles are closely related, *administration* being a major factor in *maintenance of morale*. In American military doctrine these principles are considered part of normal routine, usually justly so—though the American serviceman just as routinely finds some reason to complain.

The Russians include *annihilation* among their principles of war. Considering their operations in World War II and at present in Afghanistan, it would be well to take the term at face value. Annihilation always has been a feature of Oriental warfare. Genghis Khan, Tamerlane, and others of their ilk through the centuries routinely massacred whole populations. Mercy was hardly a characteristic of the Imperial Japanese Army, and Mao Tse-tung taught that "to preserve one's own forces and annihilate the enemy" was the "first essential of military operations."[17]

In actual fact, however, most nations have practiced the strategy of annihilation at one time or another, especially on a small scale during colonial wars. America's Indian struggles were often kill-or-be-killed affairs. One Arizona militia officer, rebuked for killing squaws and children, voiced an attitude common among American frontiersmen: "I fight on the broad platform of extermination."[18] (So did the Apaches he hunted, with preliminary torture when possible.) Some American generals have shown a no-quarter attitude, like Stonewall Jackson's "No, kill them all. I do not wish them to be brave." and "Always kill the brave officers."[19]—a degree of ferocity normally encountered only among noncombatants. These incidents, however, were minor and unusual in American history. It was strategic bombing, especially with nuclear weapons, that really introduced annihilation to Western warfare as a definite strategy. Men who would protect civilians caught in the midst of ground fighting somehow find killing them no trial to their conscience 30,000 feet above an enemy city or in a safe headquarters thousands of miles distant. On occasion, nevertheless, their strategy has had a grim justification. The atomic bombing of Hiroshima and Nagasaki killed approximately 210,000 Japanese and forced Japan's immediate surrender. Had those two bombs not been used, it probably would have been necessary to invade Japan against fanatical resistance by its armed forces and civilians alike. Probable Allied casualities were estimated at one million. Japanese losses would have been far greater.

An unrecorded British sergeant-major put the whole concept of strategy into one sentence that was almost an echo of U. S. Grant's definition. "Hit the other fellow, as quick as you can, and as hard as you can, where it hurts him most, when he ain't lookin'."

Grant himself had a final observation: " . . . in war anything is better than indecision. *We must decide.*"

NOTES

Epigraph and Introduction

1. F.J. Huddleston, *Warriors in Undress* (London: John Castle, 1925), pp. 221–22.
2. Author unknown.
3. Attributed to Alfred Machen, British author, 1863–1947.
4. *Othello,* act I, sc. 1, line 22.
5. Rudyard Kipling, *Verse: Definitive Edition* (Garden City, N.Y.: Doubleday Doran and Co., Inc., 1945), pp. 571–72.

Chapter I

1. Stephen V. Benet, *Selected Works,* vol. 1 (New York: Farrar & Rinehart, 1942), p. 102.

2. Octave Aubry, *Les Pages Immortelles de Napoleon* (Paris: Editions Correa, 1941), pp. 228–29.
3. Bruce Catton, *Glory Road* (Garden City, N.Y.; Doubleday & Company, 1952), p. 230.
4. Department of Military Art and Engineering, U.S. Military Academy. *Notes for the Course in the History of the Military Art* (West Point, N.Y.: Department of Military Art and Engineering, 1964), p. 10.
5. Carl von Clausewitz, *On War* (Princeton, N.J.: Princeton University Press, 1976), pp. 80, 86–88, 605 (edited and translated by Michael Howard and Peter Paret).
6. Corinthians 14:8.
7. Thomas R. Phillips, *Roots of Strategy* (Harrisburg, Penn.: Military Service Publishing Company, 1940), pp. 49 and 172.
8. Hans Delbruck, *History of the Art of War,* vol. 1 (Westport, Conn.: Greenwood Press, 1975), pp. 569–70 (translated by Walter Renfroe).
9. Phillips, *Roots of Strategy,* p. 172.
10. Ibid., p. 436.
11. Officer's identity now unknown.
12. Douglas S. Freeman, *Lee's Lieutenants,* vol. 1 (New York: Charles Scribner's Sons, 1942), p. 470.

Chapter II

1. "Who died far away, before his time, but as a soldier, for his country." John Bartlett, *Familiar Quotations* (Boston: Little, Brown & Co., 1941), p. 764.
2. Hans Delbruck, *History of the Art of War,* vol. 1 (Westport, Conn.: Greenwood Press, 1975), pp. 569–70.
3. F.E. Adcock, *The Greek and Macedonian Art of War* (Berkeley, Calif.: University of California Press, 1957), p. 65.
4. This was the legendary "10,000." Actually, Cyrus had another smaller Greek force operating to cover his flank. The two must have totaled over 13,000.
5. Adcock, *Art of War,* p. 76.
6. See Donald W. Engels, *Alexander the Great and the Logistics of the Macedonian Army* (Berkeley: University of California, 1978).
7. Better known as "Arrian," the Greek form of his name. He used original sources whenever possible. Made a Roman citizen by the Emperor Hadrian, he was a successful frontier commander, 131–37.
8. Ca. 200 a "Castra Peregrina" appears—possibly an imperial field headquarters.
9. J.F.C. Fuller, *Julius Caesar* (London: Eyre & Spottiswoode, 1965), pp. 298–99. Caesar was deified as "The Invincible God."
10. Delbruck, *History,* p. 568.
11. Ibid., p. 570.
12. From the poem "Soldier" by MG Charles T. Lanham.

13. Stephen V. Benet, *Selected Works,* vol. 1 (New York: Farrar & Rinehart, Inc., 1942), p. 102.
14. Fuller, *Julius Caesar,* pp. 210, 321.
15. The Rubicon River was the boundary between Cisalpine Gaul and Roman Italy.
16. Theodore A. Dodge, Civil War veteran and biographer of "Great Captains," considered Caesar "perhaps the greatest man the world has ever seen" but also believed that "more than half of Caesar's campaigns were consumed in extricating himself from the results of his own mistakes" (*Caesar,* vol. 2 [Boston: Houghton-Mifflin, 1892] pp. 691–92, 767).
17. It commonly is known as *The Military Institutions of the Romans,* the title given it by a Lt. John Clarke; his translation, published in London in 1767, is considered the standard English version.
18. Thomas R. Phillips, *Roots of Strategy* (Harrisburg, Penn.: Military Service Publishing Company, 1940), p. 102.
19. Ibid., p. 172.
20. Ibid., p. 67.
21. Henry H. E. Lloyd, *A Political and Military Rhapsody on the Invasion and Defense of Great Britain and Ireland* (London: 1795, p. 284).
22. George A. Billias, *George Washington's Generals* (New York: William Morrow & Co., 1964), p. 260.
23. William F. Rickenbacker, "The Latin Hour," *National Review,* 19 August 1983, p. 1014.

Chapter III

1. Barrow, G.W.S., *Robert Bruce and the Community of the Realm of Scotland* (Berkeley, Calif.: University of California, 1965), p. 194.
2. Eric Linklater, *Robert the Bruce* (New York: D. Appleton-Century Co., 1934), pp. 61–62.
3. Ibid., pp. 39–43, 147–49.
4. Charles Oman, *A History of the Art of War in the Middle Ages,* vol. 2 (London: Methuen & Co., Ltd., 1924), p. 99.
5. Herbert E. Maxwell, *Robert the Bruce and the Struggle for Scottish Independence* (New York: G.P. Putnam's Sons, 1909); p. 222.
6. George MacDonald Fraser, *The Steel Bonnets* (New York: Alfred A. Knopf, 1972), p. 28. The "family" was his "*familia regis,*" the king's military staff and personal escort.
7. Ibid., p. 28.
8. John Buchan, *Oliver Cromwell* (Boston: Houghton Mifflin, 1934), p. 296. This may have been a grudge fight. Leslie and Cromwell had served together during the English Civil War, during which Leslie's dash and skill probably saved Cromwell from defeat at Marston Moor, but Cromwell's report gave him little credit.
9. Oman, *History of the Art of War,* p. 101, fn. 1.

10. Edward M. Earle, ed., *Makers of Modern Strategy* (Princeton, N.J.: Princeton University Press, 1944), pp. 3, 13–14.
11. Ibid., pp. 12, 24.
12. Charles Oman, *A History of the Art of War in the Sixteenth Century* (New York: E.P. Dutton & Company, Inc., 1937), p. 396.
13. F.J. Hudleston, *Warriors in Undress* (London: John Castle, 1925), p. 161.
14. Earle, *Modern Strategy,* p. 25.
15. Fairfax Downey, *Cannonade* (Garden City, N.Y.: Doubleday & Company, 1966), p. 333.
16. From Barwick's *A Breefe Discourse Concerning the Force and All Manuall Weapons of Fire,* published in London in 1594.
17. Oman, *Sixteenth Century,* p. 497.
18. Geoffrey Parker, *The Army of Flanders and the Spanish Road, 1567–1659* (Cambridge, Mass.: Cambridge University Press, 1972), p. 132.
19. Sir James Turner, *Pallas Armata* (London: At the Rose and Crown, 1683).
20. Oliver L. Spaulding, Hoffman Nickerson, and John W. Wright, *Warfare* (Washington, D.C.: The Infantry Journal, 1937), pp. 465–66.
21. Michael Lewis, *The Spanish Armada* (New York: Macmillan, 1960), p. 102.
22. Garrett Mattingly, *The Armada* (Boston: Houghton Mifflin Company, 1959), p. 123. Not so oddly, Genghis Khan used almost this same phrase.
23. C.G. Cruickshank, *Elizabeth's Army* (Oxford: Clarendon Press, 1966), is a splendid reference here.
24. Onasander apparently was a Greek who lived in Rome during the first century A.D. He wrote a number of works on methods of training soldiers, to include sham battles.
25. Cruickshank, *Elizabeth's Army,* p. 201.
26. Henry J. Webb, *Elizabethan Military Science: The Books and the Practice.* (Madison, Wis.: University of Wisconsin Press, 1965), p. 31.
27. G. Geoffrey Langsam, *Martial Books and Tudor Verse* (New York: Columbia University, 1951), p. 26.
28. David Beers Quinn, *England and the Discovery of America* (New York: Alfred A. Knopf, 1974), p. 301. I have slightly modernized the text quoted.
29. Cyril Falls, *Mountjoy: Elizabethan General* (London: Odhams Press, Ltd., 1955), p. 159.
30. Ibid., pp. 33, 106.

Chapter IV

1. R.M. Hatton, *Charles XII of Sweden* (London: Weidenfeld and Nicolson, 1968), pp. 145, 184.
2. Callot did two series of engravings, *Les Grandes Miseres de la Guerre*

(1633) and *Les Miseres et Malheurs de la Guerre* (1633–35). Earlier panoramic prints he did for Louis XIII and Spanish officers had shown the dramatic side of war.

3. Cartridges—a musket ball and a charge of powder wrapped in a twist of paper—had been known for at least a century. This, however, was their first large-scale use.
4. Ingvar Andersson, *A History of Sweden* (New York: Praeger, 1956), p. 171 (translated by Carolyn Hannay).
5. Hatton, *Charles XII,* p. 115 fn.
6. C.H. Firth, *Cromwell's Army* (London: Methuen & Co. Ltd., 1962), p. 171.
7. Ibid., p. 177.
8. Ibid., p. 208.
9. J.F.C. Fuller, *A Military History of the Western World,* vol. 2 (New York: Funk & Wagnalls Company, 1955), p. 46.
10. Andersson, *History of Sweden,* p. 178.
11. Ibid., p. 201.
12. Electors were the ruling princes of the major German principalities, who were entitled to elect the Holy Roman Emperor. Poland was an elective monarchy, but its ''elections'' were normally won by force or large-scale bribery.
13. F.J. Huddleston, *Warriors in Undress* (London: John Castle, 1925), p. 241.
14. Hatton, *Charles XII,* p. 521.
15. Turenne became a Catholic in 1668 as a gesture of fidelity to Louis XIV.
16. Max Weygand, *Turenne: Marshal of France* (Boston: Houghton Mifflin, 1930), p. 23 (translated by George B. Ives).
17. Francis Lloyd, *Marshal Turenne* (London: Longmans, Green & Co., 1907), p. 67.
18. ''Princes of the blood'' (or ''blood royal'') were directly related to the king.
19. Octave Aubry, *Les Pages Immortelles de Napoleon* (Paris: Editions Correa, 1941), p. 202.
20. Oliver L. Spaulding, Hofman Nickerson, and John Womack Wright, *Warfare* (Washington, D.C.: The Infantry Journal, Inc., 1939), p. 502.
21. Sir Charles Oman, *A History of the Art of War in the Sixteenth Century* (New York: E.P. Dutton and Company, Inc., 1937), p. 476. *Enfants perdus* (lost children) were men detailed for especially risky duty. An arquebus was a light musket.
22. Henri Bouchot, *L'Epopee du Custume Militaire Francais* (Paris: L. Henri May, n.d.,), p. 76.
23. Quotations in this paragraph are from pp. 13 and 6 of Turpin de Cissé, *Commentaires sur les Memoires de Montecuccoli* (Paris: Lacombe Libraire, 1769).

24. Weygand, *Turenne,* p. 328.
25. Bouchot, *L'Epopee,* p. 67.
26. Weygand, *Turenne,* p. 120.
27. Ibid., p. 151.
28. Probably the latest edition of them was published in Paris in 1909, edited by Paul Marichal.
29. Lloyd, *Marshal Turenne,* pp. 29, 50, 144, 159, 365.
30. Cyril Field, *Old Times Under Arms* (London: William Hodge and Company, Limited, 1939), pp. 114–15.
31. Walter C. Horsley, ed. and trans., *The Chronicles of a Campaigner: M. de la Colonie, 1692–1717* (London: John Murray, 1904), p. 24.
32. Henri and Barbara Van der Zee, *William and Mary* (New York: Alfred A. Knopf, 1973), p. 81.
33. Thomas R. Phillips, *Roots of Strategy* (Harrisburg, Penn.: Military Service Publishing company, 1940), pp. 323–24.
34. Christopher Duffy, *The Army of Frederick the Great* (New York: Hippocrene Books, 1974), p. 181.
35. *Nouvelles Decouvertes . . .* was published in Paris, 1726. In his *Polybe,* Folard used a new translation from the original Greek by a Dom Vincent Thuiller, with his own added commentary ''enriched by critical and historic notes.'' Published in Paris, 1727–30.
36. Huddleston, *Warriors,* pp. 164–65. His full name was Hermann Moritz, but the Hermann soon was mislaid.
37. Phillips, *Strategy,* p. 182.
38. Ibid., p. 201.
39. The original edition by Arkstee and Merkus in Amsterdam was curiously illustrated. The translation in *Roots of Strategy* is clear but somewhat abridged.
40. Phillips, *Strategy,* p. 194.
41. Ibid., pp. 185, 261.
42. This probably was the weapon invented by Isaac de la Chaumette, a Huguenot, who had to flee to England in 1721. The Saxe-Chaumette relation needs more research.
43. Phillips, *Strategy,* p. 298.
44. Ibid., pp. 200–201.
45. Ibid., pp. 190–91.
46. Ibid., p. 183.

Chapter V

1. Rudyard Kipling, *Verse: Definitive Edition* (Garden City, N.Y.: Doubleday, Doran and Co., Inc., 1945), pp. 727–28.
2. J.F.C. Fuller, *A Military History of the Western World,* vol. 2 (New York: Funk and Wagnalls Company, 1955), p. 129.
3. Ibid., p. 128.

4. F. J. Huddleston, *Warriors in Undress* (London: John Castle, 1925), p. 239. The translation is my own, from Frederick's version of French.
5. Thomas R. Phillips, *Roots of Strategy* (Harrisburg, Penn.: Military Services Publishing Company, 1940), p. 356.
6. Christopher Duffy, *The Army of Frederick the Great* (New York: Hippocrene Books, Inc., 1974), p. 67.
7. Octave Aubry, *Les Pages Immortelles de Napoleon* (Paris: Editions Correa, 1941), pp. 202–3, 229–30.
8. François Jules Louis de Loyzeau de Grandmaison, *A Treatise on the Military Service of Light Horse and Light Infantry* (Philadelphia: Robert Bull, 1777), p. 4.
9. Ibid., pp. 4–5.
10. Henry H.E. Lloyd, *A Political and Military Rhapsody on The Invasion and Defense of Great Britain and Ireland* (London: 1795), p. x.
11. Ibid., p. xii. Subensign is an unusual grade—apparently something like a third lieutenant.
12. Ibid., p. xiv.
13. Huddleston, *Warriors,* p. 133.
14. Lloyd, *Rhapsody,* pp. 10–13.
15. Ibid., pp. 21–22.
16. Huddleston, *Warriors,* p. 138.
17. Ibid., p. 137.
18. Department of Military Art and Engineering, U.S. Military Academy. *Resume of the History of the Principles of War,* 1962 (prepared for use by instructors only).

NOTE: For interesting material on Lloyd, see Amos Perlmutter and John Gooch (eds.). *Strategy and the Social Sciences.* London: Frank Cass, 1981, pp. 22–24.

Chapter VI

1. Johann Ewald, *Diary of the American War* (New Haven: Yale University Press, 1979); p. 108 (translated by Joseph P. Tustin). Turpin de Crisse, a French hussar officer during the Seven Years' War, wrote several respected works on war, especially his *Commentaires sur les memoires de Montecuculi.* Grandmaison, another French officer, was the author of *A Treatise on the Military Service of Light Horse and Light Infantry,* probably the outstanding work on that subject.
2. Gen. Humphrey Bland published the first edition of this work in 1727; it went through possibly nine revisions, the last in 1762. It was practically the ''Bible'' of the British Army.
3. Christopher Ward, *The War of the Revolution,* vol. 1 (New York: Macmillan, 1952), p. 366.
4. George Washington, *Farewell Address,* September 17, 1796.
5. Charles W. Elliott, *Winfield Scott* (New York: Macmillan, 1937), p. 37.

6. Published by Magimel in Paris in 1813. A very detailed work, including sample administrative forms, it was used by the French Army until the 1830s.
7. Probably the *Ordonnance* of August 1791, an excellent, easily comprehended work, reflecting Guibert's influence.
8. Elliott, *Scott,* p. 198, fn.
9. Ibid., p. 423.
10. Both Scott and Taylor belonged to the opposition Whig party. Army officers then indulged energetically in politics; Scott was twice the (unsuccessful) Whig candidate for president.
11. Octave Aubry, *Les Pages Immortelles de Napoleon* (Paris: Editions Correa, 1941), p. 203.
12. Elliott, *Scott,* pp. 501–502.
13. K. Jack Bauer, *The Mexican War* (New York: Macmillan, 1974), p. 322.
14. Rudyard Kipling, *Verse* (Garden City, N.Y.: Doubleday, Doran & Co., 1945), p. 620.
15. Sam Houston's opinion. Carl Sandburg, *Abraham Lincoln: The War Years,* vol. 1 (New York: Harcourt, Brace & Co., 1939), p. 247.
16. Elliott, *Scott,* p. 718.
17. Tully McCrea, *Dear Belle* (Middletown, Conn.: Wesleyan University Press, 1965), p. 33 (Catherine S. Crary, ed.).
18. The original was published in Paris, 1806; O'Connor's translation in New York by J. Seymour, 1817. Opinions differ as to the accuracy of O'Connor's work, but no objections seem to have been offered then, possibly because of O'Connor's reputedly itchy trigger finger.
19. Thomas E. Griess, "Dennis Hart Mahan: West Point Professor and Advocate of Military Professionalism, 1830–1871" (Dissertation, Duke University, 1968), p. 214. This is the best review of Mahan's work.
20. Stephen E. Ambrose, *Halleck: Lincoln's Chief of Staff* (Baton Rouge, La.: Louisiana State University Press, 1962), pp. 5–6.
21. Robert G. Carter, *The Art and Science of War Versus the Art of Fighting* (Washington: 1922). A small, hilarious book.
22. Henry W. Halleck, *Elements of Military Art and Science* (New York: Appleton, 1846), pp. 140–41. A revised edition in 1859 covered lessons of the Mexican and Crimean wars; in the latter, Halleck noted the greater efficiency of the French military supply services as compared to the semicivilian English system.
23. See page 159. His translation was not published until 1864.
24. Herman Hattaway and Archer Jones, *How the North Won* (Bloomington, Ind.: University of Indiana Press, 1983), p. 239.
25. Kenneth P. Williams, *Lincoln Finds a General,* vol. 1 (New York: Macmillan Company, 1950), p. 351.
26. The first, and probably the ablest, of these was Kenneth P. Williams—see especially the appendix to volume 5, *Lincoln Finds a General.* Her-

mann Hattaway and Archer Jones, *How the North Won,* continue this work.

27. Lee was graduated from West Point in 1829, one year before Mahan returned from France. His major education in war was service on Scott's staff in Mexico as an Engineer officer, but it must be remembered that an Engineer officer of that period was expected to be expert in almost any duty.
28. Douglas S. Freeman, *Lee's Lieutenants,* vol. 2 (New York: Scribners, 1943), p. 2 and fn.
29. Jackson's observations were mostly oral; some undoubtedly had been prettied up by the time they were printed. These are from Freeman, *Lee's Lieutenants,* vol. 1, p. 470.
30. Bruce Catton, *This Hallowed Ground* (Garden City, N.Y.: Doubleday & Co., 1956), p. 314.
31. Lloyd Lewis, *Sherman: Fighting Prophet* (New York: Harcourt, Brace & Co., 1932), p. 450.
32. Ibid., p. 616.
33. Ibid., p. 648.
34. New York: Appleton, 1878.
35. *Military Policy of the United States* (Washington, D.C.: Government Printing Office, 1904). This work was based in good part on Halleck's *Elements. . . .* Part of it is definitely twisted history, but Elihu Root, Secretary of War 1899–1904, found it a major inspiration for his far-reaching Army modernization, especially the long-overdue development of an effective staff system.
36. Maj. Gen. Enoch Crowder. *Encyclopedia Americana,* 1969 ed., s.v.
37. William A. Ganoe. *MacArthur Close-Up.* Privately printed, New York, 1962, p. 30. Ganoe, better known as the author of *The History of the United States Army* (New York: D. Appleton and Co., 1924), a popular but confused work, was MacArthur's adjutant and first acolyte. His little book is all incense and tall candles but unconsciously draws a sharp portrait. The parallels between it and the MacArthur of Courtney Whitney's equally awed *MacArthur: His Rendezvous with History* (New York: Alfred A. Knopf, 1956), covering 1940–54, are amazing.
38. Stephen E. Ambrose, *Duty, Honor, Country* (Baltimore: Johns Hopkins Press, 1966), p. 267.
39. T.R. Fehrenbach, *This Kind of War* (New York: Pocket Books, Inc., 1964), p. 447.
40. Vorin E. Whan, Jr., ed., *A Soldier Speaks* (New York: Praeger, 1965), pp. 256–57.
41. Ibid., p. 75.

Chapter VII

1. Thomas R. Phillips, *Roots of Strategy* (Harrisburg, Penn.: Military Service Publishing Co., 1940), p. 432; Emil Ludwig, *Napoleon* (New

York: Garden City Publishing Co., 1926), p. 578. Technically, Napoleon should be considered ''Napoleon Bonaparte'' until 1804, when he became the Emperor Napoleon I. Some sore-nosed English and German historians still refer to him as Bonaparte throughout his life, implying that he never was a legal ruler.

2. Logistics is the military science of planning and carrying out the movement and supply of the armed forces.
3. J.F.C. Fuller, *The Generalship of U. S. Grant* (New York: Dodd, 1929), p. 1.
4. See Suggested Additional Reading.
5. Ludwig, *Napoleon,* p. 74.
6. Richard Aldington, *The Duke* (New York: Garden City Publishing Co., 1943), p. 243.
7. Octave Aubry, *Les Pages Immortelles de Napoleon* (Paris: Editions Correa, 1941), p. 106.
8. Antoine de Pas Feuquieres, *Memoirs,* 4 vols. (Paris: Rollin Fils, 1737). Clausewitz admired him; Pierre Joseph Bourcet, *Memoires Historiques . . . 1757 . . . 1762* (Paris: Maradan, 1792). Napoleon undoubtedly was also able to read his *Principes de la Guerre des Montagnes,* then available only in manuscript copy.
9. Jacques A., Comte de Guibert, *Essai General de Tactique.* The best-known edition is that printed in two volumes by C. Plonteux in Liege, 1775.
10. Edward M. Earle, ed., *Makers of Modern Strategy* (Princeton, N.J.: Princeton University Press, 1944), p. 66.
11. Aubry, *Les Pages,* p. 231
12. Ibid., p. 234.
13. Phillips, *Strategy,* p. 425.
14. George Meredith, *Napoleon,* XIII. From John Bartlett, *Familiar Quotations,* p. 576.
15. *Correspondence de Napoleon Ier,* 32 vols. (Paris: Henri Plon, 1858–70). Though Napoleon III had it screened for material that might detract from Napoleonic glory, it is the finest source of authentic Napoleonic thoughts and doctrine.
16. Arthur Chuquet, *Ordres et Apostilles de Napoleon (1799–1815),* 4 vols. (Paris: Librairie Ancienne Honore Champion, 1911–13).
17. *Commentaires de Napoleon Premier,* 6 vols. (Paris: Imprimerie Imperiale, 1867). An abridged English translation, edited by Somerset de Chair, was published in London, 1948.
18. It is included in Phillips, *Roots of Strategy,* which attributes it to a General Burnod.
19. Jacques de Chastenet de Puysegur, *Memoires, avec des instructions militaires,* 2 vols. (Paris: 1747). Puysegur was a lieutenant general under Louis XIV. Dietrich von Bulow was a minor Prussian nobleman, author of several, often contradictory, books on warfare that occasionally con-

tain bits of wisdom. His pro-Napoleon views led the Prussian government to rule him insane; he died in confinement in 1807. Karl Ludwig Charles, Archduke of Austria (1771–1847), one of the ablest generals of the period was a cautious strategist but a daring, inspiring combat leader.

20. Ferdinand Lecomte, *Le General Jomini: Sa Vie et Ses Ecrits* (Paris: Tanera, 1860), p. 152. A Swiss soldier, Lecomte was briefly an observer of our Civil War and made sensible comments. This book, however, is complete hero worship, at odds with common sense.
21. Jomini had been awarded the cross of the Legion of Honor in 1807, apparently for services in 1805. Ney now nominated him for the grade of officer of the Legion of Honor, which would bring a more ornate version of the cross and a higher pension. Jomini also had been made a baron in the imperial nobility, but this seems to have been an automatic award to colonels and brigadier generals.
22. A. du Casse, *Memoires de Prince Eugene,* vol. 4 (Paris: Michel Levy Freres, 1859), pp. 128–30.
23. Lecomte, *Jomini,* p. 365.
24. *Carnet de la Sabretache* (Paris: Le Sabretache, 1896), pp. 55–56.
25. *Traite des grandes operations militaires,* 8 vols. (Paris: 1804–16), several times revised. The title of the first volumes originally was *Traité de grande tactique.*
26. *Line of operations*—the axis of a military force's advance toward the enemy. *Interior lines*—you hold a central position and so can concentrate quickly and hit in any direction, while the enemy forces must move longer distances by *exterior* lines around your position to link up. This theory does not always work. Custer ended up occupying a central position at the Little Big Horn.
27. *Histoire des Guerres de la Revolution,* 15 vols. (Paris: Anselin et Pochard, 1820).
28. Paris: Anselin, 1827. Jomini wrote a fifth volume covering 1815 twelve years later but did not use the first-person style.
29. Paris: Anselin, 1838. The first American translation in 1854 was very unsatisfactory.
30. Napoleon, *Correspondence.* August 16, 1813, Item 20, p. 382.
31. For this often-grinding process, see Christopher Duffy, *The Army of Frederick the Great* (New York:: Hippocrene Books, 1974), pp. 30–32.
32. Ramsay W. Phipps, *The Armies of the First French Republic,* vol. 2 (London: Oxford University Press, 1929), pp. 40–41. "Citizen" was the standard Revolutionary form of polite address.
33. Napoleon used this term to describe how his whole army was under his direct control "like a battalion in the hands of a good commander, ready for anything."
34. *Notes sur la Prusse dans sa grande catastrophe: 1806* (Paris: Chapelot, 1903) (translated by A. Niessel).

35. Carl von Clausewitz, *On War* (Princeton, N.J.: Princeton University Press, 1976), p. 66 (Michael Howard and Peter Paret, editors and translators).
36. Department of Military Art and Engineering, USMA. *Jomini, Clausewitz and Schlieffen.* (West Point, N.Y.: Department of Military Art and Engineering, 1964).
37. An essay Clausewitz wrote for the crown prince on "The Important Principles of Military Leadership" was published as *Principles of War* (edited and translated by H. Gatzke) by the Military Service Publishing Company of Harrisburg, Penn. in 1942.
38. Clausewitz, *War,* pp. 66–67.
39. Published in London by N. Trubner & Co., 1873; translation by J. J. Graham.
40. Published by Princeton University Press, edited and translated by Michael Howard and Peter Paret.
41. Clausewitz, *On War,* p. 152.
42. Ibid., p. 136.
43. Ibid., p. 95.
44. Ibid., pp. 75, 87, 219, 227, 248, 258–62.
45. Ibid., pp. 119–21.
46. Phillips, *Strategy,* p. 436.
47. Clausewitz, *On War,* pp. 101, 117. This might well be applied to the American "mass media" as well as warfare!
48. The phrases used are from the Graham translation in its "revised edition with Introduction and Notes by Colonel F. N. Maude" (London, 1918), pp. 3–5, 41–45. The Howard/Paret edition is somewhat less "bloody."
49. Department of Military Art and Engineering, p. 31.
50. Published in London, 1911.
51. Clausewitz, *On War,* pp. 88–89.

Chapter VIII

1. Anthony John Trythall, *"Boney" Fuller: The Intellectual General, 1878–1966* (London: Cassell & Company, Ltd., 1977), p. 159.
2. Junker means literally "young gentleman" but was applied to any male member of the landholding aristocracy in Eastern Germany, especially in East Prussia. Looser, more modern usage extended it to all upper-class Prussians, especially army officers.
3. In 1870 Moltke was ennobled and became Count Helmuth von Moltke.
4. Jay Luvaas, *The Military Legacy of the Civil War* (Chicago, Ill.: University of Chicago Press, 1959), p. 126.
5. Carl von Clausewitz, *On War* (Princeton, N.J.: Princeton University Press, 1976), p. 30 (Michael Howard and Peter Paret, editors and translators).
6. J.F.C. Fuller, *A Military History of the Western World* vol. 3 (New

York: Funk & Wagnalls Company, 1956), p. 134.

7. F. W. von Mellenthin, *German Generals of World War II* (Norman, Okla: University of Oklahoma Press, 1977), p. 275; Kenneth Macksey, *Guderian: Creator of the Blitzkrieg* (New York: Stein and Day, 1976), p. 23.
8. Hans G. L. Delbruck (1848–1929), soldier, German politician, newspaper editor, professor of history, and author of *History of the Art of War*, attempted to rationalize and debunk early military history. An English translation of volume 1 by Col. Walter J. Renfroe was published by Greenwood Press, Westport, Connecticut, 1975. Schlieffen himself wrote *Cannae* (English translation by the Command and General Staff School Press, Fort Leavenworth, Kansas, 1936). Unfortunately, this book is full of oversimplifications.
9. Clausewitz, *On War*, p. 56.
10. The *Reichstag* was the lower house of the German parliament.
11. Walter Goerlitz, *History of the German General Staff, 1657 to 1945* (New York: Praeger, 1953), p. 142.
12. Ian Hamilton, *Gallipoli Diary* (New York: George H. Doran Co., 1920), vol. 1, pp. 132–133; vol. 2, p. 74.
13. An English translation by John N. Greeley and Robert C. Cotton was published as *Battle Studies: Ancient and Modern Battle* (New York: Macmillan 1921), but is incomplete.
14. Luvaas, *Civil War*, p. 152.
15. Philip Guedalla, *The Two Marshals* (London: Hodder and Stoughton, 1943), p. 261.
16. Edward M. Earle, ed., *Makers of Modern Strategy* (Princeton, N.J.: Princeton University Press, 1944), p. 216.
17. Charles Ardant du Picq, *Etudes sur le Combat* (Paris: Chapelot, 1914), pp. 154–155.
18. Ferdinand Foch, *The Memoirs of Marshal Foch* (Garden City, N.Y.: Doubleday, Doran and Company, Inc., 1931), p. lviii (translated by Col. T. Bently Mott). A book of somewhat dubious accuracy.
19. Foch, *The Principles of War* (New York: 1920), p. 286 (translated by Hilaire Belloc).
20. Foch, *Memoirs*, p. xxv.
21. Robert Lee Bullard, *American Soldiers Also Fought* (New York: Maurice H. Louis, 1939), pp. 20–21.
22. This was a typical General Staff operation—an exacting examination of both German and Allied combat requirements and realistic training to instill the new doctrine, with great attention to troop morale and indoctrination.
23. J. D. Hittle, *The Military Staff: Its History and Development* (Harrisburg, Penn.: Military Service Publishing Company, 1944), p. 72. A book that clearly demonstrates that "a little learning is a dangerous thing."

24. F. J. Huddleston, *Warriors in Undress* (London: John Castle, 1925), p. 208 fn.
25. Lin Piao, ''Long Live the Victory of the People's War.'' Department of Military Art and Engineering, U.S. Military Academy. *Revolutionary Warfare,* vol. 3, 1968, p. 201.
26. Walter Mills, *The Martial Spirit* (Cambridge, Mass.: The Literary Guild, 1931), p. 322.
27. *Encyclopedia Americana,* 1967 ed., s. v. ''Geopolitics.''
28. Herbert G. Wells, *The War in the Air* (London: George Bell and Sons, 1908).
29. Giulio Douhet, *The Command of the Air* (New York: Coward-McCann, Inc., 1942, translated by Dino Ferrari).
30. Earle, *Modern Strategy,* p. 492.
31. John R. Elting, Dan Cragg, and Ernest Deal, *A Dictionary of Soldier Talk* (New York : Scribners, 1984), pp. 176–77.
32. Lee Kennett, *A History of Strategic Bombing* (New York : Scribners 1982), p. 55. An extremely useful book.
33. J.F.C. Fuller, *Armament and History* (New York : Scribners 1945), p. 146.
34. Kennett, *Strategic Bombing,* p. 87.
35. William Mitchell, *Winged Defense* (New York: G. P. Putnam's Sons, 1925), p. 19.
36. Earle, *Modern Strategy,* p. 488.
37. Kennett, *Strategic Bombing,* pp. 75–76.
38. Ronald H. Bailey, *The Air War in Europe* (Alexandria, Va.: Time-Life Books, Inc., 1979), p. 65.
39. Ibid., p. 191.
40. Albert Speer, *Inside the Third Reich* (New York: Macmillan Company, 1970), pp. 346, 350, 554.
41. H. Essame, *The Battle for Germany* (New York: Bonanza Books, 1969), pp. 47–48.
42. E. T. Williams and C. S. Nicholls (eds.), *The Dictionary of National Biography, 1961–1970* (New York: Oxford University Press, 1981), p. 700.
43. Trythall, *Fuller,* p. 47. ''Slosh'' means ''thrash'' or ''beat.''
44. ''Boney'' is slang for ''Bonaparte.'' The English have trouble admitting that Gen. Napoleon Bonaparte *did* become the Emperor Napoleon I.
45. London: Sifton Praed and Co., 1931.
46. London: Murray, 1920; London: Hutchinson and Co., 1926.
47. *The Outline of History* (New York: Macmillan Company, 1922), p. 1085.
48. London: Sifton Praed, 1931 and 1932.
49. New York: Funk and Wagnalls Company, 1954.
50. New York: Dodd Mead, 1921; New York: Scribners, 1933.

51. New Brunswick, N.J.: Rutgers University Press, 1960 and 1961.
52. Fuller, *Armament and History,* p. xvii.
53. Kenneth Macksey, *The Tank Pioneers* (New York: Jane's Publishing Incorporated, 1981), p. 112.
54. For excellent pictures and descriptions of these vehicles, see Douglas Botting, *The Second Front* (Alexandria, Va.: Time-Life Books, Inc., 1978), pp. 172–81.
55. New York: Random House, 1939.
56. Brian Bond, *Liddell Hart: A Study of His Military Thought* (New Brunswick, N.J.: Rutgers University Press, 1977), pp. 43, 127.
57. Trythall, *Fuller,* p. 157.
58. *The Tanks,* 2 vols. (London: Cassell, 1959).
59. London: Collins, 1950; London: Cassell, 1948.
60. Basil H. Liddell Hart, "Sherman—Modern Warrior," *American Heritage,* August 1962, p. 105.
61. Macksey, *Guderian,* p. 41.
62. The 1922 Treaty of Rapallo between Germany and Russia, then both semioutcasts from the European system, included establishment of three German experimental training centers in Russia.
63. Heinz Guderian, *Panzer Leader* New York: Dutton, 1952.
64. Macksey, *Guderian,* p. 32 fn.
65. von Mellenthin, *German Generals,* p. 89.
66. Macksey, *Guderian,* p. xi (author's translation).

Chapter IX

1. E. D. Phillips, *The Mongols* (London: Thames and Hudson, 1969), p. 82.
2. Sun Tzu Wu, *The Art of War* (Harrisburg, Penn.: Military Service Publishing Company, 1957), p. 51.
3. Ibid., pp. 9–12. Sun Tzu Wu's correct name seems to have been Sun Wu. ("Tzu" probably was an honorific, meaning "master.") A more modern version is "Sunzi." Ch'i was generally the present-day Shantung area, across the Yellow Sea from Korea. Wu was in the Yangtze River delta, roughly inland from modern Shanghai.
4. Ibid., p. 44.
5. Ibid., pp. 40–47.
6. Ibid., p. 48.
7. Ibid., pp. 58–60.
8. Ibid., pp. 62–70.
9. Ibid., pp. 48, 50, 58–70.
10. Ibid., pp. 96–99.
11. Edgar O'Ballance, *The Red Army of China* (New York: Praeger, 1963), p. 13. I recall this, also, from boyhood reading, as "China is a sleeping giant. Do not wake her"; Thomas R. Phillips, *Roots of*

Strategy (Harrisburg, Penn.: Military Service Publishing Company, 1940). For Sun Tzu's opinions, see pp. 16–17, 41, 49; for Napoleon's, pp. 429–30.

12. The total length of this defensive system, including spurs, loops, and secondary walls would be close to four thousand miles.
13. Rene Grousset, *The Empire of the Steppes* (New Brunswick, N.J.: Rutgers University Press, 1970), p. 216 (translated by Naomi Walford).
14. Harold Lamb, *Genghis Khan: Emperor of All Men* (Garden City, N.Y.: Garden City Publishing Company, 1927), p. 66.
15. Phillips, *Mongols,* p. 41. Various translations use "God," "Heaven," and "Sky" interchangeably.
16. Lamb, *Genghis Khan,* p. 76.
17. The reason for this prohibition seems to have been in the vague animism that was all the religion Genghis Khan's Mongols originally possessed. Springs and rivers were the home of earth spirits which must not be disturbed.
18. Grousset, *Steppes,* p. 249.
19. Ibid., p. 249.
20. B. H. Liddell Hart, *Great Captains Unveiled* (Edinburgh: Blackwood, 1927). A typical Hart potboiler, attractive but more propaganda than history.
21. Vorin E. Whan, Jr., ed., *A Soldier Speaks* (New York: Frederick A. Praeger, 1965), pp. 63–64.
22. In 1936, Mao described his youth to Edgar Snow, an American journalist; we can only hope that Mao told a reasonably honest tale.
23. James E. Sheridan, *China in Disintegration* (New York: The Free Press, 1975), p. 157.
24. Samuel B. Griffith, *Mao Tse Tung: Sun in the East* (Annapolis, Md.: *Naval Institute Proceedings,* 1951), p. 92.
25. E. Clubb, *Twentieth Century China* (New York: Columbia University Press, 1964), p. 238.
26. U.S. Congress, *Congressional Record,* 79th Cong., 2d sess., vol. 92, pt. 12, pp. A 4495–96.
27. Sheridan, *China* p. 247. There are several slightly varying versions.
28. Published by the Foreign Language Press, Peking, 1960.
29. Mao Tse-tung, *Problems of Strategy in China's Revolutionary War: Selected Works,* vol. 1 (Peking: Foreign Language Press, 1965).
30. Sun Tzu Wu, *Art of War,* p. 74.
31. Ibid., pp. 92, 98.
32. Harold L. Boorman and Scott A. Boorman, "Strategy and National Psychology in China" *Annals of the American Academy of Political Science,* vol. 370 (March 1967), pp. 143–55.
33. Sun Tzu Wu, *Art of War,* pp. 51, 95, 96, 58.

Chapter X

1. Fred Majdalany, *The Fall of Fortress Europe* (London: Hodder and Stoughton, 1969), p. 11.
2. Forrest C. Pogue, *Organizer of Victory* (New York: Viking Press, 1973), p. 641.
3. F. W. von Mellenthin, *German Generals of World War II* (Norman, Okla.: University of Oklahoma Press, 1977), p. 32.
4. Robert Murphy, *Diplomat Among Warriors* (New York: Pyramid Books, 1965), pp. 192–193.
5. W. Averell Harriman and Elie Abel, *Special Envoy to Churchill and Stalin* (New York: Random House, 1975), p. 399.
6. Milovan Djilas, *Wartime* (New York: Harcourt Brace Jovanovich, 1977), p. 438. Stalin also told Djilas that Churchill would steal a penny out of his pocket but that "Roosevelt is not like that. He dips in his hand only for bigger coins" (pp. 388–389).
7. Harriman, *Envoy,* p. 226.
8. Murphy, *Diplomat,* p. 312.
9. Harriman, *Envoy,* p. 536.
10. Djilas, *Wartime,* p. 407.
11. Petro G. Grigorenko, *Memoirs* (New York: W. W. Norton and Company, 1982), p. 212.
12. Harriman, *Envoy,* p. 522 fn.
13. Glen B. Infield, *The Poltava Affair* (New York: Macmillan Publishing Company, Inc., 1973), p. 152.
14. Djilas, *Wartime,* pp. 386, 435.
15. From an admiring officer of Marshall's personal staff.
16. Pogue, *Victory,* p. 313.
17. Roberta Wohlstetter, *Pearl Harbor: Warning and Decision* (Stanford, Calif.: Stanford University Press, 1962), p. 348.
18. Ibid., p. 346. This "sphere" included all British, Dutch, French, and Portuguese holdings in the Far East; Manchuria; Korea; as much of China as possible; and the Philippines. Australia and India were to be acquired later.
19. William L. Langer and S. Everett Gleason, *The Undeclared War, 1940–1941* (New York: Harper & Brothers, 1953), p. 852. Quotation from Japanese Prime Minister Hideki Tojo.
20. There was a vague story that his company commander considered him too excitable to be a sergeant.
21. Desmond Young, *The Desert Fox* (New York: Harper and Brothers, 1950), p. 45.
22. Majdalany, *Fortress Europe,* p. 146.
23. Gordon A. Craig, *Germany, 1866–1945* (New York: Oxford University Press, 1978), p. 756.
24. Chester Wilmot, *The Struggle for Europe* (New York: Harper and Brothers, 1952), p. 700.

25. The U.S. Joint Chiefs of Staff were Marshall; Adm. Ernest J. King, chief of naval operations; Lt. Gen. Henry H. Arnold, army air forces; and Adm. William D. Leahy, Roosevelt's personal chief of staff. The Combined Chiefs of Staff consisted of the U.S. Joint Chiefs and their British opposite numbers.
26. Told by an air force officer, formerly of Marshall's staff.
27. Pogue, *Victory,* p. 74.
28. A personal observation by a former senior member of Marshall's staff.
29. Pogue, *Victory,* p. 371.
30. Omar N. Bradley, *A Soldier's Story* (New York: Henry Holt & Co., 1951); *A General's Life.* (New York: Simon & Schuster, 1983).
31. Pogue, *Victory,* pp. 371–72, 385.
32. H. Essame, *Patton: A Study in Command* (New York: Charles Scribner's Sons, 1974), p. 17.
33. George S. Patton, *War As I Knew It* (Boston: Houghton-Mifflin Co., 1947), p. 201.
34. Charles W. Elliott, *Winfield Scott* (New York: The Macmillan Company, 1937), p. 718.
35. Department of Defense. *The Armed Forces Officer* (Washington, D.C.: U.S. Government Printing Office, 1950), pp. 82, 129.
36. W.G.F. Jackson, *Alexander of Tunis as a Military Commander* (New York: Dodd, Mead & Company, 1972), p. 227.
37. The author heard him address the Corps of Cadets at West Point in the early 1950s. Bradley's *Soldier's Story,* filled with savage jabs at Montgomery, had been popular reading before his appearance, and the atmosphere in the lecture hall was icy. "Monty" beamed at the cadets, began talking of his service as a second lieutenant in 1914, when, armed only with a swagger stick, "as was our silly custom at that time, I found myself confronted by a large German with a fixed bayonet. (Short pause) Lacking any other weapon, I threw myself at him. (Long pause) After getting out of the hospital I put in for transfer to the staff." Thereafter, he had them!
38. Russel F. Weigley, *Eisenhower's Lieutenants* (Bloomington, Ind.: Indiana University Press, 1981), p. 506.
39. Mellenthin, *Generals,* p. 83.
40. Young, *Desert Fox,* p. 7.
41. Published as *Infantry Attacks* (Washington, D.C.: Combat Forces Press, 1956).
42. Dwight D. Eisenhower, *Crusade in Europe* (Garden City, N.Y.: Doubleday & Co., 1948), p. 156.
43. Young, *Desert Fox,* pp. 232–233.
44. Ibid., p. 256.
45. Majdalamy, *Fortress Europe,* p. 306.
46. Wilmot, *Europe,* pp. 215–16.
47. Young, *Desert Fox,* p. xii.

Chapter XI

1. From memory, from one of the many versions of the ancient *Ballad of Otterburn.* At Otterburn, in 1388, Sir James Douglas defeated a larger English force under Sir Henry ''Hotspur'' Percy but was killed at the battle's climax.
2. Jacques A., Comte de Guibert, *Defense du systeme de guerre moderne,* vol. 4 (Paris: 1779), p. 213.
3. Finagal's Laws are a collection of pithy axioms concerning the innate cussedness of inanimate objects and human nature, very popular in the 1950–60s. (Another example: ''Interchangeable parts aren't.'') Sometimes called Murphy's Laws.
4. John M. Collins, *American and Soviet Military Trends Since the Cuban Missile Crisis.* (Washington, D.C.: Georgetown University Press, 1978), p. 116.
5. Conventional forces are those that do not employ nuclear, chemical, or biological weapons—in other words, similar to the armies of World War II.
6. Bill Mauldin, *Up Front* (New York: Henry Holt and Company, 1945), p. 100.
7. A common remark, variously phrased, of that period.
8. Robert F. Kennedy, *Thirteen Days* (New York: Signet Books, 1969), p. 95.
9. A silo is an underground launching site for ballistic missiles. A ''hardened'' silo is one specifically constructed to provide substantial protection against nuclear explosions.
10. A cruise missile probably is best described as a robot ''kamikaze'' aircraft.
11. *US News and World Report,* 14 January 1985, pp. 22–23.
12. Edward P. Hamilton, ed., *Adventures in the Wilderness* (Norman, Okla.: University of Oklahoma Press, 1964), p. 197. St. Joseph River was a French post near modern Niles in southwest Michigan.
13. Francis Parkman, *The Conspiracy of Pontiac* (New York: Collier Books, 1962), pp. 297–98.
14. These were the 1925 Geneva Protocol, instituted by the League of Nations, and the 1972 Biological Warfare Convention.
15. David T. Twining, ''The Sverdlovsk Anthrax Outbreak,'' *Airforce Magazine,* March 1981, pp. 124–128.
16. Charged-particle beams aim and project atomic particles at the speed of light.
17. Listed in a speech, January 8, 1918, these outlined a reasonable peace settlement ''openly arrived at.'' Sad to say, the Germans liked them, but England, France, and Italy didn't.
18. KGB—Committee for State Security—a national political police force under the direct control of the Russian Communist party, responsible for espionage, counterintelligence, the ideological purity of Russian citi-

zens, propaganda, subversion, sabotage, disinformation, psychological warfare, and the organization and training of terrorists. It has wide powers of arrest, torture, and trial; it also controls the paramilitary Border Guard, some 160,000 strong. Its propaganda/subversive work is coordinated with that of the Communist party's Department of Propaganda and Education and the Academy of Sciences.

19. Lewis W. Walt, *The Eleventh Hour* (Ottawa, Ill.: Caroline House Publishers, Inc., 1979), p. 76.
20. John Barron, *The KGB Today* (New York: The Reader's Digest Press, 1974), p. 174.
21. MVD—Ministry of Home Affairs—predecessor of the KGB, with much the same functions. It, in turn, was a successor to the NKVD (1935–43). It included military formations much like the German SS but seldom risked in actual combat.
22. Harold Lamb, *The March of Muscovy* (Garden City, N.Y.: Doubleday & Company, Inc., 1948), p. 176.
23. Richard M. Watt, *Dare Call it Treason* (New York: Simon and Schuster, 1963), p. 293.

Appendix

1. Chester Wilmot, *The Struggle for Europe* (New York: Harper & Brothers, 1952), p. 142.
2. Department of Military Art and Engineering, U.S. Military Academy, *Resume of the History of the Principles of War* (West Point, New York, 1962).
3. H. Esseme, *Patton,* (New York: Scribners, 1974), p. 239.
4. Yorck von Wartenburg, *Napoleon as a General,* vol. 1 (West Point, N.Y.: Department of Military Art and Engineering, 1955), p. 50.
5. B. H. Liddell Hart, ed., *The Letters of Private Wheeler* (Boston: Houghton Mifflin Co., 1952), p. 18. Chatham was the army commander; Strachan, the admiral.
6. Forrest C. Pogue, *George C. Marshall, Organizer of Victory* (New York: Viking Press, 1973), p. 168.
7. J. W. Cunliffe, ed., *Century Readings for a Course in English Literature* (New York: Century Company, 1923), p. 39.
8. Lenoir Chambers, *Stonewall Jackson,* vol. 2 (New York: William Morrow & Co., 1959), p. 330.
9. Mark M. Boatner III, *The Civil War Dictionary* (New York: David McKay Co., 1959), p. 329.
10. Wartenburg, *Napoleon,* vol. 2, p. 54.
11. Oral history—told by a member of Patton's staff.
12. Obviously stolen from Shakespeare's *King Henry VI, Part II,* and thereafter abused.
13. Robert S. Henry, *"First With the Most" Forrest* (New York: Bobbs-Merrill Co., 1944), p. 424.

14. Thomas R. Phillips, *Roots of Strategy* (Harrisburg, Penn.: Military Service Publishing Company, 1940), pp. 409–10.
15. Carl von Clausewitz, *On War* (Princeton, N.J.: Princeton University Press, 1976), pp. 119–20 (edited and translated by Michael Howard and Peter Paret).
16. Phillips, *Strategy,* pp. 323–24.
17. Mao Tse-tung, *Selected Works,* vol. 2 (Peking: Foreign Language Press, 1965), p. 9.
18. Robert M. Utley, *Frontiersmen in Blue* (New York: Macmillan, 1967), p. 256.
19. Douglas S. Freeman, *Lee's Lieutenants,* vol. 1 (New York: Scribners, 1942), pp. 424 and 464.
20. Bruce Catton, *Grant Takes Command* (Boston: Little, Brown and Company, 1968) p. 105.

SUGGESTED ADDITIONAL READING

Chapter II

Adcock, F. E. *The Roman Art of War Under the Republic*. Cambridge, Mass.: Harvard University Press, 1940.

Barker, John. *The Superhistorians*. New York: Charles Scribner's Sons, 1982.

Cook, J. M. *The Persian Empire*. New York: Schocken Books, 1983.

Fuller, J. F. C. *The Generalship of Alexander the Great*. New Brunswick, N.J.: Rutgers University Press, 1960.

Kohn, Richard H. *The Eagle and the Sword*. New York: Macmillan, 1975.

Luttwak, Edward N. *The Grand Strategy of the Roman Empire*. Baltimore, Md.: Johns Hopkins University Press, 1976.

McNeill, William H. *Plagues and Peoples*. Garden City, N.Y.: Doubleday, 1976.

Norman, A. V. B. *The Medieval Soldier*. New York: Crowell, 1971.

Oman, Charles. *A History of the Art of War in the Middle Ages*, 2 vols. London: Methuen & Co., 1924.

Rogers, William L. *Greek and Roman Warefare*. Annapolis, Md.: Naval Institute Press, 1964.

Spaulding, Oliver L.; Nickerson, Hoffman; and Wright, John W. *Warfare*. Washington, D.C.: Infantry Journal, 1937.

Webster, Graham. *The Roman Imperial Army*. London: Adam and Charles Black, 1969.

Chapter III

Aussaresses, F. *L'Armee Byzantine a la Fin du VI Siecle*. Bordeaux: Feret & Fils, 1909.

Braudel, Fernand. *The Wheels of Commerce*. New York: Harper & Row, 1982.

Erickson, Carolly. *The First Elizabeth*. New York: Summit Books, 1983.

Falls, Cyril. *Elizabeth's Irish Wars*. London: Methuen, 1950.

Jenkins, Elizabeth. *Elizabeth the Great*. New York: Coward-McCann, Inc., 1958.

Jenkins, Romilly. *Byzantium: The Imperial Centuries*. London: Weidenfeld and Nicolson, 1966.

Norman, A. V. B. *The Medieval Soldier*. New York: Thomas Y. Crowell Co., 1971.

Perroy, Edward. *The Hundred Years War*. New York: Capricorn Books, 1965.

Powicke, Michael. *Military Obligation in Medieval England*. Oxford: Clarendon Press, 1962.

Runciman, Steven. *The Fall of Constantinople, 1453*. Cambridge, England: Cambridge University Press, 1965.

Verbruggen, J. F. *The Art of War in Western Europe During the Middle Ages*. Amsterdam: North Holland Publishing Company, 1977 (translated by Sumner Willard and S. C. M. Southern).

Chapter IV

Anderson, R. C. *Naval Wars in the Baltic, 1522–1850*. London: Francis Edwards, Ltd., 1969.

Duffy, Christopher. *Fire and Stone*. London: David & Charles, 1975.

Reichel, Daniel. *Davout et l'Art de la Guerre*. Neuchatel: Centre d'Histoire, 1975.

Roberts, Michael. *Gustavus Adolphus: a History of Sweden, 1611–1632*, 2 vols. London: Longman, 1953–58.

Sawczynski, A. *Les Institutions Militaires Polonaises au XVII[e] Siecle*. Paris: *Revue Internationale d'Histoire Militaire*, Tome III, 1952, No. 12.

Thane, Elswyth. *The Fighting Quaker*. New York: Hawthorne Books, 1972.

Chapter V

Churchill, Winston S. *Marlborough: His Life and Times*. London: Harrap & Co., Ltd., 1938.

Corvisier, Andre. *Armies and Societies in Europe, 1494–1789*. (Bloomington, Ind.: Indiana University Press, 1979 (translated by Abigail T. Siddall).

Emmerich, Andreas. *The Partisan War*. London: Reynell, 1789.

Ewald, Johann. *Diary of the American Revolution*. New Haven, Conn.: Yale University Press, 1979 (Joseph P. Tustin, ed. and trans.).

Luvaas, Jay. *Frederick the Great on the Art of War*. New York: Macmillan, 1966.

Quimby, Robert S. *The Background of Napoleonic Warfare*. New York: Columbia University Press, 1957.

Savory, Reginald. *His Britannic Majesty's Army in Germany During the Seven Years War*. Oxford: Clarendon Press, 1966.

Scouller, R. E. *The Armies of Queen Anne*. Oxford: Clarendon Press, 1966.

Chapter VI

Adams, Henry. *The War of 1812*. Washington, D.C.: The Infantry Journal, 1944.

Catton, Bruce. *Grant Moves South*. Boston: Little, Brown and Co., 1960.

———. *Grant Takes Command*. Boston: Little, Brown and Co., 1968.

Fuller, J. F. C. *The Generalship of Ulysses S. Grant*. New York: Dodd, Mead & Co., 1929.

———. *Grant and Lee*. London: Eyre & Spottiswoode, 1933.

Freeman, Douglas S. *George Washington*, 5 vols. New York: Charles Scribner's Sons, 1948–52.

———. *R. E. Lee, A Biography*, 4 vols. New York: Charles Scribner's Sons, 1949. (*Note:* In case of conflict between this work and the author's *Lee's Lieutenants*, use the latter.)

Jacobs, James R. *Tarnished Warrior: Major General James Wilkinson*. New York: Macmillan, 1938.

Pogue, Forest C. *Organizer of Victory, 1943–1945*. New York: The Viking Press, 1973.

Chapter VII

Alger, John I. *Antoine-Henri Jomini: A Bibliographical Survey*. West Point, N.Y.: USMA, 1975.

de Courville, Xavier. *Jomini, ou le Devin de Napoleon*. Paris: Librairie Plon, 1935.

Cronin, Vincent. *Napoleon Bonaparte*. New York: William Morrow, 1972. This I especially recommend.

Esposito, Vincent J., and Elting, John R. *A Military History and Atlas of the Napoleonic Wars*. New York: Praeger, 1964 (2nd ed., 1965, recommended).

Fuller, J. F. C. *Armament and History*. New York: Charles Scribner's Sons, 1945. An inaccurate book but an excellent picture of Clausewitz's influence, as seen by Fuller.

Markham, Felix. *Napoleon*. New York: New American Library, 1963.

Perlmutter, Amos, and Gooch, John, eds. *Strategy and the Social Sciences* Totowa, N.J.: Frank Cass and Company, Limited, 1981.

Quimby, Robert S. *The Background of Napoleonic Warfare*. New York: Columbia University Press, 1957.

Wilkinson, Spenser. *The Rise of General Bonaparte*. Oxford: Clarendon Press, 1930.

Chapter VIII

Colby, Benjamin. *'Twas a Famous Victory*. New Rochelle, N.Y.: Arlington House, 1974.

Craig, Gordon A. *The Politics of the Prussian Army, 1640–1945*. New York: Oxford University Press, 1956.

Craven, Wesley F. and Cate, James L., eds. *The Army Air Forces in World War II*, 7 vols. Chicago, Ill.: University of Chicago Press, 1948–58.

Esposito, Vincent J., ed. *The West Point Atlas of American Wars*, vol. 2. New York: Frederick A. Praeger, 1959.

Harris, Sir Arthur. *Bomber Offensive*. New York: Macmillan, 1947.

Hurley, Alfred F. *Billy Mitchell: Crusader for Air Power*. New York: Franklin Watts, 1964.

Macksey, Kenneth. *Invasion*. New York: Macmillan, 1980.

———. *The Tank Pioneers*. London: Jane's, 1981.

Perlmutter, Amos, and Gooch, John. *Strategy and the Social Sciences*. London: Frank Cass, 1981.

Puleston, W. D. *The Life and Works of Captain Alfred Thayer Mahan*. New Haven, Conn.: Yale University Press, 1939.

Ritter, Gerhard. *The Schlieffen Plan*. New York: Frederick A. Praeger, 1958.

Slessor, John C. *Airpower and Armies*. New York: Oxford University Press, 1936.

Stone, Norman. *The Eastern Front, 1914–1917*. New York: Charles Scribner's Sons, 1975.

Watt, Richard M. *Dare Call It Treason*. New York: Simon and Schuster, 1963.

Webster, Sir Charles, and Frankland, Noble. *The Strategic Air Offensive Against Germany, 1939–1945*, 4 vols. London: Her Majesty's Stationery Office, 1961.

Weigley, Russell F. *Eisenhower's Lieutenants*. Bloomington, Ind.: Indiana University Press, 1981.

Chapter IX

Fehrenbach, T. R. *This Kind of War: Korea: A Study in Unpreparedness.* New York: Macmillan, 1963.

Hucker, Charles O. *China's Imperial Past.* Stanford, Calif.: Stanford University Press, 1975.

Liu, F. F. *A Military History of Modern China.* Princeton, N.J.: Princeton University Press, 1956.

Martin, H. Desmond. *The Rise of Chingis Khan and His Conquest of North China.* Baltimore, Md.: The Johns Hopkins Press, 1950.

Oberdorfer, Don. *Tet!* Garden City, N.Y.: Doubleday, 1971.

Palmer, David R. *Summons of the Trumpet.* San Rafael, Calif.: Presidio Press, 1978.

Romanus, Charles F., and Sunderland, Riley. *United States in World War II, China-Burma-India Theatre: Time Runs Out in the CBI.* Washington, D.C.: Department of the Army, 1959.

Zewen, Luo, Wenbao, Dai, Wilson, Dick, Drege, Jean Pierre, and Delahaye, Hubert. *The Great Wall.* New York: McGraw-Hill Book Company, 1981.

Chapter X

Blumenson, Martin, ed. *The Patton Papers.* 2 vols. Boston: Houghton Mifflin, 1972–1974.

Hamilton, Nigel. *Monty: The Making of a General, 1887–1943.* New York: McGraw-Hill, 1983.

———. *Master of the Battlefield: Monty's War Years, 1942–1944.* New York: McGraw-Hill, 1984.

Liddell Hart, B. H., ed. *The Rommel Papers.* New York: Harcourt, Brace and Co., 1953.

von Manstein, Erich. *Lost Victories.* Chicago: Henry Regnery Co., 1958.

Hoffmann, Peter. *The Story of the German Resistance, 1933–1945.* Cambridge, Mass.: MIT Press, 1977.

Ruge, Friedrich. *Rommel in Normandy: A Reminiscence.* San Rafael: Presidio Press, 1979.

Slim, William. *Defeat Into Victory.* London: Cassell and Co., 1956.

Speer, Albert. *Inside the Third Reich.* New York: The Macmillan Company, 1970.

Chapter XI

Dyson, Freeman. *Weapons and Hope.* New York: Harper & Row, 1984.

Ebon, Martin. *Psychic Warfare: Threat or Illusion.* New York: McGraw-Hill, 1984.

Fehrenbach, T. R. *This Kind of War.* New York: The Macmillan Company, 1963.

Henze, Paul B. *The Plot to Kill the Pope.* New York: Charles Scribner's Sons, 1983.

Holst, Johan J., and Nerlich, Uwe, eds. *Beyond Nuclear Deterrence*. New York: Crane Russak and Company, Inc., 1977.

Palmer, David R. *Summons of the Trumpet*. San Rafael, Calif.: Presidio Press, 1978.

Perlmutter, Amos, and Gooch, John, eds. *Strategy and the Social Sciences*. Totowa, N.J.: Frank Cass and Company Limited, 1981.

Pipes, Richard. *Survival Is Not Enough*. New York: Simon and Schuster, 1985.

Sterling, Claire. *The Terror Network*. New York: Berkley Books, 1982.

Suvorov, Viktor. *Inside Soviet Military Intelligence*. New York: Macmillan, 1984.

Ulam, Adam B. *The Rivals: Russia and America Since World War II*. New York: The Viking Press, 1971.

INDEX